Springer Series in

MATERIALS SCIENCE 82

Springer Series in
MATERIALS SCIENCE

The Springer Series in Materials Science covers the complete spectrum of materials physics, including fundamental principles, physical properties, materials theory and design. Recognizing the increasing importance of materials science in future device technologies, the book titles in this series reflect the state-of-the-art in understanding and controlling the structure and properties of all important classes of materials.

63 **Fundamentals of Ion-Irradiated Polymers**
By D. Fink

64 **Morphology Control of Materials and Nanoparticles**
Advanced Materials Processing and Characterization
Editors: Y. Waseda and A. Muramatsu

65 **Transport Processes in Ion-Irradiated Polymers**
By D. Fink

66 **Multiphased Ceramic Materials**
Processing and Potential
Editors: W.-H. Tuan and J.-K. Guo

67 **Nondestructive Materials Characterization**
With Applications to Aerospace Materials
Editors: N.G.H. Meyendorf, P.B. Nagy, and S.I. Rokhlin

68 **Diffraction Analysis of the Microstructure of Materials**
Editors: E.J. Mittemeijer and P. Scardi

69 **Chemical-Mechanical Planarization of Semiconductor Materials**
Editor: M.R. Oliver

70 **Applications of the Isotopic Effect in Solids**
By V.G. Plekhanov

71 **Dissipative Phenomena in Condensed Matter**
Some Applications
By S. Dattagupta and S. Puri

72 **Predictive Simulation of Semiconductor Processing**
Status and Challenges
Editors: J. Dabrowski and E.R. Weber

73 **SiC Power Materials**
Devices and Applications
Editor: Z.C. Feng

74 **Plastic Deformation in Nanocrystalline Materials**
By M.Yu. Gutkin and I.A. Ovid'ko

75 **Wafer Bonding**
Applications and Technology
Editors: M. Alexe and U. Gösele

76 **Spirally Anisotropic Composites**
By G.E. Freger, V.N. Kestelman, and D.G. Freger

77 **Impurities Confined in Quantum Structures**
By P.O. Holtz and Q.X. Zhao

78 **Macromolecular Nanostructured Materials**
Editors: N. Ueyama and A. Harada

79 **Magnetism and Structure in Functional Materials**
Editors: A. Planes, L. Manõsa, and A. Saxena

80 **Ion Implantation and Synthesis of Materials**
By M. Nastasi and J.W. Mayer

81 **Metallopolymer Nanocomposites**
By A.D. Pomogailo and V.N. Kestelman

82 **Plastics for Corrosion Inhibition**
By V.A. Goldade, L.S. Pinchuk, A.V. Makarevich and V.N. Kestelman

V.A. Goldade L.S. Pinchuk
A.V. Makarevich V.N. Kestelman

Plastics for Corrosion Inhibition

With 166 Figures

Springer

Prof. Victor A. Goldade
Prof. Leonid S. Pinchuk
Dr. Anna V. Makarevich
Metal Polymer Research Institute
National Academy of Sciences of Belarus
32a Kirov St.
246050 Gomel, Belarus
E-mail: vgoldade@tut.by

Prof. Vladimir N. Kestelman
KVN International
632 Jamie Circle
King of Prussia
PA 19406, USA
E-mail: kvnint@earthlink.net

ISSN 0933-033x

ISBN-10 3-540-23849-2 Springer Berlin Heidelberg New York

ISBN-13 978-3-540-23849-2 Springer Berlin Heidelberg New York

Library of Congress Control Number: 2004114849

Springer is a part of Springer Science+Business Media.
springeronline.com

Printed in Germany

Typesetting by TechBooks using a Springer LaTeX macro package
Cover concept: eStudio Calamar Steinen
Cover production: *design & production* GmbH, Heidelberg

Printed on acid-free paper SPIN: 10892988 57/3141/jl 5 4 3 2 1 0

Preface

One of the key problems of failure-free operation of machinery is prevention of corrosion. The global scale of modern production makes this problem even more critical. At the beginning of the 21st century industrial contamination and the corrosion-active nature of the environment reached a level such that corrosive damage of materials became commensurate with their production volume and expenditure on anticorrosion protection of machines became comparable with investments in basic production. Anticorrosion techniques changed from being an auxiliary service to industrial enterprises into a developing, scientifically intensive and generously financed branch of production.

Polymers occupy a very specific place amongst anticorrosion techniques. Polymers combine good chemical resistance with impermeability to different media and unusual deformation characteristics. The main principle of their application as anticorrosion means is the creation of a tight barrier that insulates metal machine parts or constructions from corrosion agents. The advantages of polymers allow the creation of such a barrier at minimal cost, providing protection of the working machines from corrosion, combining their manufacture with preservation and decreasing the cost of anticorrosion. This is one of the main reasons why world production of polymer materials increased by almost 50% in the past decade.

Films and coatings are the simplest and most economical type of polymer anticorrosion elements. Their specific material consumption (the ratio of the element's mass to the protected metal hardware volume) is 5 to 20 times less than that of any other kind of polymer anticorrosion means. Application of polymer film elements meets the demands of economy of hydrocarbon raw materials resources, which are restricted and not renewable. Specialization of films and coatings to their fields of application in anticorrosion protection influenced their production technologies and design of the technological equipment. Almost all modification methods, including filling, plastification, gluing, orientation stretching and the effect of physical fields, are used for polymer formation. During recent decades there has been a significant increase in the production of anticorrosion films and coatings based on high-tech, multifunctional and comparatively cheap thermoplastics such as PVC, ethylene copolymers with vinyl acetate, high-molecular PE and conducting polymers (organic metals). The list of anticorrosion film elements has broadened and includes multilayered, shrinking and stretching films, those with regulated adhesion, damping air layers as well as gas-filled (foamed) and liquid-filled plastics.

The global production of polymer films and coatings is now approaching stable growth. Behind the stabilization or decrease of the production of traditional engineering materials (cast iron, steel and nonferrous materials) is generally recognized and convincing proof of the increasing role of anticorrosion polymer film elements in industrial production.

The corrosion-active nature of the environment in combination with frictional effects can initiate the most dangerous type of corrosion damage – corrosion-mechanical wear of metals. Wear by friction alone presents an acute problem for machinery. To solve this problem, large capital investments are spent on the repair of machines and production of spare parts. The cost of the working machines is constantly being reduced due to depreciation. The efficiency of the majority of machines is far below 50% because of, first of all, losses due to friction. Corrosion-mechanical wear of metals significantly aggravates the tribological problem. It is sufficient to consider the following examples.

Worn out internal combustion engine (due to wear caused by friction and high-temperature corrosion) increase contamination of the environment by 25–30%. Depressurization of industrial apparatus with movable seals greatly increases the danger to staff as well as for the environment. It is socially very important to increase the lifespan of implanted endoprostheses containing friction joints operating in chemically active biological fluids. The application of antifrictional polymer materials enabled a decrease in the acuteness of the tribological problem.

The efficiency of polymer materials as an anticorrosion mean of protection of metal parts from corrosion can be significantly raised by modification of polymer binders using corrosion inhibitors for metals. In this case it becomes possible to realize both the barrier and inhibition mechanisms for metal protection from corrosion in the anticorrosion element. Anticorrosion polymer elements as a source of electric field can hamper corrosion processes and corrosion-mechanical wear of metal parts by the electrochemical mechanism. Realization of the barrier, inhibition and electrochemical mechanisms of anticorrosion protection with the help of polymers not only allows a profound improvement in the anticorrosion protection of metal parts but comes close to the creation of "smart" anticorrosion plastics and anticorrosion systems.

Today, mechanical engineering has at its disposal a broad range of polymer materials containing contact and volatile corrosion inhibitors, as well as inhibitor complexes. Inhibited plastics are more efficient means of protecting metals against corrosion than traditional polymer materials. Their composition, molding procedures and design of technological equipment for the treatment of inhibited polymers are commonly patented. Meanwhile, eminent scientific and commercial centers worldwide investigate the issues of scientific development, production and applications of these materials. Significant expense on this research is quickly recouped as a result of abated corrosion damage. The methodology and technological concepts of developing inhibited polymer films, coatings and structural plastics and the assimilation of polarized polymer elements into anticorrosion systems are discussed in the present book. A special chapter is devoted to inhibited engineering materials such as glues, sealants, plastic carriers of corrosion inhibitors, inhibited plastic containers and microencapsulated corrosion inhibitors (CI).

The production and application of inhibited plastics affect many aspects of ecology. Polymer modification by various CI results in the emission of considerable amounts of toxic substances. Up to 85% of inhibited polymer films and plastic containers are intended for one use only and become industrial and domestic waste. Recycling or annihilation of these materials are serious technical problems complicated by the necessity to collect waste and other management problems. Nevertheless, the efficiency of anticorrosion techniques depends heavily on the solution of these problems. A number of countries in the world have adopted long-term programs for the recycling of polymer wastes, increased the funding of extraction of valuable components and elaboration of low-waste plastic technologies.

The statistics of the global production and consumption of anticorrosion materials bears witness to the fact that inhibited polymer materials are one of most promising approaches to decrease the losses from metal corrosion. Their production and application technologies are related to non-traditional fields of processing polymer materials and many of their specific characteristics are not reflected in technical literature, for various reasons. These questions are described in this book devoted to the means of anticorrosion protection of metals. The authors hope that the book will be of interest to professionals engaged in the spheres of protecting metals from corrosion, materials science, plastics processing technology and engineering ecology.

The authors express their gratitude to Prof. A.S. Neverov and Ph.D. I.V. Tsarenko for valuable ideas concerning modification technologies of polymer films by metal corrosion inhibitors, to S.Ya. Liberman and E.I. Parkalova for industrial use of the technology of inhibited plastics and to S.V. Zotov and L.S. Pushkina for their help in the preparation of this book.

Gomel, Belarus – King of Prussia, PA
January, 2005

V.A. Goldade
L.S. Pinchuk
A.V. Makarevich
V.N. Kestelman

Contents

List of Abbreviations

AFM	atomic force microscopy
AKOR *(CI trademark)*	mineral nitrated oil
ATet	5-aminotetrazole hydrate
ATR	aatenuated total reflectance
ATr	3-amino-1,2,4-triazole
BMEA	benzoate of monoethanolamine
BMP *(CI trademark)*	neutralization product of urea-sulfated mineral oil
BTA	benzo-1,2,3-triazole
CAC	cyclohexylamine chromate
CEVA	copolymer of ethylene and vinyl acetate
CFD	copolymer of formaldehyde with 1,3-dioxolane
CI	corrosion inhibitor
CM	composite material
CTD	copolymer of trioxane with 1,3-dioxolane
DEL	double electric layer
Dichan	dicyclohexylamine nitrates
DMSO	dimethyl sulfoxide
DOP	dioctyl phthalate
DSC	differential scanning calorimetry
DSMO	*(CI trademark)*
DTA	differential thermal analysis
EIS	electrochemical impedance spectroscopy
ESCD	efficient surface charge density
FO	fuming-oxides
G-2 *(CI trademark)*	hexamethylenimine m-nitrobenzoate
GRM *(CI trademark)*	tar of vegetable oil and inedible fats
HDPE	high-density polyethylene
IFHAN-1 *(CI trademark)*	1-diethylamine-2-methylbuthanone-3
IFHANGAZ-1 *(CI trademark)*	dialkylaminoacrylonoitryle + dialkylamine
INHAM	inhibited antifrictional materials
IPF	inhibited polymer films
IRS	infrared spectroscopy
KSK *(CI trademark)*	sulfated acidic tar

LDPE	low-density polyethylene
LLDPE	linear low-density polyethylene
M-1 *(CI trademark)*	salt of cyclohexylamine and SFA
M1–P–M2	metal 1 – polymer – metal 2 systems
MNI-3 *(CI trademark)*	oxidized petrolatum
MNI-7 *(CI trademark)*	oxidized ceresin
MO	mineral oil
MPC	maximum permissible concentration
MPE	metal-polymer electret
MPJ	metal-polymer joint
MSDA-1 *(CI trademark)*	salt of dicyclohexylamine and SFA
MTet	5-methyltetrazole
NDA	nitrate of dicyclohexylamine
NG-203 *(CI trademark)*	oil solution of potassium sulfonate and oxidized petrolatum
PA	polyamide
PAni	polyaniline
PAN	polyacrylonitrile
PC	polycarbonate
PE	polyethylene
PET	polyethylene therephtalate
PFR	phenol-formaldehyde resins
PHC	polyheterocyclic compounds
PhTet	5-phenyltetrazole
PI	polyimide
Pl	plasticizer
PMMA	polymethyl methacrylate
PP	polypropylene
PPl	pentaplast (penton) – [poly-[3,3-bis-(chlormethyl)oxetane]
PPy	polypyrrole
PS	polystyrene
PTFCE	polytrifluorinechlorine ethylene
PTFE	polytetrafluoroethylene
PU	polyurethane
PVA	polyvinyl acetate
PVAL	polyvinyl alcohol
PVB	polyvinyl butyral
PVC	polyvinyl chloride
PVDC	polyvinyliden chloride
PVF	polyvinyl formal
PVFE	polyvinyl formal ethylal
RH	relative humidity
SB	sodium benzoate
SCC	stress corrosion cracking

SFA	synthetic fatty acid
SEM	scanning electron microscopy
SIM *(CI trademark)*	condensation product of alkenyl succinic anhydrate and urea
ST	selective transfer
Tet	tetrazole
Tr	triazole
TSC	thermally stimulated current
TSD	thermally stimulated depolarization
UHMWPE	ultrahigh molecular weight polyethylene
Ukramin *(CI trademark)*	a product of neutralizing fatty acids by monoethanol amine
VCI	volatile corrosion inhibitors
VITAL *(CI trademark)*	a mixture of potassium and triethanolamine salts of unsaturated fatty acids of tall oil
VNH-5 (*CI trademark*)	benzo-1,2,3-triazole subjected to heteroalkylation
VNHL-20 *(CI trademark)*	dimorpholinephenyl-methane
VNHL-49 *(CI trademark)*	N-cyclohexylphenylimine
WI	wear inhibitors
XPES	X-ray photoelectron spectroscopy
XSA	X-ray structural analysis

1 Design Principles of Plastics Containing Metal Corrosion Inhibitors

Design of plastics incorporating corrosion inhibitors (CI) presents an intricate scientific and technical problem. Knowledge from different scientific domains, including materials science, physical chemistry, electrochemistry, physics of dielectrics, physics of polymers, plastics processing technologies and tribology, is necessary to solve the problem. This interdisciplinary problem challenges the specialists in various fields with unexpected results. Any solution will have a profound commercial impact since: (i) polymers are used in the majority of products in combination with metals, (ii) contacts of metals with other materials either in the atmosphere or technical media are a potential hotspot for corrosion and, (iii) inhibited plastics have proved to be the most efficient means of eliminating corrosion available today.

Plastic containing CI is in its essence a composite material. Inhibitors and special additives, which also have a function regulating physical-chemical interactions between the polymer and inhibitor, are spread within the polymer component of the composite at a given degree of cohesion with the polymer. Compatibility of the CI with the polymer binder should not be too high as the inhibitor would not be released from the composite. In the case of low compatibility the release rate of the inhibitor will be too high and the life of the plastic as an inhibitor carrier would be unacceptably low.

Production of such composite materials encounters a serious technological obstacle in the low thermal stability of both polymers and CI. Moreover, there is always a danger of impairing the inhibitor's properties during joint thermo-mechanical processing with the polymers, in which the inhibitor loses its volatile components when heated to the polymer softening point. Therefore, technological procedures for combining polymers and inhibitors are, as a rule, at the level of inventions and often constitute the "know-how" of the developers of the inhibited plastics.

These materials are highly efficient as a means of corrosion inhibition due to their ability to realize almost all inhibition mechanisms of metal corrosion, namely: the barrier mechanism connected with the impenetrability of polymers for most corrosion media; the inhibition mechanism induced by a specific action of the inhibitors on the corrosion process kinetics; the protecting mechanism related to the effect of the polarizing charge formed in the plastic upon distribution of electrode potentials within the corrosive system, among others. Due to the diversity of these mechanisms the methods for estimating efficiency of plastics containing CI often results in incommensurate results. The methodology of selecting test methods for inhibited plastics involves a number of factors; the key is approximation of the testing scheme to the real operation conditions of an article along with simplification and acceleration of the experiment.

Materials of the considered class are used in corrosion-proof goods in the form of films, coatings and structural products. The films are commonly used in packaging, the coatings are used for protection of machinery and buildings with large specific area, and the structural products are predominantly used in fIt follows from equation It follows from equation riction joints operating in hostile media. The form of inhibited plastics depends on their processing technique, structure and characteristics.

The present chapter, which is *per se* an introduction to this book, is devoted to the topics above.

1.1 Underlying Notions

As the subject of the present book lies at the junction of several sciences, it will be expedient to give some principle terms to be used in further narration. There are three groups of terms related to metal corrosion, composite materials and tribology.

1.1.1 Corrosion of Metals

Metal corrosion is a physical-chemical interaction between a metal and a medium leading to degradation of service properties of the metal and the medium or a technical system they constitute. A chemical reaction between a metal and a medium, or their components, at their interface lies at the base of metal corrosion.

The mechanism for metal corrosion depends, first of all, on the type of hostile medium. *Gas corrosion* occurs in metal contact with an active gas. A layer of solid corrosion products (scale) is formed on the metalware surface in dry oxidative gases at elevated temperatures. Metal corrosion in electrolyte solutions, even if the solution is in the form of a thin film on a metal surface, follows the reaction

$$\mathrm{M} + \mathrm{Ox} = \mathrm{M}^{z+} + \mathrm{Red}\ , \tag{1.1}$$

where M is the metal, Ox is the oxidant particle, Red is its reduced form; Ox acquires a charge $+ze$, and Red acquires a charge $-ze$ (where e is the electron charge). In the majority of cases this reaction follows the electrochemical route, i.e. an atom M and a particle Ox are not in direct contact but the electrons are transferred from M to Ox through M's conduction zone (Fig. 1.1).

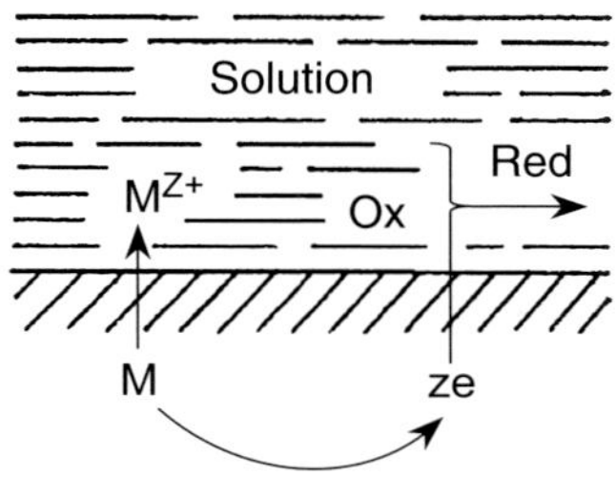

Fig. 1.1. Scheme of electrochemical corrosion mechanism of metals [1]

Electrochemical corrosion of metals follows the scheme indicated in Fig. 1.1 in two reactions, i.e. anodic dissolution of the metal and cathodic reduction of the oxidizer:

$$M = M^{z+} + ze \,, \tag{1.2}$$

$$Ox + ze = Red \tag{1.3}$$

These reactions are totally independent but interrelated through electric neutrality of the metal-electrolyte solution system.

Metal corrosion is a spontaneous process accompanied by a decrease in Gibbs' energy in the corrosion system. The Gibbs' energy varies in metal corrosion in a similar way to in any other spontaneous chemical reaction. Thermodynamic instability of the metal in contact with an electrolyte solution is judged from the magnitude of its electrode potential. Let us consider its mechanism.

Transfer of charge carriers from the metal to the electrolyte solution as well as movement of ions in the electrolyte are hampered by the corrosive system. The electrolyte solution induces the formation of a surface layer in the vicinity of the metal electrode surface consisting of spatially separated positive and negative charge carriers, which is called a double electric layer. Spatial separation of charges is always accompanied by a potential difference, therefore the double layer exerts a perceptible influence on the rate of electrode processes. The double layer consists of two parts: a compact layer and a diffusive layer (Fig. 1.2).

The compact, or Helmholtz, layer is the closest to the surface, in which the charge distribution and potentials change linearly with distance from the electrode surface. The potentials change exponentially when the more diffused outer Gouy–Chapman layer is formed.

The corroding metal potential is inconstant since it is affected by corrosion factors varying in time. As soon as the electrochemical corrosion becomes stationary, the electrode potential acquires a constant value E_{corr}. Once the stationary potential or the corrosion potential has stabilized, velocities of the anodic (1.2) and cathodic (1.3) reactions on the electrode equalize. E_{corr} measured in standard conditions relative to some standard reference electrode

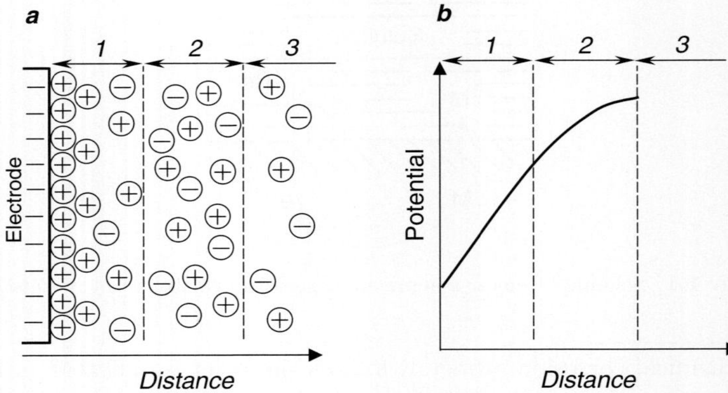

Fig. 1.2. (**a**, **b**) The double layer. Distribution of ions as a function of distance from the electrode behaving as an anode (**a**) and variation of potential with distance (**b**) for the model shown. (1) Helmholtz layer, (2) Gouy–Chapman layer, (3) bulk solution

is called the standard electrode potential E_0. Metals are placed according to ascending electrode potentials in Table 1.1 [2, 3].

The electric current passing through the corrosive system changes the electrode potential. *Polarization* is the deviation of the electrode potential from the equilibrium state when electric current is passed through the electrode-electrolyte solution interface. The unequal rates at which separate stages of the total electrode process are running are the reason for this polarization. The polarization is called concentration polarization when an insufficient amount of the reacting substance is fed to the electrode surface and its concentration near the electrode is less than its volume content in the corrosive system. The polarization is called electrochemical polarization when the limiting stage of the electrode process turns out to be the charge transfer through the electrode-solution interface (Fig. 1.2). There are also the anodic and cathodic polarization depending on the electrode potential bias for direct current passage. The phenomenon that counteracts the electrode polarization is called *depolarization*. Polarization can be lowered or averted by adding special substances called depolarizers.

It is convenient to use polarization diagrams, sometimes called Evans' diagrams, to forecast metal corrosion. These are the graphs of electrode potential versus the logarithm of current or the logarithm of current density [4]. One such diagrams is presented in Fig. 1.3. Metal M corrodes in acidic solution, in which hydrogen atoms are reduced. The inhomogeneous structure of the metal promotes the formation of anodic and cathodic areas on its surface. The point of intersection of the anodic and cathodic curves corresponds to the maximum density of the corrosive current i_{corr} on the abscissa axis and to corrosive potential E_{corr} on the ordinate axis.

Table 1.1. Standard potentials of electrode reactions of metals in aqueous medium at 25°C

Electrode Reaction	E_0, V	Electrode Reaction	E_0, V	Electrode Reaction	E_0, V
$Li^+ + e^- = Li$	−3.045	$Zr^{4+} + 4e^- = Zr$	−1.53	$Sn^{2+} + 2e^- = Sn$	−0.136
$Rb^+ + e^- = Rb$	−2.925	$Ti^{3+} + 3e^- = Ti$	−1.21	$Pb^{2+} + 2e^- = Pb$	−0.125
$K^+ + e^- = K$	−2.925	$V^{2+} + 2e^- = V$	−1.18	$W^{3+} + 3e^- = W$	−0.11
$Cs^+ + e^- = Cs$	−2.923	$Mn^{2+} + 2e^- = Mn$	−1.18	$Bi^{3+} + 3e^- = Bi$	+0.226
$Ca^{2+} + 2e^- = Ca$	−2.87	$Nb^{3+} + 3e^- = Nb$	−1.10	$Sb^{3+} + 3e^- = Sb$	+0.24
$Na^+ + e^- = Na$	−2.714	$Zn^{2+} + 2e^- = Zn$	−0.763	$Re^{3+} + 3e^- = Re$	+0.30
$La^{3+} + 3e^- = La$	−2.52	$Cr^{3+} + 3e^- = Cr$	−0.74	$As^{3+} + 3e^- = As$	+0.30
$Mg^{2+} + 2e^- = Mg$	−2.356	$Ga^{3+} + 3e^- = Ga$	−0.53	$Cu^{2+} + 2e^- = Cu$	+0.340
$Am^{3+} + 3e^- = Am$	−2.32	$Fe^{2+} + 2e^- = Fe$	−0.44	$Cu^+ + e^- = Cu$	+0.520
$Pu^{3+} + 3e^- = Pu$	−2.07	$Cd^{2+} + 2e^- = Cd$	−0.402	$Hg_2^{2+} + 2e^- = 2Hg$	+0.796
$Np^{3+} + 3e^- = Np$	−1.86	$In^{3+} + 3e^- = In$	−0.342	$Ag^+ + e^- = Ag$	+0.799
$Be^{2+} + 2e^- = Be$	−1.85	$Tl^{3+} + 3e^- = Tl$	−0.336	$Rh^{3+} + 3e^- = Rh$	+0.80
$V^{3+} + 3e^- = V$	−1.80	$Mn^{3+} + 3e^- = Mn$	−0.283	$Pd^{2+} + 2e^- = Pd$	+0.987
$Al^{3+} + 3e^- = Al$	−1.67	$Co^{2+} + 2e^- = Co$	−0.277	$Ir^{3+} + 3e^- = Ir$	+1.000
$Ni^{2+} + 2e^- = Ni$	−0.257	$Mo^{3+} + 3e^- = Mo$	−0.20	$Pt^{2+} + 2e^- = Pt$	+1.19
$Ti^{2+} + 2e^- = Ti$	−1.63	$Ge^{4+} + 4e^- = Ge$	−0.15	$Au^{3+} + 3e^- = Au$	+1.52

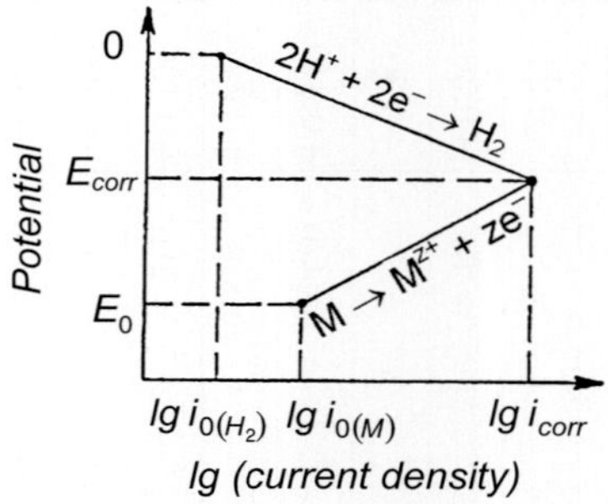

Fig. 1.3. Evans' diagram

The parameters necessary to construct the diagram are usually measured by the potential. The instrument automatically maintains the desired potential between the working and reference electrodes by passing the appropriate current between the working and counter electrodes. This current is controlled, and the potential of the working electrode is measured.

The corrosion process rate and mechanisms depend upon internal and external factors. The innternal factors are the nature of the metal, its chemical composition, structure, electrode surface state, and the presence of residual stresses. The external factors are chemical composition of the corrosive medium, and conditions of the process, including temperature, pressure, flow velocity. The competing effects of these factors determine which direction of the corrosion process dominates.

Anodic dissolution of metals may run together with formation of oxides, hydroxides and hydrogen ions. Typical cathodic processes of metal corrosion are electrochemical reduction reactions of hydrogen (hydrogen depolarization), oxygen (oxygen depolarization) or oxygen-containing ions.

The metal can enter into interactions with environment components and form compounds that are stable to corrosion during anodic dissolution. *Passivity* is the thermodynamic state of reaction in metals at which their corrosion retards abruptly. The transfer of a metal into the passive state is called *passivation,* and the layer formed on the metal electrode surface is called the passivating layer. Oxide and salt passivation of metals are distinguished by the composition of passivating layers.

M. Pourbaix [5] has devised a compact summary of thermodynamic data of potential – pH diagrams related to corrosion behavior of any metal in water. These diagrams are now available for most common metals. Diagram (Fig. 1.4) shows specific conditions of potential and pH under which the metal either does not react (immunity) or is able to react to form specific oxides or complex ions. These data indicate the conditions under which diffusion-barrier films may form on the electrode surface. The diagram outlines the nature of stoichiometric compounds into which any less stable compounds

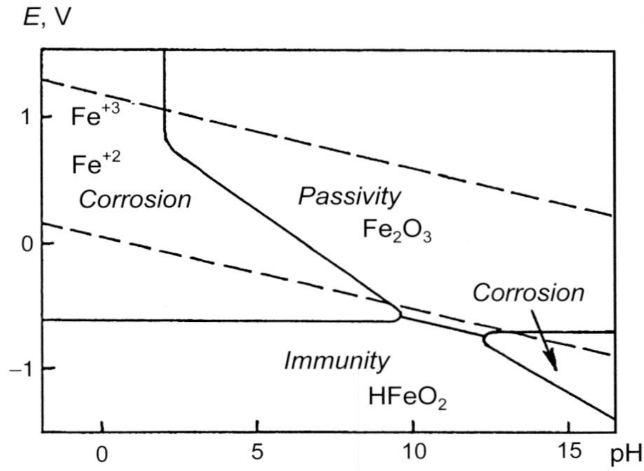

Fig. 1.4. A simplified Pourbaix' scheme for iron at 25°C

may transform whenever equilibrium is achieved. Pourbaix' diagrams describe the equilibrium state of a metal when immersed in an acid or alkali.

Corrosion inhibitors are substances whose introduction in small amounts into a corrosion system (aggressive medium, polymer coating, lubricating or packing material, etc.) produces noticeable corrosion abatement in metals. CI are subdivided into adsorption and passivating CI by their action mechanism. The former protect metals by affecting the kinetics of the electrode corrosion processes. The latter promote the formation of an oxide or other films on a metal product to transfer the metal into the passive state.

Metal corrosion is classified by the type of the hostile medium and conditions at which it runs. Some aspects of atmospheric, marine, underground and biological types of corrosion are considered in this book. Heterogeneity of a metal or medium, specific conditions of oxidizer supply and withdrawal of corrosion products, which do not intrude into the equipotential state of the corroding metal surface, may bring about the areas where either anodic or catholic reactions dominate. *Local corrosion of* metals without passivation may mean the intensive local dissolution of the metal in anodic regions. *Crevice corrosion* is local corrosion of metals in a structure having holes and narrow clearances. The electrolyte solution contacting open areas of a metal structure gains limited mass exchange with the solution trapped in the voids and clearances. The main reason for crevice corrosion in atmospheric conditions is moisture condensation on the article surface. In liquid media there is a hampered access of O_2 and removal of corrosion products from the gaps, where the latter leads to violation of metal passivity. As a result, the electrode potential of the metal in a clearance shifts to the region of negative values. Hence, a galvanic element appears in the article, experiencing crevice corrosion with a potential difference of 0.5–0.75 V. In the element, the metal in the clearance acts as an anode and the metal on the open surface of the article as a cathode. Formation of such galvanic elements results in corrosive-mechanical failure of metals in friction joints, which is called mechano-chemical wear.

1.1.2 Composite Plastics

Composite materials are multiphase systems consisting of two or more components that preserve their structure and properties within the composite formulation.

The present work is devoted to polymer composites whose *matrix*, i.e. uninterrupted component within the composite structure, consists of a polymer material. The matrix imparts a given shape to the composite product, distributes mechanical loads in the material and protects other components from external effects. The matrix material predetermines the range of working temperatures, resistance to corrosion, electrical properties, aging regularities of the composite, as well as processing technique of the composite into articles.

The materials under study are filled composites, which means that the matrix contains dispersed *fillers* – particles of organic or inorganic solid matter, liquids or gases. The filler particle size varies in a wide range. A distinct feature of a composite is the presence of heterogeneous inclusions visible without a microscope. These components tend to become finer as the technology perfects and may reach 1 μm in size, necessitating a microscope to identify them. By the end of the twentieth century there is a new trend in manufacturing of composite materials in which particles of nanometer size (10^{-9} m) are used. A large portion of the volume of these fine particles occupies the surface layer, so their properties differ from those of the constituent substance, especially in the macrovolume. Surface phenomena play a paramount role in technological operations of manufacturing a composite consisting of nanoparticles [6]. A *nanocomposite* is a composite material whose components have the dimensionality of colloidal particles (10^{-9}–10^{-7} m).

Classification of composites by the phase inclusion size bears a philosophical aspect: how small should a component in the matrix be not to make the term "composite material" so universal as to include in fact all materials? Interatomic distances in molecules and crystals are of $1.5 \cdot 10^{-10}$ m dimensionality, distances between iterative elements of the crystalline structure are 10^{-10}–10^{-9} m, while the size of the smallest intermolecular voids in polymers is 10^{-8} m. Note that mean nanoparticle size (plastic pigments are 10^{-8}–10^{-5} m in size, the diameter of monocrystalline fibers or whiskers is 10^{-7}–10^{-5} m, glass microspheres are 10^{-6}–10^{-4} m) is commensurate with parameters of monolithic simple materials. This means that in the totality of engineering materials, nanocomposites occupy a place at the boundary between composite and simple materials.

Particles of the dispersed phase can be spread within the matrix randomly, although they commonly tend to arrange in a certain order (Fig. 1.5).

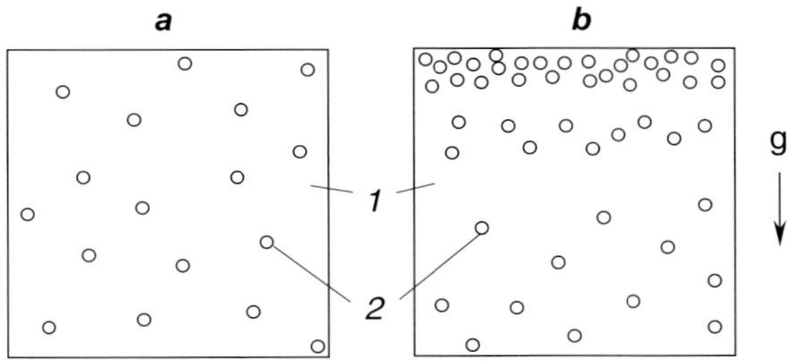

Fig. 1.5. Structural schemes of dispersion-filled composites: (**a**) isotropic, (**b**) anisotropic. (1) matrix, (2) dispersed phase

When the particles are chaotically scattered, the materials are called *isotropic*, i.e. their properties are identical in all directions. When the particles are distributed evenly within the matrix at a concentration gradient g (g – the vector indicating the direction of fastest change of particle concentration in the matrix), the composite is *anisotropic* and its properties depend upon the direction. The purpose of manufacturing isotropic composites is to place a respective component exactly in the site where properties of the composite article are to be realized most fully.

Formation of the composite is accompanied by mechanical, physical and chemical interactions of the components. This leads to phases and interfaces between them. Interfacial interactions of components are a function of their thermodynamic, kinetic and mechanical compatibility.

Thermodynamic compatibility is a property of the composite matrix and other components to be in the state of thermodynamic equilibrium at temperatures corresponding to its formation regimes or the operation conditions of the material.

Kinetic compatibility is a property of the composite components to be in a metastable equilibrium state (as opposed to a stable one) regulated by electric interactions between the components, by conditions of adsorption, diffusion or chemical reactions between them. Components that are compatible in a thermodynamic sense (temperature, pressure and ambient effects) might become incompatible kinetically, which can lead to the release of one of the components from the matrix.

Mechanical compatibility is reached if deformation characteristics and heat-expansion coefficients of components (which can be compatible neither thermodynamically nor kinetically) are similar, so that the composite integrity will not be disrupted during operation and mechanical properties will remain stable.

Thus, the components form an integral composite material when they are interrelated through mechanical, physical and chemical bonds. The mechanical bonds arise under the action of friction forces when the components become engaged mechanically via the natural or purposefully created asperities on contacting surfaces. These bonds define the properties of polymer overcoats on metal parts as well as triboengineering characteristics of metal-polymer friction joints. Physical bonds are the result of molecular interactions on the surface layers of components. Gravitation, electrostatic and magnetic interactions of the composite ingredients belong to physical bonds. Chemical bonds represent joining of atoms and ions of different contacting components into molecules and crystals. The work needed to break a chemical bond is much larger than that to disrupt physical intermolecular bonds.

Among the advantages of the composites is the possibility of assembling them from required components via the necessary type of links. The term "design of compounds" used in engineering means the choice of the form and dimensions of the compound with allowance for the magnitude and time of the load action, part configuration and material strength. This term in materials science will justify speaking about design composites with preset type of binder and other components, their distribution within the matrix, matching of the technological procedures using the mechanical, physical or chemical bonds and, bearing in mind their strength and stability, with account of operation conditions, the cost of and specific demands made of the composites.

Composition optimization is the process of selecting the best correlation of componential concentration in the composite. The best composition meets the optimum criterion, which is the qualitative index expressing the threshold of composite efficiency. It assists in comparing the alternative compositions, aimed at choosing the best one. Optimization of the composition is based on experimental determination of its characteristics through testing samples of various compositions. Optimization of the manufacturing technology of composites embraces numerous factors, requiring a creative approach. the most reliable optimization is the fabrication of samples of an optimum composition proceeding from all alternatives of the technology. Testing of such samples and comparison of their technical and economic characteristics can be the basis for selecting an optimum technology. This path is, however, time consuming and requires significant expenditure. The composite processing technology is usually chosen on the basis of experimental results and the availability of traditional production equipment.

Modern trends in the design of composite materials presuppose the development of intelligent or smart composites [7]. These materials adapt to service conditions either through varying their charge state or physical-chemical structure, or regulation of this adaptation degree by a feedback system [8]. The principle of operation of such smart materials is illustrated in Fig. 1.6.

This material contains structural elements functioning as [9]: (i) a sensor – a sensitive element or a meter responding to ambient variations; (ii) an actuator – a structure with which the material exerts an effect on the environment; (iii) a processor – structural element controlling the actuator functions based on signals incoming from the sensor. Material 2 contains structures

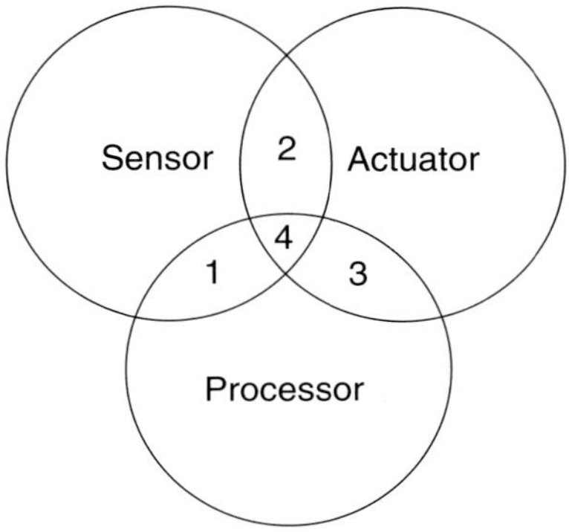

Fig. 1.6. Scheme of smart material structure. (1) adaptive structure, (2) controlled, (3) active, (4) intelligent structure

functioning as a sensor and actuator simultaneously. This controlled material may influence the environment in the form of mechanical loading, electrical or magnetic field, etc. Material 3 combines the functions of the processor and actuator. This active material is programmed to respond to variations in the environment properties. The majority of polymer packaging films containing CI belong to this group. Material 1 is fit with structural elements serving as a sensor and processor. This is an adaptive material able to adapt to varying ambient conditions. Intelligent or smart material 4 incorporates the elements functioning as a sensor, actuator and processor. Along with exerting an active influence of physical, chemical or biological nature upon the environment it also regulates its own activity, i.e. the level of this influence depending on the degree of variation of the environment.

Composite materials consisting of a few components, each of which is intended for individual functions within the composite, are optimum for imparting smart properties.

1.1.3 Friction of Inhibited Plastics

Tribology is a science studying friction, wear and lubrication, along with interactions of the contacting physical bodies during their mutual motion.

Friction is the resistance to relative movement of two contiguous bodies arising in the zones of their contact and is effective in the contiguous plane. The friction pair is a system of two elements whose contacting surfaces are moving relative to each other. One of the elements is a frictional material; the other is a counterbody (rider). The friction surface is a solid body surface participating in friction. Wearing is the process during which the material is separated from the friction surface and residual deformation is accumulated in the solid body being abraded that is exhibited in a gradual change of the body size and shape. Wear is the result of the wearing process. A lubricating material is fed to the friction surface to reduce wear rate and friction force. Lubrication is the effect of the lubricating material. Triboengineering is an applied sphere of tribology that represents the final stage of elaboration of the friction joint.

Mechanical energy generated by the movement of the contacting bodies is converted by friction into other types of energy, leadings to triboelectric and tribochemical phenomena.

In the case that the frictional material is a polymer belonging to the class of dielectrics, it becomes electrified byt friction. *Electrization* means the formation of excess and similar charges on the surface or in the bulk of the body participating in friction. High electrization of dielectrics is attributed to their low electroconductivity. When a polymer is rubbed against a metal, the electrization potential of the polymer part reaches its stationary value in a few seconds, after which it remains invariable. *Electrets* are dielectrics that preserve their electrization for a long time and are, in fact, analogues of a permanent magnet.

When metal parts rub in an electrolyte, it is possible to form short-circuit galvanic microelements (Fig. 1.7). Potentials ${}_1E_3$ and ${}_2E_3$ appear at the metal-electrolyte interface and contact potential difference ${}_1E_2$ in the contact sites of the parts. The electromotive force of these elements promotes electrode processes on the friction surfaces. The processes appear even though ${}_1E_3 = {}_2E_3$ because of the galvanic elements resulting from crevice corrosion in the friction zone.

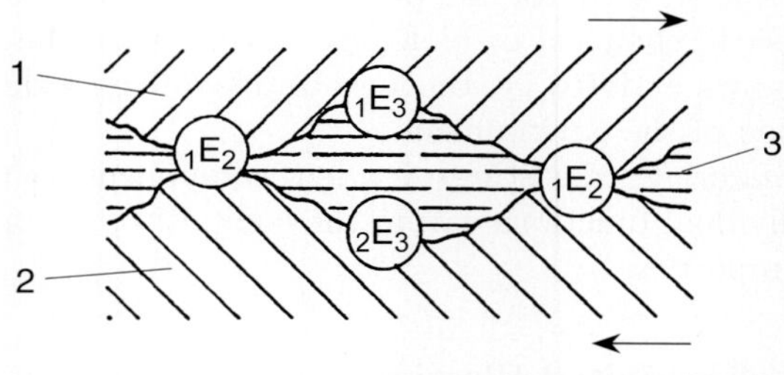

Fig. 1.7. Formation of galvanic elements in the contact of metal parts. (1) and (2) friction surfaces of metal parts; (3) electrolyte

Friction and corrosion mutually affect one another. Friction violates integrity of the passivating layers on the friction surface of metal parts. Movement of the friction surfaces relative to each other accelerates supply of corrosion matter to the freshly formed surface areas of the mechanical counterbody and removal of the products of electrode reactions. These stages of the electrode process are less intensive in static conditions and follow the mechanism of diffusion or convection. As can be seen, corrosion serves as a catalyzer of mechanical wear during metal friction in electrolytes. *Mechano-chemical wear* is the wearing of a material during friction accompanied by its chemical interaction with the environment.

Tribochemistry is a domain of chemistry that studies chemical and physical-chemical variations in solid bodies under the action of frictional energy. Wear of metal counterparts in polymer-metal friction pairs is far higher than in metal-metal pairs [10]. To explain this phenomenon the following model has been proposed [11]. Microasperities on the friction surface experience high-intensity impacts. A microasperity is transferred into an excited state called a triboplasma at the moment of impact. It is characterized by a high local temperature (above 10^4 K) and a short lifetime ($<10^{-7}$ s). Various physical processes leading to photon and electron emission take place in the triboplasma state of the microasperity. Energy accumulated on microasperities is not fully dissipated; a part of it remains in the surface layer and initiates chemical reactions on the friction surface. Free radicals are formed in the polymer-metal friction pair. These radicals are the products of macromolecular destruction and act as surfactants towards the metal counterbody [10].

Polymer-metal friction pairs are also characterized by frictional *transfer of material*. This implies the material displacement from the polymer part friction surface onto that of the metal counterbody. Macrotransfer is realized either as a fatigue detachment of polymer particles sticking to the counterbody or as galling, i.e. sticking of the viscous-flow fragments from the polymer surface layer, their extension and movement in the friction direction.

All the mentioned phenomena occur during friction of the polymer-metal pairs in electrolyte solutions, including crevice corrosion and formation of galvanic pairs in the friction zone, triboelectrization of the polymer frictional material and formation of electrets, superposition of the electrode potential of the metal counterbody and that of the polymer triboelectrization potential, tribochemical reactions yielding macroradicals and their interaction with the newly formed areas of the metal friction surface, as well as frictional transfer of the polymer material to the counterbody. Participation of the inhibited plastic in friction makes the situation still more complex due to liberation of CI in the friction zone and their effect on the kinetics of electrode processes, on lubrication and formation of fresh sites of passive films on the friction surface of the metal counterbody. The interrelationship between these processes is reviewed in Chap. 4.

1.2 Corrosion Processes in Metal-Polymer Contacts

Local corrosion of metals occurs because of many factors, the most critical of which are inhomogeneity of the metal phase and difference in composition of the corrosive medium, i.e. electrochemical heterogeneity on different areas of the metal surface.

The most widespread corrosion types are point corrosion or pitting, filiform, crevice, contact and intercrystalline corrosion. The highest danger for the metal-polymer system is presented by the crevice type of corrosion.

Crevice corrosion is often observed in clearances between metals and in places of loose contact between the metal and dielectric (including corrosion resistant materials). Passivating materials and alloys, such as stainless chromium and nickel-chromium steels, aluminum and magnesium alloys are most inclined to crevice corrosion [2, 12, 13].

Crevice corrosion starts because of hampered access of electrolyte and oxygen in the space of the clearance. This changes the pH of the electrolyte solution in the clearance, spurs the anodic and retards the cathodic processes. As a result, the metal electrode potential shifts to the negative with respect to the potential on the open surface. These processes bring about electrochemical microelements of the slot-open-surface type, in which the metal in the clearance serves as an anode [14].

Let us consider some peculiarities of local corrosion of metals in contact with polymers stimulated by changes (inhibition or acceleration) of corrosion processes at the metal–polymer interface. These changes can be attributed either to a limited velocity of movement of the substances participating in the corrosion process or to chemical and electrochemical effects of the polymer on the metal and their influence on the corrosive media activity.

1.2.1 Corrosion Specifics at the Metal-Polymer Interface

Abnormal local corrosion of metals in contact with plastics induced by formation of concentration elements [14] was related in the 1930s to specific properties of plastics. Indeed, plastics contain monomer remainders, stabilizers, plasticizers, and other corrosion-active components. For example, testing of metal screws screwed into a threaded polymer plate in a thermal humid chamber showed [15] that corrosion of the screws was caused by migration of ammonium, hydrochloric and acetic acids from the polymer material. Analogous results were obtained when studying aluminum corrosion in a closed volume containing emissions of enamels, lacquers and different plastics [16]. It was recognized for more than a century that rubber initiated corrosion of contacting metal parts. This is the reason for protecting metals in contact with rubber by applying special coatings.

A convenient model for studying crevice corrosion specifics at the metal–plastic interface turned out to be the "polymer coating on a metal substrate" system.

Anticorrosion properties of coatings are largely based on chemical resistance of the polymer used for the coating, its permeability for corrosive media and adhesion to the object being protected. One of the major criteria of the protecting durability is the coating continuity. Any defect on the coating may disable the article. The defects may occur because of technological conditions at coating formation, variations in the material properties under operation in hostile media, etc. A single defect on the coating does not impair its protective ability at once, and at the forefront in such conditions is the stability of adhesive bonds.

When a polymer is brought into contact with a low-molecular-weight liquid, it can adsorb onto the polymer surface, diffuse into its bulk and, finally, cause its swelling and dissolution. Named processes bring about in the first place changes in molecular interactions of the polymer, which affect itsphysical-mechanical, electrical and other properties. Corrosion resistance of the metal-polymer joints is affected significantly by electrochemical interactions between the components due to the presence of the areas of dissimilar electrode potential on the metal substrate [17]. Physical-chemical interactions between components of the metal-polymer system influence the two main factors that condition reliability and protective ability of the polymer coatings: firstly, penetrability of aggressive media initiating corrosion processes and metal failure, and secondly, physical-mechanical barrier properties and adhesive strength of the polymer coatings.

Delamination of the polymer coating under the action of aggressive media is accompanied by a shift of the substrate electrode potential into the region of negative values, so-called downgrading. Variation of the substrate potential at coating peeling proves that either the cathodic reaction of oxygen under the coating is retarded or the process of metal ionizing is accelerated [18]. It is possible to control both anodic and cathodic processes via regulation of the electrode potential of the metal substrate.

To investigate the coating adhesion stability one can use a method during which the coating delamination rate is recorded around a handmade point defect (puncture) reaching the substrate [19].

Kinetics of electrochemical processes in the electrode-coating-electrolyte systems has been examined by the potentiostatic method within the ± 4 V exponentiation range. The samples were polarized in a three-electrode electrochemical cell installed with a thermostat to maintain a constant temperature regime. The working electrode was a sample in the form of a metal (or graphite) plate with a one-sided coating.

The diagrams proving that the delamination kinetics of polymer coatings is dependent upon the potential value and sign are presented in Fig. 1.8. The difference in the location of minima on the curves corresponding to graphite and steel means that the delamination process is interrelated with redox reactions on the substrate surface inducing the formation and failure of oxide films. Thus, by alternating the substrate electrode potential, one can regulate the protective capability of the polymer coatings in electrolytes.

Moreover, corrosion processes may run on metals covered with varnish – and paint materials in an atmosphere and lead to the formation of numerous tortuous filiform areas of damage. This type of failure, called *filiform corrosion,* is a specific form of crevice corrosion. The filiform type of corrosion does not, as a rule, lead to metal failure but only impairs its appearance. Filiform

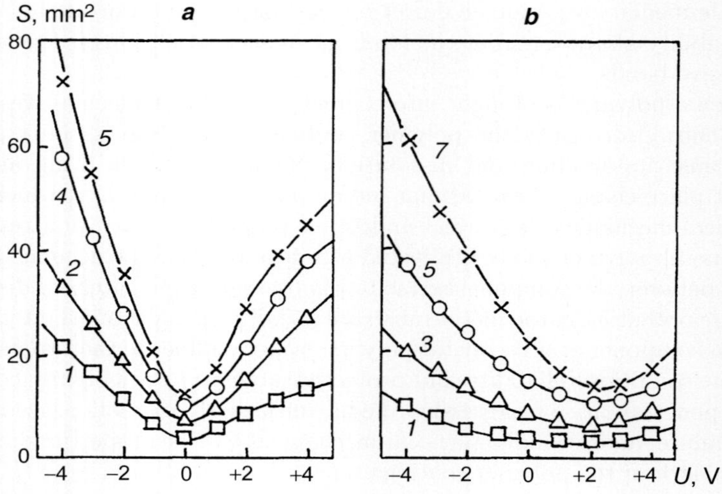

Fig. 1.8. Delamination kinetics of polyethylene coatings from graphite (**a**) and steel (**b**) substrates in a 5% solution of H_2SO_4 versus potential value U fed to the substrate and exposure time in electrolyte (numbers on curves indicate exposure time in hours)

corrosion on steels is exhibited as a network of 0.1–0.5-mm-wide thread-like corrosion products [2]. The color of the threads is typical of Fe_2O_3, while their heads are green or blue, which indicates the presence of the bivalent iron.

The filiform corrosion is induced by the generation of electrochemical elements of differential aeration in the corrosion system. The filiform corrosion usually occurs in a medium of high relative humidity (65–95%). At 100% humidity the threads widen and can swell the coating. Phosphating of the metal surface and priming with zinc chromate assists in diminishing the filiform corrosion, although not averting its emergence.

To combat the crevice corrosion one should seal gaps and clearances by metal materials not provoking corrosion of the basic material and apply electrochemical protection and corrosion inhibitors [13, 20].

1.2.2 Polymer Materials as a Source of Corrosion Agents

Comprehending corrosion as failure of materials under the action of the environment, it would be justified to relate this term to fracture of polymer materials under the action of ambient factors, including elevated temperatures, ionizing radiation, UV radiation, ozone, mechanical loads, chemical and biological matter.

However, the term "corrosion" is not accepted in relation to polymer materials. Such terms as aging, decomposition and destruction are used for characterizing polymer fracture under external effects, where the first term is used in the most analogous sense to "metal corrosion". Usually, two types of processes are running during polymer aging: destruction and macromolecular cross-linking. Time makes the aging mechanism and the increase of intermediate and final products more complicated.

The aging reaction is often classified by the nature of its inducing agent and the character of its effect on macromolecules (thermal, oxidative, radiation, mechano-chemical). As a result of aging, polymer materials change their properties and composition. These changes are caused by physical-chemical reactions leading to the release of low-molecular-weight products, oxidation of macromolecules, formation of active fragments (macroradicals), macromolecular destruction and syneresis of liquid-phase components from the polymer composition. We will dwell upon just a few characteristic examples for different polymer classes out of the great variety of failure processes in polymer materials.

Destruction of macromolecules as a result of *chemical interactions* with the environment leads to the fracture of polymer materials with the formation of active components. Participation of individual macromolecules in destruction processes is dependent upon the amount of the constituent reactive groups. Introduction of substituents of different compositions into the polymer chain changes their stability in thermal oxidation reactions and in response to chemical reagents and moisture effects.

The presence of multiple bonds in the macromolecule backbone chain greatly reduces its resistance to external effects. Introduction of heteroatoms (nitrogen, oxygen, sulfur) impairs polymer resistance to chemical agents.

The polarity of macromolecules and the solvent determine the effect of organic solvents on polymer materials. Polymers containing polar groups are resistant to non-polar substances but can swell and be dissolved in polar substances. The polymer permolecular structure exerts a considerable effect on its stability under aging as well. Crystalline polymers dissolve slower than non-crystalline ones, which is due to different diffusion velocities of low-molecular-weight components in the crystalline and amorphous polymers.

A critical factor that influences significantly aging of polymer materials is *temperature*. The temperature effect leads to changes in the chemical structure of macromolecules and their permolecular packing. Destruction of a macromolecule is accompanied by formation of low-molecular-weight products like water, hydrogen chloride, carbonic oxide and other. The temperature effect in air results in a thermal oxidative destruction of macromolecules. More low-molecular-weight compounds are formed than in thermal decomposition in an inert gas or vacuum during thermal oxidative destruction of polyethylene. Thermal oxidative destruction is accompanied by formation of hydroxyl, peroxyl, carboxyl and ester groups. Saturated low-molecular-weight hydrocarbons along with unsaturated ethane-, ethylene-, propylene- and butane-type hydrocarbons, as well as carbon oxide, water, formaldehyde, fatty acids and ketones were found in the products of the polymer thermal oxidative destruction. This failure regularity is characteristic of other types of polymer materials as well.

Polymers containing halogens, including chlorine and fluorine, release highly active products like hydrogen halide and some other low-molecular-weight compounds during thermal oxidative destruction of polyethylene. For instance, the thermal effect in air on polyvinyl chloride leads to the release of hydrogen chloride, while the polymer sample changes icolor and becomes cross-linked. Polymer materials with side substituents (polyvinyl acetate, polyvinyl formal) decompose under heating in oxygen and release acetic acid, ketones and carbon dioxide. They form cyclic compounds, carbonic oxides, water and ammonium during fracture of polyamides. Decomposition of polybutylene terephthalate in air is accompanied by formation of carbonic oxides, terephthalic acid, acetic aldehyde and other low-molecular-weight compounds.

Silicone-organic polymers are more resistant to elevated temperatures. As a result of thermal-oxidative destruction, they release low-molecular-weight products like water and formaldehyde.

In fact all polymer materials release hydrogen during thermal-oxidative destruction.

Thermal-oxidative destruction of polymer materials is accompanied by the formation of corrosion-active components that may essentially change corrosion processes in the polymer-metal contact.

An important factor leading to alteration of polymer characteristics is the *mechanical effect*, e.g. milling, attrition, rolling, mixing or ultrasoound. It is possible to break atomic links by mechanical action to form active macromolecular fragments and release low-molecular-weight products. Mechanical destruction follows the radical, ionic or mixed mechanisms depending on the ambient composition. A specific fracture mode of polymer materials is *stress-induced aging*. The effect of mechanical stresses intensifies oxidation processes, thermal oxidative destruction and crack propagation in materials.

Photochemical effects on polymer materials activate macromolecules through the adsorption of energy quanta followed by their dissociation into active components. The activated molecules are capable of various interactions with other molecules.

Photochemical destruction of polymers runs together with formation of oxide compounds, low-molecular-weight and structurized products. These results in the generation of cracks on the surface of polyolefin articles, and their mechanical and dielectric properties are noticeably impaired.

As a result of photochemical destruction, the halogen-containing polymers split off low-molecular-weight products having analogous composition to those isolated during thermal-oxidative destruction. For example, as polyvinyl chloride breaks down, the release of hydrogen chloride is observed.

Polyimides show an increasing content of the monomer and raised brittleness under the effect of photochemical factors.

Splitting of hydrogen is observed under exposure of polymer materials to *ionizing radiation* along with formation of multiple bonds, release of low-molecular-weight hydrocarbons, destruction and cross-linking. Depending on the dominant process, polymer materials can be subdivided in their relation to ionizing radiation into two groups: (i) those in which the polymers form a spatial cross-linked structure, (ii) those in which the polymers are destroyed.

The ionizing radiation results in active macroradicals, oxide compounds and low-molecular-weight products. Thus, exposure of polyolefins in air is accompanied by generation of low-molecular-weight compounds and cross-linking. Fluorine is released under the action of ionizing radiation on polytetrafluoroethylene.

The corrosion-active products can be released from polymers together with technological low-molecular-weight additives like plasticizers and lubricants. Low-molecular-weight substances on the surface of polymer parts can enter into interactions with atmospheric oxygen, active components of the lubricating medium or dissolve in the working fluids, thus changing their composition and properties, particularly those that are corrosive.

Polymer materials can also breakdown under *the action of microorganisms.* Along with impairing the appearance, microorganisms worsen hygienic characteristics of materials based on polymers, lower their mechanical strength, wear resistance, dielectric and other properties. Products of microorganism vital functions released on the polymer surface change corrosion processes during the contact of biologically damaged polymers with metals.

Proceeding from the above, it can be concluded that polymers and compositions on a polymer base can become, under the action of operating conditions, the sources of products changing corrosion kinetics in the polymer–metal systems.

1.2.3 Polymer Effect on Electrochemical Polarization Kinetics of Metals

The efficiency of protecting systems based on polymers is conditioned largely by the barrier effect of polymer elements insulating metals from the environment. The electrochemical barrier, which is dependent upon the substrate polarization, can be far more efficient than the diffusion-mechanical one governed by the material permeability. It is feasible to attain a low rate of coating peeling and negligible corrosion damage of the substrate via its polarization even in the case of continuity violation of the polymer coating (crack, puncture).

It is recognized that once the barrier action of structural polymers is broken, they cannot impose any effect on metal corrosion in electrolyte solutions, although the linear-selective dissolving of the metal is attributed to specific properties of dielectrics in individual cases. Investigations of the polymer-wetting liquids are of high practical importance for forecasting electrochemical stability of metal-polymer systems.

The authors have studied the effect of polymer powders and films on corrosion rate of steels and their physical-chemical properties in aqua solutions of electrolytes. The experimental method was as follows.

One of two platinum electrodes was immersed into an NaCl solution (10^{-4} mol/l) to a fixed depth. The immersion depth of the second electrode was a variable factor. Kinetics of the second electrode potential variation towards that of the silver chloride was registered and compared to the case when a PVB powder was placed on the solution surface (dispersion degree 160–200 μm, powder layer density 4 mg/cm^2).

Application of polymer powders on the solution surface leads to the generation of an electrode potential gradient in solution depth (Fig. 1.9). The strongest reduction of the potential in time was recorded at a minimal (15 mm) distance of the electrode from the powder layer on the surface. With increasing immersion depth, the potential variation becomes less prominent and is not registered at 60 mm depth.

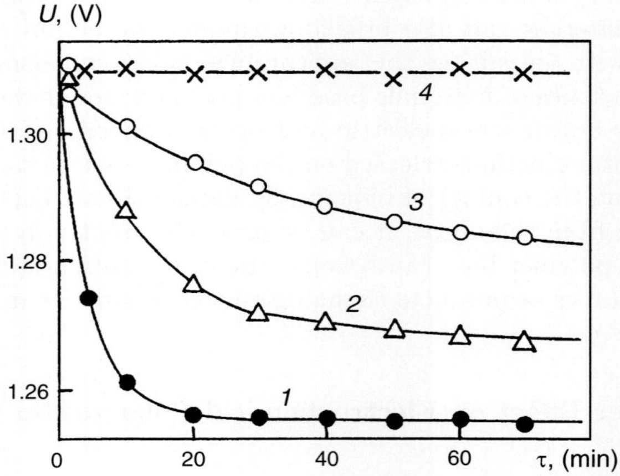

Fig. 1.9. Dependence of platinum electrode potential (U) in NaCl solution with PVB powder on its surface versus time (t) and immersion depth of the electrode (mm): (1) 15; (2) 30; (3) 45; (4) 60. Polarization current density is +0.1 mA/cm^2

Based on the known representations of the electrode processes, the derived data can evidently be related to a considerable change of the local depolarizer concentration in the vicinity of the upper electrode as compared to that at a certain depth; this is because of the polymer powders placed on the solution surface. This phenomenon can apparently be attributed to diffusion of water-soluble additives from the powder into solution as well as adsorption of the depolarizer on the polymer particles.

Potentials of the electrodes immersed into aqua solutions are changed significantly during cooling down to freezing in the presence of polymer powders on the liquid surface [21]. It was established that PE, PVB and PPl powders did not in fact change polarization kinetics in distilled water but their effect was perceptible in diluted aqua solutions of electrolytes (10^{-4} mol/l). The polarization stress varies insignificantly in more concentrated solutions (0.1 mol/l and higher) when a powder is placed on the electrolyte surface. This fact speaks in favor of the adsorption mechanism of the polymer effect on metal polarization in electrolyte solutions.

The contacting metals are subject to a specific effect due to the polymer materials. This effect should be considered during the design of anticorrosion metal-polymer parts and joints and when predicting their service parameters.

1.2.4 Permeability of Polymers to Electrolytes

One of the reasons for local corrosion at the metal-polymer interface is sorption of electrolytes by polymers and permeability of the polymer barrier towards electrolytes. Sorption of electrolytes (acid solutions, bases and salts) leads to essential variation in the service characteristics of the protecting polymer coatings and anticorrosion packaging films under mechanical loads. These variations under mechanical loads, especially in seals and friction joints, are much deeper and can affect mechanisms of contact interactions.

Transfer of aqua electrolytes in polymers has characteristics [22] that distinguish it from the transfer of other low-molecular-weight matter. The presence of water in the medium sorbed may result in formation of aqua associates and electrolyte in the polymer. Depending on the amount of sorbed water, the polymers are subdivided into three groups: (i) hydrophobic, i.e. slightly swelling in water (water concentration in the polymer below 0.5%), (ii) mildly hydrophilic (water concentration 0.5–10%), (iii) hydrophilic (water concentration above 10%). This subdivision is very conventional since the permeability mechanism depends not only on water concentration in the polymer but on the nature of their physical-chemical interactions with electrolyte ions and macromolecular active centers as well [23].

Hydrophobic polymers (polyolefins, polystyrene, polyethylene terephthalate, pentaplast, fluoroplasts) contain small amounts of polar groups in their macromolecular composition able to interact with water and electrolytes. To that extent, the permeability mechanism of electrolytes in hydrophobic polymers is similar to that of gases.

The comparison of permeability values of low-molecular-weight substances diffusing into hydrophobic polymers from water solutions to those from the dry gaseous phase (e.g., HCl, NH_3, CO_2, SO_2) showed that these values were commensurate, providing that the vapor elasticity over the solution and its partial pressure in the gaseous phase were equal. It is believed that volatile electrolytes are transferred in hydrophobic polymers in the form of non-dissociated molecules devoid of hydrate shells.

It is important for hydrophobic polymers that their permeability to nonvolatile electrolytes is at least three orders of magnitude lower than that for volatile electrolytes. Low permeability for nonvolatile electrolytes is related to their low sorption by the non-polar polymers.

Liquid electrolytes display significant differences in permeability values. For instance, permeation of hydrogen chloride through polyethylene film when diffusing from concentrated hydrochloric acid is detected in a few minutes, whereas permeation of potassium chloride is not recorded even after three months. Permeation of nitric acid through fluoroplastic film is recorded in a few tens of minutes, while it is necessary to wait more than a year for sulfuric acid to be detected [22]. With increasing concentration of the electrolyte its permeation rises. Polymers whose electrolyte diffusion factor is $D < 10^{-17} \mathrm{m}^2/\mathrm{s}$ are considered to be practically impermeable. Polyolefins, fluoroplasts and polyesters are easily permeable to hydrochloric, hydrofluoric, nitric, acetic and fluorosilicic acids, and ammonium diffusing form aqua solutes. They are less permeable for sulfuric and phosphoric acids, salts and caustic alkali. Phosphoric acid easily diffuses into PVC as well. In this case [23], diffusion of the acid is conditioned by the presence of a plasticizer in the polymer.

The effect of technological additives on permeability of polymers is connected with variations in their sorption capacity, formation of defects and interactions of the electrolyte and additives. Impregnation of fillers improves, as a rule, permeability of polymers and intensifies clusterization of water and the penetrant. When polyethylene is filled by talc, HCl and H_2O clusters formed in the polymer can be observed in microscope. Water and HCl sorption increases proportionally to the volume content of talc up to 17% concentration. Further increase in talc concentration does not result in sorption growth because of filler particle aggregation in the polymer binder.

The permolecular structure of polymers exerts a perceptible effect on permeability. Thermoplasts whose structure is not spherulitic and materials with a coarse-spherulitic structure are highly permeable because of different development levels of the microcapillary system of permolecular formations.

Mildly hydrophilic polymers (weakly swelling in water) are characterized by a limited amount of polar groups in macromolecules. Their typical representatives are aliphatic polyimides (water absorption at room temperature is 2–12%), epoxy resins, collagen and polyester resins. Swelling, dilution and destruction processes take place in mildly hydrophilic polymers during sorption and electrolytic diffusion. During sorption, geometrical parameters of samples vary continuously in these systems and affect clearances, seals and tribojoints. Moreover, surface concentration of the electrolyte, water absorption, crystallinity degree and molecular mass (at chemical destruction) also vary. Diffusion of electrolytes in polymers of the above class will be exemplified on PA [22, 23].

Acid dissociates at low acid concentrations inside the polymer and its ions start to bond with the charged end groups of macromolecules. Concentration dependencies of the diffusion factor of the electrolyte and self-diffusion factors of individual ions can be explained by saturation of the bonding centers. As a result of the diffusing ions' interaction with the polymer polar groups (centers) this dependence is observed at the initial stages of sorption of the ionic mobility on the filling degree of the adsorption centers. The sorbed ions start to bond with polyamide end groups $-NH_3^+$ and $-COO^-$ with increasing electrolyte (acid) concentration in the polymer. Their electrostatic and chemical interaction with the macromolecule attenuates with further diffusion of ions. This results in an abrupt increase of the diffusion factor of the acid on transition through the critical region (equal concentration of the bound ions and binding centers).

Volatility of the electrolyte exerts an insignificant effect on permeability. The penetration depth of acidic solutions into epoxy resin diminishes as follows: $H_2SO_4 > HNO_3 > HCl$. The permeation depth of alkali is significantly (about ten times) less than that of the acids. The concentration dependence of the diffusion factor of an acid depends on its and the polymer's origins. Thus, with growing concentration of HCl in a solute its permeation depth grows too. The diffusion factor of H_2SO_4 in a polyester resin reduces at 60°C from $3 \cdot 10^{-7}$ to $4 \cdot 10^{-18}\ m^2/s$ as the acid concentration in the solute increases from 1 to 25%. Diffusion of H_2SO_4 into epoxy resin set by amine hardener is accompanied by binding and the permeation depth changes by more than a factor of 10 when the hardener content doubles.

The permeability of salt solutes (NaCl, KCl) in epoxy resins corresponds roughly to that of nonvolatile solvents: $10^{-15}-10^{-19}\ g/(m \cdot s \cdot Pa)$ in hydrophobic polymers.

Hydrophilic polymers contain a commensurate amount of hydrophilic and hydrophobic fragments in macromolecular composition, including proteins, some polyurethanes and acrylate copolymers. Water is a good solvent for hydrophilic polymers, and therefore it often violates the original structure of the polymer on sorption of aqua solutions. Non-polar fragments, due to their macromolecular flexibility on sorption of aqua solutions, can associate and form areas of elevated hydrophoby to hamper the electrolyte motion. Hydrophilic sections of the chain encircled by the solvent are responsible for the electrolyte transfer. The majority of polymers cannot form fully hydrophobic areas because their polar and non-polar sections in macromolecules are frequently varying.

Films of hydrophilic polymers are the traditional objects of research into electrolyte transfer in polymers. For a long period this process was described as diffusion of water molecules through pores in the films. Selectivity of membranes towards the penetrants was attributed to the presence of charges on the pore walls impeding diffusion of similarly charged ions.

In the 1960s the concept of electrolyte transfer was recognized as a process of diffusive penetrability consisting of sorption, diffusion in the polymer and desorption stages [24]. The main idea contends that the electrolyte should dissolve in the polymer matrix to exercise the transfer. Its concentration in the polymer is defined by the distribution constant of the electrolyte between the polymer phases and the contacting solution. By analogy with gas penetrability, the transfer of electrolytes can be characterized by the permeability ratio P

$$P = DK\ , \tag{1.4}$$

where D is the diffusion factor and K is the distribution constant.

In its physical sense the permeability ratio of electrolytes is the amount of a substance transferred per unit time through a unit surface of a polymer film of a unit thickness with the electrolyte activity in the outer solution equal to a unit. The dimensionality of the permeability ratio coincides with the diffusion factor since the distribution constant is a dimensionless quantity.

Since solubility of electrolytes in hydrophilic polymers essentially depends on the presence of water, an important characteristic of transfer is the volume fraction of water contained in the polymer [20]. Localized water inclusions, sufficient to solvate the electrolyte and raise its solubility, are formed in the polymer matrix. In the case of a high water concentration in the polymer, the concentration of the electrolyte in water can become equal to its concentration in the ambient solution.

The diffusion factor of the electrolyte is a function of water content in the polymer. To describe the dependence of the diffusion factor on the volume fraction of water in the polymer, the representation of the free volume theory is successfully used [25, 26]. Providing that the free volume of the swollen polymer is proportional to the volume fraction in water, a simple equation relating the diffusion factor of the electrolyte in the polymer D and water D_0 will be

$$\mathrm{Lg}\, D = \lg D_0 - K\,(1/H - 1)\ , \tag{1.5}$$

where H is the volume fraction contained in the water.

Equation (1.5) describes satisfactorily the experimental data. Penetrability of hydrophilic films of cellulose acetate towards NaCl increases sixfold from $H = 0.08$ to 0.56 when wetted [27].

Hence, polymer structure and properties change under the effect of sorbed electrolyte. Consequently, permeability of polymers towards electrolytes can be chosen as a critical factor visualizing the intensity of corrosion processes in the polymer–metal contact.

1.3 Components of Inhibited Plastics

Inhibited plastics are composite materials (CM) consisting of inhomogeneous components (phases) separated by interfaces. They contain thermoplastics or thermosetting plastics, or their blends as a matrix (binder) representing an uninterrupted component within the CM. The matrix adds the necessary configuration and continuity to CM, transfers and redistributes loads over the material and defines its major physical and chemical properties [28–31]. The polymer matrix of inhibited plastics is a carrier of the functional component, a CI, which protects conjugated metal parts from corrosion. The degree of CI binding with the polymer carrier depends on the thermodynamic compatibility of CI, the binder and other components as well as on the material structure. Both the micro- and macrostructure of CM can be varied within a wide range by altering their formulation and production process. The inhibited plastics contain, along with CI, ingredients that impart certain service characteristics and alleviate processing of the composites [32–34].

When choosing the components of inhibited CM, one should take into account:

- operation conditions and corresponding anticorrosion, deformation and strength, diffusive and other characteristics contributing to durability of the composite article,
- process behavior of components making possible their joined processing in the material,
- economical aspects, such as demands, anticipated production volume and scheduled service life, raw materials and their cost, available processes for using secondary materials and wastes, presence of special equipment and rigging and necessity of their retrofitting,
- social factors connected with conditions and safety of production and application of the material and its products, process impact on the environment and means of recycling industrial and domestic waste.

1.3.1 Binders for Inhibited Plastics

Plastics are materials based on polymers that are found in the plastic or viscous-flow state during processing into articles and are solid, glassy and/or in a crystalline state when they are final products [29, 30, 35].

The thermoplastic binder is chosen for the purpose of acting as an inhibited CM and is aimed at reaching the required service characteristics of the final product.

The range of plastics used as (matrix type) binders for inhibited CM is extremely broad and includes practically the whole list of commercial thermal and thermosetting plastics [36–45].

Thermoplastics are based on linear or branched polymers, copolymers or their blends that are reversibly transferred at heating into a plastic or viscoplastic state as a result of melting of the crystalline and/or softening of the amorphous (glassy) phase [29]. Inhibited CM based on thermoplastics are largely adopted in anticorrosion techniques. Most thermoplastics are produced at a large scale in petrochemical enterprises and are comparatively inexpensive. The impressive list of thermoplastic brand names whose properties can be varied depending on the polymer nature furnishes in fact an unlimited choice of materials for specific applications. Thermoplastic materials are highly effective since goods made from them are manufactured by simple, highly productive, wasteless and economical methods (injection molding, extrusion, thermal forming, etc.). Thermoplastic articles and semis can be further machined, e.g. cut or chopped, welded or glued. Thermoplastics, with rare exceptions, are low toxicity, safe, easily modified and may be subjected to secondary processing.

Polymerizable (polymerization-synthesized) thermoplastics based on flexible chain (mainly carbochain) polymers have become very popular. They include polyethylene (PE), polypropylene (PP), poly-4-methyl-1-pentene and their chlorinated derivatives–chlorosulfonized PE, vinyl polymers (polyvinyl chloride (PVC)), polystyrene (PS), polyvinyl acetate (PVA), polytetrafluoroethylene (PTFE), as well as polyacrylates – polymethacrylate, polymethyl methacrylate (PMMA) and ethylene copolymers with *a*-olefins, such as copolymers of ethylene and vinyl acetate (CEVA), copolymer of ethylene with vinyl alcohol, and so on. As for the polymerizable heterochain plastics, the most applicable binder of inhibited CM is apparently polycaproamide (PA6).

Both flexible and rigid heterochain polymers subjected to *polycondensation* (polycondensation synthesis) are used today in rather large volumes for manufacture of inhibited plastics. The list includes polyethers (polyalkylene and polyphenylene oxides, pentaplast) and polyesters (polyethylene terephthalate, polycarbonate (PC)), polycondensated polyamides (PA6, 6), polyvinyl acetals (polyvinyl formal (PVF), polyvinyl butyral (PVB), polyvinyl formal ethylal (PVFE), polyimides (PI), polyurethanes (PU) and other.

A separate group of thermoplastic binders for inhibited CM form artificial polymers, in particular plastics produced by chemical processing of cellulose. They are largely cellulose ethers (etrols), namely ethylates, acetobutyrates, acetates.

Thermosets are based on liquid or solid active oligomers (resins). The resins are able to convert into a viscous-flow state on heating and transforming into densely netted glassy polymers when set at elevated temperatures and/or in the presence of hardeners and irreversibly lose their ability to transfer into the viscous-flow state. Hardened thermosets become infusible and insoluble [29, 30]. The advantages of thermosets as CM binders over thermoplasts lie in their ability to regulate the viscosity, wettability and saturation level of the fillers. As for the drawbacks, they mainly arise from exothermal effects, volume shrinkage and liberation of volatile matter on setting, leading to defects, shape instability and elevated brittleness of the articles made from them. Processing techniques of thermosetting plastics are rather complex; they are energy and labor intensive, inefficient and often unsafe.

Thermosets are incapable of remolding and welding at the stage of hardening. Salvaging of thermosetting products brings difficulties too since their parts are usually glued to each other or bonded by mechanical methods. That is why thermosets are used far less than thermoplastics as CM binders. They commonly serve for fabricating articles for long-term usage to ensure durable characteristics.

Depending on the type of active oligomers that form the base of thermosets, they are subdivided into phenoplasts (based on phenolformalfdehyde resins (PFR)), aminoplasts (based on urea and melamine-formaldehyde resins), epoxyplasts (on the base of epoxy resins (ER)), etheroplasts (based on acrylic acid derivatives – oligomers), imidoplasts (based on oligoimides or blends of imide-forming monomers), organosilicone polymers (based on organosiloxane oligomers), alkyd plastics (based on alkyd resins, such as pentaphthalic, glyptal).

The main process and service characteristics of polymer binders (thermoplasts and thermosets) of inhibited CM as well as the list of goods of anticorrosion purposes are presented in Table 1.2.

Properties of the binders for inhibited plastics are also described in Sects. 2.1, 3.5, 4.3, and Chap. 5.

1.3.2 Corrosion Inhibitors

CI are chemical compounds or their blends able to retard metal corrosion even in small additions in aggressive media. CI change the kinetics of electrochemical reactions that bring about corrosion so that the corrosion process rate retards significantly.

Commercial CI include more than three thousand brands differing in their purpose, physical-chemical properties and mechanisms [12, 46–48].

By their ***designation*** CI are subdivided into five classes: inhibitors in solutions of acids, alkalis, water and neutral water media, in non-aqueous liquid media and the inhibitors of atmospheric corrosion.

By their ***mechanism of influencing the velocity of electrochemical reactions*** CI are divided into three groups: inhibitors hampering only anodic (anodic CI) or only cathodic (cathodic CI) reactions and mixed CI that abate both reactions.

By their ***physical-chemical properties and protection principle*** there are volatile, contact and creep inhibitors of atmospheric corrosion. *Contact* CI are the substances whose protecting effect is exhibited only on direct application on the metal surface. Many of them migrate via diffusion and undergo creeping. *Volatile* CI with relatively high volatility can easily transfer into vapor and be adsorbed onto metal surfaces to protect them from corrosion.

As for the ***chemical structure*** (the most universal criterion) CI are divided into organic and inorganic.

Table 1.2. Types of polymer binders for inhibited CM and their chief processing and service characteristics

Plastics	Processing Characteristics of Plastics	Service Characteristics of Plastic Products	Range of Rust-Inhibiting Products
Thermoplastics			
PE	Processed into articles at 120–280°C. Readily film forming, hot-sealed, modified and reprocessed.	Mechanical strength, elasticity, chemical and frost resistance as well as resistance to oxidation and moisture are satisfactory. Non-toxic, good dielectric parameters.	Packing films, insulation adhesive tapes, coats, lining, melt-blown vessels, containers, structural products, electroinsulating seals, self-lubricating antifrictional materials.
PP	Injection-molded into structural goods and extruded into films, tubes, fibers.	Low specific mass, high physic-mechanical characteristics (impact viscosity, resistance to multiple bending, wear resistance), satisfactory thermal, chemical and water resistance, low resistance to frost and oxidation, good dielectric properties.	Packing films, sheets, coats, fibrous materials, pipes, machine parts, containers, electric insulation.
PVC	Can be extruded and rolled. Is insufficiently heat resistant with flow temperature – 150–220°C, can be dehydrochlorinated at 150–180°C, should be combined with thermal stabilizers, can be easily plasticized to acquire elasticity.	High mechanical strength, water repellency, solubility in organic solvents. Resistant to acids, alkalis, salts, to atmospheric conditions and fungi, combustible with difficulty, has good dielectric properties.	Films, coatings, lining, sheets, continuous profiles, pipes, melt-blown containers, insulation.
PS	Perfectly injection-molded or extruded at 190–230°C, stuff	Perfect dielectric, resistant to moisture and radiation, acids	Structural purpose products, films, coatings, containers, electric insulation

	colored, flow index varies from 2 to 30 depending on the technique, is film forming. Requires light stabilization and combination with rubbers to elevate impact and atmospheric resistance (impact resistant PS).	(except for concentrated nitric and acetic acids) and alkalis. Subject to photo-aging at UV radiation in air (turns yellow, undergoes embrittlement and cracking). Insufficiently heat resistant, looses shape at above 60°C), is brittle. PS modified by rubbers is sufficiently resistant to impact.	and decorative materials for radio and electronics, instrument making and engineering branches, foam materials.
PTFE	Processed by powder sintering, fusing of suspended particles, injection molding, molding, extrusion, welding, machining.	High mechanical and electrical resistance, wide operation temperature range (from 260 toi +260°C), low friction coefficient. Surpasses all known materials in chemical resistance. Resistant to atmospheric, corrosion and radiation effects.	Veils, films, coats, fibrous materials, pipes, insulation, rustproof containers, facing of chemical reactors, structural and antifrictional machine parts, seals.
PVA	Processed in water dispersions (films, coats) and solutions of organic solvents (glues), easily plasticized.	Cold-flow, resistant to atmospheric effects, is hydrolyzed by alkali solutions, and decomposes at above 180°C, isolating acetic acid.	Water paints, spackling, glues
Polyacrylates (PMMA and other)	Film-forming properties, low flow when softened. Easily plasticized and cross-linked at above 150°C, isolating volatiles.	Perfect transparency, is highly elastic, resistant to light, oxygen, diluted acids and alkalis. Does not withstand alkali attack under elevated temperatures (hydrolyzes till polyacrylic acid). Satisfactory mechanical resistance, low thermal resistance (70–80°C).	Sheets, films, paint coats (enamel, primer, filling), glues, impregnation agents, organic glass.

(*continued*)

Table 1.2. cont.

Plastics	Processing Characteristics of Plastics	Service Characteristics of Plastic Products	Range of Rust-Inhibiting Products
Polyformaldehyde (polymethylene oxide)	Injection molded at 190–240°C, extruded and molded. The polymer with –OH end groups undergoes thermooxidative destruction at 100°C, with blocked end groups at 160°C. Can be combined with thermal stabilizers, quickly crystallizes (the molds need heating)	Restricted heat and chemical resistance (swells in organic solvents, unstable to strong acids and alkalis). Inflammable, resistant to alternating loads. Its products display shape stability, low creep, high wear resistance, and low friction coefficient, mechanical and dielectric properties are insensitive to humidity.	Structural goods for engineering, automobiles and instrument-making, electric engineering, friction parts.
Phenol ether resin	Injection molded at 320–340°C and extruded at 240–300°C into articles, and films by calendering and casting.	Resistant to temperatures and chemical agents, including water, overheated vapor, solutions of alkali, mineral acids and salts, microorganisms and radiation. Decomposes in aliphatic hydrocarbons, has high mechanical and dielectric strength, stiffness, dimensional stability, resistance to sparking, self-extinction when removed from fire.	Pipes, profile articles, structural, antifrictional and electroinsulating units, automobile parts, electronic and radio engineering products, paint coats, heat-resistant foam plastics.
PETP	High thermal resistance (~ 290°C), mainly extruded and injection molded into products	High strength and dielectric characteristics. Frost-resistant (–60°C), heat- (170°C), bio- and chemically resistant, withstands moisture, acids and diluted alkalis, difficulty soluble in organic solvents, is optically transparent with low light stability.	Packaging films, containers, pallets, coats, fibrous materials

PC	Low melt flow index. Processed into products by injection molding at 230–310°C, extrusion, pressing, solution casting.	High deformation and strength characteristics (impact viscosity, hardness, bending resistance, and elasticity), dimensional stability and low creep. Resistant to heat (115–130°C), frost (–100°C) and chemical agents (water, salt solutions, diluted alkalis, hydrocarbons, alcohols). Transparent, self-extinctive, has satisfactory dielectric properties.	Films, sheets, structural articles.
Aliphatic PA	Injection molded, pressed, calendered or extruded into products. High melting point and narrow thermal plasticity range. Inclined to thermo-oxidative destruction, good at cold drawing, poorly compatible with plasticizers and stabilizers.	High strength (resistance to impact loads) and dielectric properties, low friction coefficient, high heat, oil, gasoline and biological resistance. Can be hydrolyzed at elevated temperatures by acids and alkalis. Unstable to UV and oxidizer effects.	Films, coatings, glues, electric insulation, varnish, fibrous materials, sealing, structural and antifrictional parts.
Polyvinyl acetals (PVF, PVB, PVFE)	Extruded, injection molded and cast in solution. Perfect film-formers, easily plasticized.	Elastic, perfectly soluble in organic solvents (except for PVF). Resistant to sunlight, oxidizers, low thermal resistance, elevated adhesion to metals, glass, and plastics.	Glues, adhesive layers of laminated films, easily peelable coats, electroinsulation enamels, profile articles.

(continued)

Table 1.2. cont.

Plastics	Processing Characteristics of Plastics	Service Characteristics of Plastic Products	Range of Rust-Inhibiting Products
PU (linear)	Are processed by reactive or reactive-injection molding, injection molding, rolling (thermoelastoplasts), solution and dispersion casting, form stable foams (foam materials are produced from honeycomb PU, i.e. thermosets)	Highly strong, hard and elastic. High wear, water, oil, gasoline, atmospheric and radiation resistance. Unstable at elevated temperatures and to alkali effects. High adhesion to the majority of materials. Physical-mechanical characteristics are dependent on temperature differences, accumulation of residual deformation at prolonged loading, are combustible and expensive	Coatings, films, glues, sealants, electric insulation, structural articles; packing, heat and noise-insulating foams.
Ethyl cellulose	Injection molded, molded and extruded into articles. Readily plasticized, machined, and glued by solvents.	Operation temperature range is within –60 to 80°C. Resistant to the effects of water, acids, alkalis, petrol, oil products, and atmospheric conditions. High impact resistance, electric stability, adhesion to different substrates and gas permeability.	Films, paint and extruded coats, glues, electric insulation, structural products.
Cellulose acetates	Film and fiber-forming. Mainly extruded and cast into articles. Low thermal stability (thermal destruction initiation interval is within 190–210 °C). Require combining with thermal stabilizers. Compatible with plasticizers.	High strength, luster, transparency, low deformability, shape stability. Stable to the effects of water, diluted acids, gasoline, oils, and alcohols, are saponified by alkalis, and are incombustible.	Packaging films, paint coats, fibrous materials, structural articles.

Thermosets			
Phenolic resins (phenol formaldehyde, phenol furfural)	Processed into articles by curing at heating. High adhesion to fillers, well impregnated and soluble.	Good mechanical strength, wear resistance, low friction coefficient. Perfect dielectric properties, resistance to heat, fire, oil, gasoline, acids and corrosion. High adhesion to metals.	Engineering goods operating in hostile environment, antifrictional machine parts, elements of radio equipment, facing, construction units, paint coats.
Epoxy resins (epoxydian, epoxy furfural, epoxyethereal)	Processed into products by hot and cold cure using hardeners (di- and polyamines, polyatomic acids, alcohols, PFR), have good wettability, adhesion to fillers.	High adhesive, physic-mechanical (strength, wear resistance) and triboengineering (antifrictional) characteristics, thermal and chemical resistance and stability in humid atmosphere.	Paint coats, glues, sealants, potting compounds, friction parts, electric insulation, very strong structural articles
Organosilicon liquids (poly-organosiloxanes)	Cold set.	High elasticity, thermal and atmospheric resistance, hydrophobic, good dielectric properties.	Structural articles, glues, paint coats, lining.

The inhibiting effect of *inorganic* compounds is attributed to the presence of some cations (Ca^{2+}, Zn^{2+}, Ni^{2+}, As^{2+}, Sb^{2+}) and anions (${CrO_4}^{2-}$, ${Cr_2O_7}^{2-}$, ${NO_2}^{-}$, ${PO_4}^{3-}$, ${SiO_3}^{2-}$) [43,47,48].

Inorganic cathodic CI are able to restrict corrosion rate of metals via diminishing oxygen concentration in the corrosive medium (sodium sulfite, hydrazine), formation of screening insoluble residues (calcium bicarbonate, zinc sulfate and chloride for steel) on cathodic sites of the metal or by raising overstrain of the cathodic process as a result of reduction of metal cations and their deposition on the metal surface that is being protected (salts of heavy metals, such as antimony, bismuth, arsenic).

Inorganic anodic and mixed CI present water-soluble salts of ammonium, alkali (sodium, potassium), alkaline-earth (calcium, strontium, magnesium, barium) and other (zinc, lead) metals containing active anions in their molecules (nitrates, nitrites, chromates, carbonates, phosphates, molybdates, silicates).

Inorganic anodic passivators impregnated into a hostile medium retard the anodic reaction of metal solvation. *Anodic passivators* represent salts containing oxidizing anions (nitrites, chromates) that replace the anions of aggressive electrolytes (sulfates, chlorides) from metal surfaces in the course of competing adsorption and enrich the metal electrode potential, shifting it to the region of anodic passivation. Thus, adsorbed anions impede anodic solution by oxidizing the metal to largely insoluble oxides, forming an impermeable protecting film on the metal surface. *Mixed passivators* are inorganic salts of weak acids and strong bases (carbonates, phosphates, silicates, borates) subjected to hydrolysis in aqua media. The hydrolysis generates hydroxyl ions, promoting the formation of passivating oxides on cathodic and anodic sites of the metal.

Film-forming anions (phosphates, polyphosphates, and carbonates) can also form sparingly soluble salts on the metal surface that hamper corrosion propagation. These anions can become an active part of rust converters that transform corrosion products into insoluble compounds. For instance, orthophosphoric acid transforms rust into insoluble iron phosphates.

The *protecting* CI form a separate group (e.g. metal pigments like zinc, aluminum, copper) that promotes cathodic polarization of the metal with a more positive potential. These CI liberate electrons while dissolving in electrolytes and shifting the metal electrode potential up to a negative value, at which the anodic reaction is suppressed fully.

Inorganic CI are commonly used as a functional component of inhibited polymer coatings (see Chap. 3), and less often as inhibited films (sodium nitrate, protecting metal pigments, sodium-borosilicate layers of films).

Most *organic* CI realize their inhibiting effect through their sorption on the metal surface and the formation of adsorbed protecting layers that insulate the surface from the hostile medium. This is why organic compounds are usually mixed effect CI. Functional derivatives of aliphatic, alicyclic, aromatic and heterocyclic compounds having adsorption-active atoms of nitrogen, oxygen, sulfur and multiple bonds also display CI properties. Molecules of organic CI frequently contain not one but a few functional groups. The efficiency of the corrosion-inhibiting effect of organic CI depends on the electronic structure of the functional groups (defining the character of chemisorptive bonds with the metal) and the size and conformation of their molecules responsible for the screening mechanism of adsorbed molecules.

Many organic CI are in fact salts consisting of organic cations (e.g. amine cations) that contain heteroatoms along with inorganic (CrO_4^{2-}, $Cr_2O_7^{2-}$, NO_2^-, CO_3^{2-}) and organic ($C_6H_5COO^-$, $NO_2C_6H_4COO$) anions. These compounds dissociate, as a rule, in aqua solutions of electrolytes forming active cations and anions capable of affecting the kinetics of electrode reactions in the necessary direction.

The CI as components of inhibited plastics can be of different organic brands [37–39, 41, 43, 44, 49–51], e.g.:

- aliphatic, cyclic, aromatic and heterocyclic amines, their derivatives and salts,
- amines, their derivatives and salts,
- morpholine, its derivatives and salts,
- carbonic acids, their derivatives and salts,
- nitric compounds,
- organic phosphates and phosphonates,
- different natural matter.

Cyclic amines are the most applicable functional components of inhibited plastics. They are first of all cyclohexylamine, dicyclohexylamine and their derivatives (nitrosodicyclohexylamine, cyclohexylamine), salts of cyclic amines with inorganic (nitric, chromic and carbonic) acids and organic (synthetic fatty, benzoic, nitric and dinitrobenzoic) acids as well as amines (N-cyclohexyl phenylimine) and their salts (methanitrobenzoate of hexamethylenimine).

The class of *aliphatic and aromatic amines* is presented by primary (octadecyl, isopropyl and allyloamines), secondary (tributyl, triisopropyl amines) and other amines as well as by benzyl amine, salts of aliphatic amines (diisopropylamine nitrite, octadecylamine salicylate, hexamethylene diamine chromate, ethyl and diethyl amine benzoate, 2-ethyl-hexyl-heptyl amine molybdate), heteroalkylated lower amines (aminoketones, aminonitriles CI of Russian grade IFHAN; amine alcohols).

Heterocyclic amines form a class of CI incorporating piperidines, pyridines, quinolines, imidazolines, pyrrolidines, azoles, oxazoles, triazoles and others. They display high thermal stability making possible their processing jointly with plastics. The most promising representatives of this class, due to a combination of service and toxicological properties, are the derivatives of the five-membered nitrogen-containing heterocycles of imidazole,1,2,3- and

1,2,4-triazoles, and tetrazole [52–59]. Benzo-1,2,3-triazole (Russian trademark BTA) was used as an inhibiting component of films and other plastic goods intended for protection against atmospheric and water corrosion for predominantly nonferrous metals [46]. Recently anticorrosion properties of a broad series of polynitrous heterocyclic compounds of imidazole, 3-R-5R'-1,2,4-triazoles and 5-R-tetrazoles were studied and modification methods of anticorrosion PE films by these CI were proposed [50, 51, 60].

There are numerous examples among the *carboxylic acids* of using synthetic fatty acids (SFA) of various fractions, as well as caproic, pelargonic, oleic, benzoic acids, nitrous and ammonium salts of benzoic and sebacic acids, salts of dicarbonic, nitrous and dinitrobenzoic acids as components of inhibited plastics.

The majority of CI of amine and organic acids are strongly volatile (Table 1.3) and are referred to as mixed volatile CI of atmospheric corrosion of adsorptive–activation and activation-blocking types.

Nitrobenzoate amines and nitrous derivatives of azoles form a separate group of CI. Owing to their high oxidizing capacity, they accelerate the cathodic process, which leads to shifting of the electrode potential of the corrosive system to values maintaining the passive state of the metal [47, 61]. Contact CI of nitrated paraffins, SFA, alkyl phenols, mineral oils, oxidized and nitrated petrolatum and ceresin are also widely used today [62].

Organic phosphorus-containing compounds (mono and dialkyl esters(C_2-C_{10}) of phosphoric acid, hydroxyethylidene-diphosphonic acid, nitrile-tris (methylene phosphonic acid) are mainly used as ingredients of polymer coatings. They are related to CI of the anodic-passivating type [63].

Natural CI working as plastic modifiers include oxidized petrolatum, ceresine, tannins and tall oil.

Characteristics of the most used CI as components of inhibited plastics are given in Table 1.3.

A broad spectrum of *synergetic mixtures* of CI is used to strengthen their inhibiting action and expand the range of metals protected [39, 46, 47]. To extend the life of inhibited plastics, they are modified by CI mixtures of different volatility. Highly volatile components of inhibitive mixtures intensify the formation of adsorptive CI layers on hardware surfaces prior to the onset of the corrosion process. Low-volatility CI are able to prolong the anticorrosion effect of the plastics.

The most important requirement imposed on CI as a component of inhibiting plastics is the thermal stability of the inhibitor at joint processing with the polymer. The upper temperature threshold of such processing depends on the phase transition temperature, that of thermal and thermooxidative destruction of CI, and intensive evaporation of volatile components of the inhibitive mixture. The chief criteria for CI adoption expediency as a plastic modifier is minimal loss of the volatile fraction and retained protective

Table 1.3. CI as components of inhibited plastics and their properties

Main classes and brands of CI (Russian and other trademarks)	Chemical formula	Appearance under normal conditions	Density g/cm^3	Pressure of saturated vapor at 20°C, Pa	T_m or T_b, °C	Solvents	Protected Metals
Inorganic CI							
Sodium nitrate	$NaNO_2$	White hygroscopic powder	2.17	–	271 (decomposition 320)	w, e, eal	Fe, steel, Al and its alloys
Ammonium carbonate	$(NH_4)_2CO_3$	White powder	1.73	– –	169 (decomp. 190)	w, eal, ac	steel
Strontium chromate	$SrCrO_4$	Lemon-yellow powder	3.75	–	decomp. 1000	w	Fe, steel, Al, Mg and Ti alloys
Zinc orthophosphate (hydrate)	$Zn_3(PO_4)_2{\cdot}8H_2O$	White powder	3.11 (15°C)	–		insoluble	Fe, steel
Sodium metasilicate	Na_2SiO_3	Colorless crystalline powder	2.61	–	1088	w	steel, Al and alloys, Pb, brass, Cu
Organic CI *Cyclic amines, imines, morpholine, their derivatives and salts*							
Cyclohexylamine	$C_6H_{11}NH_2$	Liquid	0.82	133	(134)	w, eal, e	steel
Dicyclohexylamine	$(C_6H_{11})_2NH$	Cold-crystallizable liquid	0.92 (13°C)		20 (254–255)	w, eal, bz	steel, Ni

(continued)

Table 1.3. cont.

Main classes and brands of CI (Russian and other trademarks)	Chemical formula	Appearance under normal conditions	Density g/cm^3	Pressure of saturated vapor at 20°C, Pa	T_m or T_b, °C	Solvents	Protected Metals
Salt of dicyclohexylamine and SFA (MSDA-1)	$(C_6H_{11})_2NH\cdot HOOCR$ (R = $C_{10}H_{21}$-$C_{13}H_{25}$)	Paste-like brown substance	0.91–0.93	0.19 (40°C)	40–50	eal, bz, benzene, mineral oils	steel, cast iron, Zn, Cu and alloys, Al, Ag, Sn, babbitt
Salt of cyclohexylamine and SFA (M-1)	$(C_6H_{11})NH_2\cdot HOOCR$ (R = $C_{10}H_{21}$-$C_{13}H_{25}$)	Paste-like brown substance	0.91–0.93	0.13	40–50	w, eal, bz, mineral oils	steel, cast iron, Al, Cu, alloys, Zn, babbitt
Dicyclohexylamine nitrate (NDA, Dichan[a], VPI-260[a])	$(C_6H_{11})_2NH\cdot HNO_2$	White yellowish powder		$1.3\cdot 10^{-2}$	165–180	w-eal mixture, eal, ac, ethylene glycol	steel, Al and its alloys, Ni, Cr, Co
Cyclohexylamine chromate	$[(C_6H_{11})NH_2]_2\cdot H_2CrO_4$	Bright yellow powder		$9.3\cdot 10^{-4}$	127	w, eal	steel, cast iron, Cu, brass, Ni, Sn, Ag
Cyclohexylamine carbonate	$[(C_6H_{11})NH_2]_2\cdot H_2CO_3$	White powder	0.5 (poured)	21.3	106–112	w, eal, ac, ch, bz, CCl_4	steel, cast iron, Pb, Al, Ni, Zn, Sn
Cyclohexylamine metanitrobenzoate	$C_6H_{11}NH_2\cdot HOOCC_6H_4NO_2$	White powder		$1.8\cdot 10^{-4}$ (40°C)	134	eal	Fe, Cu, brass, Zn, Cd, Mg, Al, Sn, Ni, Ag

Hexamethylenimine m-nitrobenzoate (G-2)	$(CH_2)_6NH\cdot HOOCC_6H_4NO_2$	White powder		$1.8\cdot10^{-4}$ (40°C)	132–133	w, eal	Steel, coated steel, Ni, Cu, Sn, Pb, Al, Ag and their alloys
N-cyclohexyl-phenylimine (VNHL-49)	C_6H_5–CH=N–C_6H_{11}	Light-yellow liquid	0.97	0.57 (24°C)	(305)	w (weak), eal	Steel, Al, Cu, their alloys, Ni, Zn, Sn, Cd, Cr, Ag
Morpholine	C_6H_8ONH	liquid	0.999	$1.76\cdot10^{3}$	(128–130)	w, eal, e	Steel, Al, Ni, Ag, brass
Dimorpholino-phenylmethane (VNHL-20)	$C_6H_5CH(C_6H_8ON)_2$	white powder	1.098	$3.5\cdot10^{-4}$	98–99	eal, ac in hexane	Steel, cast iron, Al, Cu and their alloys, Pb, Ni, Zn, Sn, Cd, Cr, Ag
Aliphatic and aromatic amines, their derivatives and salts							
2-ethyl hexyl heptylamine molybdate	$[RR'NH_2]_2MoO_4$ ($R = 2\text{-}(C_2H_5)C_6H_{13}$, $R' = C_7H_{15}$)	White powder					Fe, steel, Cu
1-diethylamine-2-methylbutanone-3 (IFHAN-1)	$(C_2H_5)_2NCH_2CH(CH_3)C(O)CH_3$	Yellow liquid	0.86	13–27	(210)	eal, toluene	Steel, Cu, brass, Zn, Cd, Mg, Al, Sn, Ni, Ag

(*continued*)

Table 1.3. cont.

Main classes and brands of CI (Russian and other trademarks)	Chemical formula	Appearance under normal conditions	Density g/cm^3	Pressure of saturated vapor at 20°C, Pa	T_m or T_b, °C	Solvents	Protected Metals
Dialkylamino-akrylonitrile + dialkyamine (IFHANGAZ-1)	RR'N–$(CH_2)_2$–C≡N (80%) + RR'NH (20%) (R, R' =C_7H_{15}-C_9H_{19})	Yellow liquid	0.84–0.86	1.4	–40÷–60 (190 at 1.33 kPa)		Steel, Cu, brass, Al and alloys, Zn, Ni, Cd, Sn, Mg, Ag,
Monoathanolamine benzoiate (BMEA)	$NH_2C_2H_4OH \cdot HOOCC_6H_5$	Viscous yellowish liquid		$2.9 \cdot 10^{-2}$ (40°C)	(171–172)	w, eal	steel, cast iron
Heterocyclic compounds, their derivatives and salts							
Imidazole	$C_3N_2H_4$				88–90 (257)	w, eal, e	Steel
Benzo-1,2,3-trizole (BTA)	$C_6H_4N_3H$	White powder		$1.0 \cdot 10^{-3}$	98–100	w, eal, e, bz, ac	Steel, cast iron, Cu, its alloys, Ag
3-Amino-1,2,4-triazole (ATr)	$NH_2C_2N_3H_2$	White powder		at BTA level	146–148	eal, ac	Steel, Cu, its alloys
5-phenyltetrazole (PhTet)	$C_6H_5CN_4H$	White powder			213 (decomp.)	DMSO, morpholine	Steel, Cu, brass, bronze

5-Phenyltetrazole piperidine salt (T-4)	$C_6H_5CN_4H \cdot HNC_5H_{10}$	White powder			116–117	eal	Cu and alloys
5-Aminotetrazole (hydrate) (ATet)	$NH_2CN_4H \cdot H_2O$	White powder			200–204 (decomp.)	w, eal	Steel, Cu and its alloys
Carboxylic acids and their salts							
Oleic acid	$C_{18}H_{17}CH{=}CH(CH_2)_7COOH$	Brown liquid	0.895		16 (286 Pa)	eal, e, bz, ch	Fe, steel
Pelargonic (nonanoic) acid	$CH3(CH_2)_7COOH$	Liquid	0.90	Moderate volatility	12.5 (253–255)	eal, ch, bz, ac	Steel
Sodium benzoate (SB)	C_6H_5COONa	White powder	0.96			w, eal, pyridine	Steel, alloys, Al, Sn, bronze
Nitrocompounds							
Mineral nitrated oil (AKOR)	–	Oily liquid of dark brown to black color			44–45	mineral oils, organic solvents	Steel, cast iron, Cu and alloys, Al
Natural compounds							
Oxidized petrolatum (MNI3)	–	Viscous brown substance			60	mineral oils, organic solvents	Steel, cast iron, Cu and alloys, Al
Oxidized ceresin (MNI-7)	–	Viscous yellow-brown substance				mineral oils, organic solvents.	Steel, cast iron, Cu and alloys, Al

wa – water; eal – ethyl alcohol; ac – acetone; e – ether; bz – benzole; ch – chlorophorm; DMSO–dimethylsulfoxide ability of the inhibitors after combining with the polymer carrier. Just a few CI respond to the named criteria in full.

The *aggregate state* of the inhibitor determines its impregnation procedure into the polymer matrix.

Solid powder CI are processed jointly with polymer melts by extrusion, injection molding, compression molding, etc. or are bonded to the polymer carrier via gluing or hot lamination.

Liquid CI are used as saturation components of porous and fibrous polymer matrixes. The CI must have sufficient fluidity and wetability in this case. The heating stage aimed at reduction of CI viscosity exerts certain limitations on the use of highly volatile inhibitors. Application of vacuum techniques for impregnation of the polymer porous structures is also restricted by the resultant losses of the most valuable and difficult to preserve volatile fractions of CI.

Some liquid CI can be plasticizers (Pl) of the polymer binder at the same time. Many oil-soluble CI show this property towards polyoilefins. The CI soluble in Pl, solvents and other CM expand their limits of blending with thermoplastic binders.

There are also the requirements imposed on CI as components of polymer glue compositions. CI should not deteriorate the colloidal stability of the glue composition and adhesive strength of glued joints with metals in this case.

An important property of CI as the inhibited plastic ingredients is their toxicological indices (see 6.2).

It is strongly advisable for formulations of inhibited plastics to make allowance for the effect the CI exerts on the polymer binder properties during processing and service life of the final products. Some CI may activate thermooxidative destruction and atmospheric aging of polymers, while others vice versa, are able to retard these adverse processes.

1.3.3 Other Ingredients of Inhibited Plastics

In fact, all inhibited plastics along with CI contain the ingredients that facilitate their processing and define the performance of the obtained materials to a large extent [32, 33, 45]. Their content may vary in plastics within a wide range from portions to tens of percents. The main ingredients of inhibited CM are fillers, Pl, stabilizers, dies, cross-linking agents (for thermosets), structural formers, cellulating agents (porephores) lubricants, fire retardants and antistatics. The demands imposed on the ingredients are their compatibility with other components of the polymer CM, the stability of properties at storage, processing and operation and safety for man.

Thus, impregnation of *fillers* turns out to be one of the most useful methods of creating polymer materials with preset process and service parameters [28, 31]. The filler content does not, as a rule, surpass 45–50% by weight, as its excess evokes difficulties in processing of thermoplastic blends into goods. There are, however, highly filled plastics in which concentration of the filler is three times that of the polymer.

Fillers exert a specific effect on plastics properties, including mechanical strength, hardness, chemical stability, thermal, electrophysical characteristics and frictional parameters (see Table 1.4). The fillers are often impregnated to cut the cost of CM. The fillers are classified according to their

- origin: organic or inorganic, natural (vegetable or mineral), artificial or synthetic,
- aggregate state: solid, liquid and gaseous (in foam plastics),
- structure: dispersed or powder (grainy, needle-like, flaky and other) and continuous reinforcing of either monolithic or fibrous structure, including unidimensional (braids, threads), 2D (porous tapes, films, fabric, non-woven veils, paper) and skeleton (molded cellular and solid-woven carcasses, 3D cloths and so on).

Table 1.4. Effect of solid fillers on plastics properties

Properties of CM Regulated by Fillers	Fillers
Tensile strength	Wood powder, cellulose, inorganic (glass, carbon) and organic (viscous, cotton, polyester, polyacrylontrile (PAN), PA) fibers, graphite
Impact viscosity	Cellulose, organic fibers, glass fiber, asbestos, aerosil, betonite, vermiculite, calcium silicate.
Stiffness, hardness	Chopped cellulose, jute, inorganic (asbestos, glass) and inorganic (PA, PAN, etc.) fibers, powder PTPE, graphite, soot, bronze, aluminum oxide, kaolin, vermiculite, chalk, dolomite, calcium silicate, silicon dioxide, mica, talc.
Heat resistance and dimensional stability	Inorganic (asbestos, glass, carbon) and organic (PA, PAN, and other) fibers, commercial carbon, lamellar aluminum, kaolin, betonite, vermiculite, chalk, calcium silicate, mica, talc.
Heat and electric conductivity	Graphite, commercial carbon, carbon fibers, Al, Cu, Ag, bronze and other metal powders.
Electric insulating properties	Wood powder, inorganic (glass, asbestos) and organic fibers, Al oxide, kaolin, asbestos, talc, mica, silicon oxide.
Chemical resistance	Organic fibers (PA, PAN), glass fibers, graphite, coke, lamellar aluminum, asbestos, kaolin, vermiculite, mica, talc.
Processability	Soot, kaolin, lignin, organic fibers, aluminum oxide.

Plasticizers (Pl) are impregnated into plastics to impart or raise elasticity and/or plasticity at processing and during service life [30,32,33]. Pl facilitate dispersion of loose ingredients in the polymer binder and lower the flow point, brittleness and glass transition of polymer materials. Along with this, Pl usually impair heat stability. Some types of Pl are able to improve resistance to fire, light and temperature of the plastics. Pl should be compatible with other components of inhibited CM, be low volatile, inert chemically and devoid of any smell at the same time. Many Pl are inclined to exudation (syneresis), a property of the liquid phase to be spontaneously released from the polymer composition as a result of relaxation of inner stresses and reduced volume of the polymer matrix [30]. Exudation occurs as a result of physical-chemical processes induced by thermodynamic non-equilibrium of the polymer – Pl system. Exudation is traditionally considered as a drawback of plasticized CM. As applied to inhibited plastics, this phenomenon has been made the basis for a series of useful solutions [38,41,44]. Their essence consists in using Pl as a CI carrier in the inhibited CM. During operation the Pl transports CI via the exudation mechanism from the polymer matrix volume to the surface of the protected article. Some CI, e.g. oil-soluble contact CI, are good plasticizers themselves.

Pl may be natural or synthetic. Petroleum (mineral) and vegetable oils, resins and paraffins are the most well-known natural Pl. Among synthetic Pl there are esters of aromatic carboxylic acids and aliphatic alcohols (dialkyl, dimethyl, dibutyl, di(2-ethyl hexyl) and other phthalates, tri and pyromellitates), esters of aliphatic acids and aliphatic alcohols (dialkyl adipinates, sebacates, azelates), ethers of orthophosphoric acid and alcohols or phenols (tricrezyl, triphenyl, tri(2-ethyl hexyl) and other phosphates). This group also includes polyesters produced by polycondensation of dicarboxylic acids with glycols followed by interactions with hydroxyl or carboxyl-containing oligoesters with monocarbonic acid or monohydric alcohol (e.g. dibutyl ester of polyethylene glycol adiminate, dibutyl ester of polyethylene glycol adipinate sebacate). Physical, chemical and toxicological properties of Pl as well as the principles of their selection are reviewed in [30,32,33,64,65].

Stabilizers are employed to protect polymer materials from aging, which may result in loss of a set of useful properties due to chemical and physical transformations occurring in polymers during operation, processing or storage under the effect of heat, light, oxygen and other factors. The main types of stabilizers of inhibited plastics are antioxidants, which are inhibitors of thermo-oxidative destruction, and light stabilizers, which are inhibitors of photo-oxidative destruction.

Antioxidants improve the stability of plastics to oxygen. Such substances as oxygen acceptors (fine-dispersed metals and oxides of transient metals of lower valence) break the chains of radical reactions on thermo-oxidative destruction of polymers or destroy intermediate products of these reactions, e.g. hydroperoxides (sulfides, metal salts of variable valence, derivatives of secondary aromatic amines, quinoline, phenols).

The ability of *light stabilizers* is based on the absorption of the photochemically active component of sunlight, extinguishing excited states (deactivation of excited molecules that have absorbed a light quantum), retardation of light-induced dark reactions as well as destruction and deactivation of impurities and products of photoreactions catalyzing photochemical processes in polymers. As for light stabilizers, some inorganic pigments are often used (soot, TiO_2, ZnS), along with organic chromophores (ethers of salicylic acid, o-oxy-p-alkoxybenzophenon, benzotriazole), organometallic compounds (phosphonates and nickel dithiocarbamates) and stable radicals (nitroxyl and their analogues). Some CI of amine type areparticularly volatile (Russian trademark NDA, G-2, BTA, 5-Ph-Tet and other) and can also be polymer stabilizers.

Dyes are used to manufacture colored inhibited plastics. Both organic dyes of different brands and inorganic pigments may suit this purpose [66]. Some pigments displaying the properties of metal CI are discussed in Sect. 1.3.2.

Cross-linking agents (hardeners) are introduced into inhibited CM based on thermosets to create cross-links between macromolecules at a certain stage in their processing. They form a honeycomb structure of the polymer matrix imparting specific properties to a hardened product (strength, insolubility). Different polyfunctional compounds, like diamines, glycols, dicarboxylic acids and their anhydrides, amine alcohols and isocyanates, may be used as hardeners. Cross-linking agents include siccative additives that accelerate drying of paint coats.

Structural formers (oxides, carbides, nitrides, salts of organic acids, fine-dispersed powders, surfactants) exert an effect on the permolecular structure formation and assist in yielding materials with a given microstructure. A number of powder CI can be related to active structuring agents. In particular, impregnation of small amounts of fine-dispersed CI (NDA, G-2, 5-PhTet) in the polymer film materials improves their strength and barrier properties (see Sect. 2.3).

Porephores assist in the formation of the porous structure in polymer materials by foaming their gas-vapor phase. This phase appears in the polymer matrix during porephore evaporation (e.g. the low-molecular hydrocarbon) under the action of heat or chemical reactions with participation of the porephore accompanied by generation of gaseous products. Many volatile CI actively transfer into the vapor state under certain conditions (e.g. at joint processing with the polymer melt) and are capable of foaming the polymer binder. This is how inhibited plastics with CI in the vapor phase are obtained.

Lubricants are used to decrease tackiness and avert sticking of polymer materials to the forming surfaces of the processing equipment. Lubricants can be made of paraffin, wax, salts of higher fatty acids and other matter apt to migration to the polymer surface. They also improve dispersion of loose ingredients in the polymer binder at the same time.

Fire-retardants are additives (halogen-derivative of hydrocarbons, phosphoric compounds, antimony, isocyanates) that hamper ignition and restrict inflammability of the polymer materials. Some Pl of inhibited plastics as well as pigment CI introduced into anticorrosion polymer coats may display fireproofing properties [43].

Antistatics impede the generation and accumulation of static electricity in polymer goods. This group of additives includes various surfactants like amines, quaternary ammonium bases and some other substances. Many organic CI are typical surfactants and are therefore good antistatics. Antistatic properties are also intrinsic to polymer films and coatings containing CI in the form of metal-protecting pigments (zinc, aluminum, copper, and etc.) as well as anticorrosive film materials based on conducting polymers.

1.4 Thermodynamic and Process Compatibility of the Components of Inhibited Plastics

Thermodynamic compatibility of the components of a polymer material implies their capability for mutual dissolution under certain conditions with the formation of a stable single-phase blend [30]. This state of the material is stable in time, during which no matter or energy is transferred through it [38].

In contrast, inhibited plastics represent *thermodynamically non-equilibrium systems* that experience relaxation changes in their composition and structure, and are continuously exchanging substances with the ambient medium during their lifetime.

This non-equilibrium chemical and physical (mechanical) state of inhibited plastics is caused by incorporated CI and their carriers (Pl) in concentrations exceeding the thermodynamic compatibility threshold with the polymer. The thermodynamic non-equilibrium is a useful property of inhibited plastics since it expands their anticorrosion functional features. The greater the deviation from the non-equilibrium state of such systems, the quicker the relaxation processes in them accompanied by the release of excess CI into the ambient medium and to the object being protected. The structure and properties of the components of inhibited plastics CI depend on their composition and can be carried to the mated part by either diffusion in the gaseous (thanks to CI volatility) or liquid phase via its exudation, its mixture with a Pl, or washing of water-soluble CI and diffusion through the material moisture. These processes bring about mechanical relaxation of the polymer matrix. The point at which the inhibited anticorrosion material moves into a state of thermodynamic equilibrium is usually correlated with its protective ability.

Process compatibility of CM components means their capability to form a blend with satisfactory mechanical characteristics that preserves their structure and properties within a necessary period [30, 67].

The state of thermodynamic equilibrium (implying the formation of a single-phase blend of components) can rarely be reached due to the high viscosity of polymer composite systems. Polymer CM are most often dispersed systems whose composition varies with time and individual components may form phase areas of different sizes. One of their components (the polymer phase) is a continuous dispersed medium, i.e. a matrix, in which all other components are spread as a dispersion of spatially separated particles called a dispersed phase [31].

Polymer composite systems usually exist in a metastable (unstable) state of mechanical equilibrium. This is because a mixture of mutually insoluble components separates extremely slowly due to the very low diffusion coefficients of the polymer matrix of the ingredients. This state of polymer systems is sometimes defined as *kinetic compatibility* [28].

The term m*echanical compatibility* is used for polymer CM that consist of components that are thermodynamically incompatible but correlated within certain limits by their elastic and plastic characteristics, thermal expansion ratios and other physical-mechanical properties.

Thermodynamic compatibility of polymer CM components is observed rather infrequently as there is not commonly necessary to reach it. Polymer materials can be produced with preset properties even in conditions of limited compatibility between components. It is desirable to attain a thermodynamic non-equilibrium state for these materials for specific purposes, e.g. corrosion-proofing by inhibited plastics. Process compatibility of components, on the other hand, should be treated as a requisite quality of CM.

1.4.1 Phase Equilibrium in Polymer-Solvent Systems

An ideal inhibited plastic material combines considerable phase and diffusion mobility of CI with mechanical strength and high barrier characteristics. These contradictory requirements are met when a polymer-solvent system is used as the inhibited plastic base (with high polymer content). The required physical-mechanical parameters in the described systems are promoted by the polymer, while the solvent is either a liquid CI or its solution (dispersion) in Pl.

Polymer solutions differ from those of low-molecular-weight matter by specific properties of macromolecules characterized by large size, broad molecular-mass distribution, flexibility range, numerous conformations and capability of conformal recombinations in response to temperature fluctuations and solvent type [29].

True polymer solutions in low-molecular-weight liquids are probable only within a certain temperature range. Mutual solubility of components beyond this range becomes confined and the solution starts to separate with cooling into phases coexisting in some equilibrium.

Phase equilibrium in polymer-solvent systems is subdivided into the following five types [41,65,68]:

1. *Crystalline equilibrium*, which is characteristic for polymer solutes separating into phases when transferring into the temperature and concentration zone of confined compatibility of components: (i) crystalline polymer and (ii) non-saturated (low-concentration) polymer solution in a solvent.
2. *Amorphous (liquid) equilibrium*, taking place when the system decomposes at a temperature higher or lower than a certain limit into two equilibrium amorphous phases having dissimilar polymer content. One of the phases is the solvent and the other is a superconcentrated polymer solute.
3. *Liquid-crystalline equilibrium* is typical of solutions decomposing into two phases with increasing polymer content, i.e. isotropic (low-concentration) and anisotropic (highly concentrated) polymer solutions converting further into a single-phase liquid-crystalline system.
4. *The equilibrium in the cross-linked polymer-solvent systems* sets during absorption of the solvent till some critical degree of polymer swelling. The cross-linked polymer gel coexists in the equilibrium state in the heterogeneous region with a pure solvent.
5. *A combination of the crystalline and amorphous equilibrium* is characteristic for readily crystallizable polymer of PE type. This type of equilibrium is of the strongest interest from the viewpoint of wide adoption of crystallizing polymers as the base of inhibited plastics.

Since polymer solutions obey the rule of phases, it is expedient to use a traditional thermal analysis of the cooling curves in studying their phase state. A peculiarity of the polymer-solvent system is its propensity to form solid solutes under definite ratios of components, which has been confirmed by X-ray diffraction analysis [69].

The experimental cooling curves (*a*) and a graph of a bicomponent HDPE-mineral oil system (*b*) are shown in Fig. 1.10. Mineral oil is a perfect PE plasticizer and can be the CI carrier in inhibited plastics. The presented graph is characteristic for dual systems of the peritectic kind with limited solubility of components in the solid state [65].

The system is in the state of an isotropic colloidal solution L under elevated temperatures (above the liquids line). Vertical lines I, II and III on graph (*b*) correspond to compositions, whose cooling curves are shown in Fig. 1.10.

The compositions of type I with less than 5% of oil content (the thermodynamic compatibility threshold of the oil and PE) release only a single equilibrium β-phase (solid PE solution in oil) on cooling of the homogeneous melt.

During crystallization of type II melts with 5–30 wt% of Pl, initially a homogeneous solid β-solution is formed (enriched PE), which becomes unstable

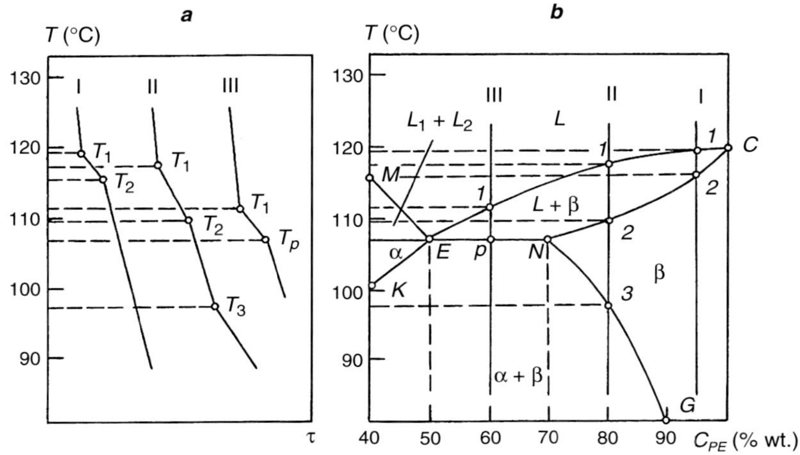

Fig. 1.10. Cooling curves (**a**) and a portion of the diagram of state (**b**) of HDPE-vacuum oil system: L, α and β – zones of liquid phase existence (melt), solid solutions of PE in oil and oil in PE, respectively; L $+\beta$, $\alpha+\beta$ – zones of coexisting liquid phase, solid solution and two solid solutions; L_1+ L_2 – stratification zone of two liquid phases (saturated polymer solution in oil and vice versa); ELC and ENC – liquids and solids lines

as the T_3 temperature and reaches and isolates some amount of oil-enriched α-phase, whose composition corresponds to the EK line (its continuation is not shown in Fig. 1.10).

In the course of cooling of melts with 30–50 wt% of Pl (type III) below the liquids line (below T_1), the melt releases a solid oil solution in PE (β-phase) with up to 30% oil content. Further cooling to the peritectic temperature (T_{m}) brings about the third α-phase (PE solid solution in oil). When the liquid phase L disappears fully upon the peritectic reaction $L+\beta \leftrightarrow \alpha$, the system transfers into a biphasic field α and β. As the temperature goes down, the composition of phases α and β changes in accordance to the lines EK and NG.

PE-oil systems with low PE concentration (oil – above 90%) are characterized by liquid phase equilibrium [69].

Structure formation of polymer CM with low polymer content in the PE-Pl melt apparently takes the following path. Equilibrium phases α and β are nucleated during melt crystallization, although do not stratify fully with formation of a single interface because of high viscosity of the polymer system. Thus enriched by the polymer, the β-phase approaches a solid body in its properties and forms a continuous porous matrix with the polymer-depleted α-phase. As a result, a gel structure typical of the biphasic polymer with characteristic deformation properties and inclination to syneresis is formed [70].

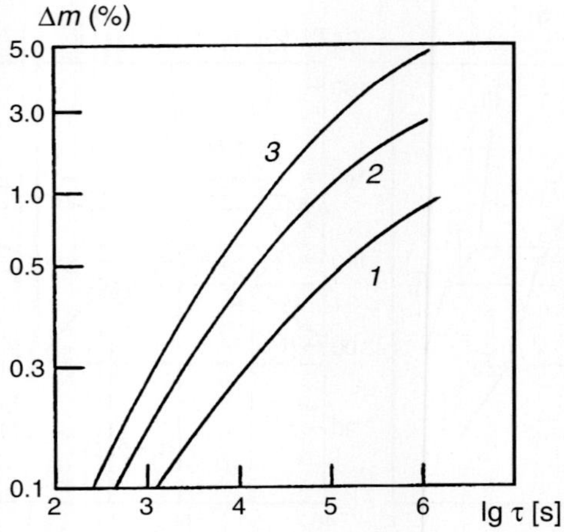

Fig. 1.11. Exudation kinetics of Pl from PE film – vacuum oil samples in response to oil concentration in CM: (1) 20 wt%; (2) 40%; (3) 60 wt%

The syneresis rate of the liquid phase increases as its concentration in the polymer composition augments (Fig. 1.11).

Synergetic processes in the biphasic polymer gels are commonly attributed to stress relaxation because of unbalanced phase decomposition during crystallization of the plasticized polymer melt. The liquid liberates from the gel simultaneously with contracting geometrical parameters (shrinkage) [70].

Experimental kinetic curves of shrinkage of PE-based film samples plasticized by the vacuum oil are illustrated in Fig. 1.12.

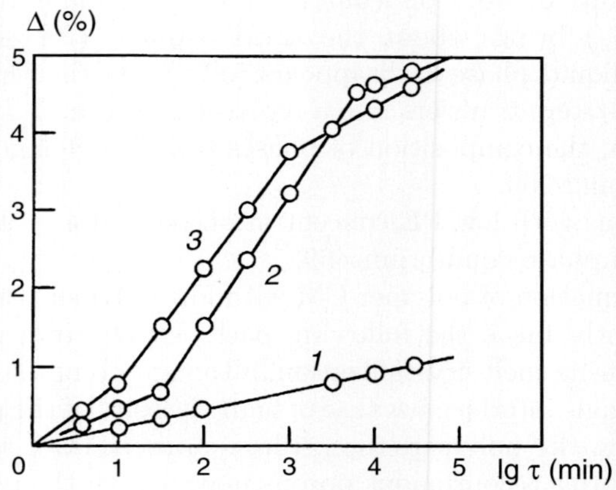

Fig. 1.12. Kinetic shrinkage curves across thickness (Δ, %) of PE + vacuum oil film samples. Initial oil concentration: (1) 30, (2) 40 and (3) 50 wt%

When oil concentrations in the films are about 40–50%, the dependencies Δ-(lg τ) (curves 2 and 3) form a section where the shrinkage rate is rather high. This section is not presented in the diagram of the sample with below 30% oil content (line 1). Proceeding from the analysis of the diagram of the state of the system under consideration (Fig. 1.10), the accelerated shrinkage of samples 2 and 3 can evidently be attributed to intensified relaxation processes accompanying phase breakdown of the system into the matrix – β and low-viscous α phases. This is one more piece of evidence of the similarity of the processes lying in the base of syneresis and shrinkage brought about by phase transformations in the polymer-solvent system.

1.4.2 Compatibility of Components of Inhibited Plastics and Its Regulation

To characterize the *thermodynamic compatibility* of components of inhibited plastics, the following empirical parameters can be used:

(1) optical transparency of the plastic,
(2) Pl or CI + Pl mixture content in the material above which syneresis begins (for plasticized polymers),
(3) lowering of T_m of the crystalline or T_g of the amorphous polymer binder on mixing with Pl (or CI solution in Pl),
(4) CI solubility in Pl under different temperatures (when the CI carrier is Pl) [30,65].

Optical transparency as a criterion of componential compatibility is often used for films and coatings [41]. However, some films and coatings based on incompatible components can be optically transparent. For instance, very fine films consisting of two layers of a material laminated into phases may turn out to be transparent [71]. In a number of cases, optical transparency of heterogeneous polymer materials is conditioned by the equality of the refraction parameters of their components (i.e. the difference is below 0.01) or particle size of the dispersed phase (below 0.1 μm) [71]. Hence, if the temperature dependencies of refraction parameters of components differ, heterogeneous CM can be transparent at some temperature because of the equality of the refraction parameters becoming blurred on heating or cooling.

It is very important for formulations of the liquid-filled inhibited plastics to estimate the limits of thermodynamic compatibility (C_{lim}) of Pl and CI + Pl blends with a polymer binder. For this purpose, the presence or absence of syneresis in the studied system judged from the liquid phase separation [65,70] can be a convenient empirical criterion for estimating C_{lim}.

Since inhibited plastics, especially films, are mostly based on PE (see 1.3), the value of C_{lim} estimated towards PE of such Pl as esters of aromatic and aliphatic dicarboxylic acids and mineral oils reaches 5 wt%. Polyatomic alcohols and vegetable oils are less compatible with PE, C_{lim} is about 2% (Table 1.5).

Table 1.5. Thermodynamic compatibility parameters* of LDPE and Pl at 25°C

Pl	Pl Content in PE Composition (wt %)	Presence or Absence of Syneresis
Dioctyl phthalate	7.5	+
	4.7	no
Di(2-ethylhexyl) sebacate	7.2	+
	4.5	no
Vaseline oil	7.0	+
Glycerin	4.9	insignificant
	2.1	no
Castor oil	3.0	+
	2.6	no

*According to investigation results of the films produced by hot-pressing of PE + Pl blends

CI and Pl mixtures are characterized by lower $C_{\lim}$ values in contrast to the initial Pl [50, 51]. There are, however, cases at a certain liquid-phase ratio when three-component systems like polymer – liquid 1 – liquid 2 display a critical compatibility concentration of the mixture of liquids with the polymer, which exceeds that of the corresponding paired systems [65].

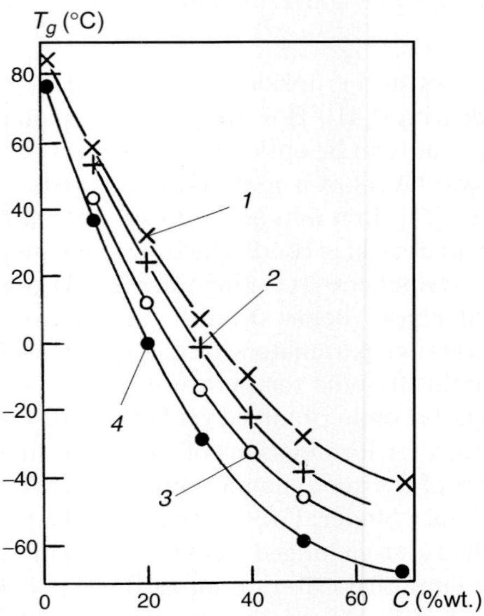

Fig. 1.13. Glass transition temperature (T_g) of PVC versus dioctyl phthalate content (C). T_g determined by: (1) dynamometric balance; (2) thermo-mechanical method; (3) DTA; (4) DSC

Introduction of a Pl that is perfectly compatible with the polymer leads, as is known, to a negligible lowering of the melting, glass transition, brittleness and flow temperatures of the polymer [65,72]. Note that the T_g of PVC drops significantly if this polymer is combined with the esters of aromatic (dioctyl phthalate) (Fig. 1.13) and aliphatic (dioctyl sebacate) dicarboxylic acids [72].

The decrement of ΔT_m of PE at its plastification by mineral oils reaches 15 °C and more. The plasticizing ability of vegetable oils towards PE is much worse ($\Delta T_m \sim 3-5$ °C) (Table 1.6). This conclusion on the degree of compatibility of PE with different classes of Pl is in good agreement with investigation results of synergetic processes in analogous systems.

Syneresis plays a positive role since it alleviates transportation of the liquid-phase CI to the protected hardware. The CI content in the material is however restricteded so as not to impair strength and barrier properties of CM that are highly filled by liquids.

Table 1.6. Melting point (T_m) and temperatures of thermal destruction initiation (T_{di}) and intensive thermal destruction initiation (T_{idi}) of PE, Pl, CI and their blends according to derivatographic analysis

Samples	T_m, °C	T_{di}, °C	T_{idi}, °C
LDPE	107	225	335
Plasticizers:			
Vaseline oil	–	185	220
Vacuum oil	–	200	220
Glycerin	–	140	–
Di(2-ethylhexyl)sebacate	–	180	270
Castor oil	–	185	–
Sunflower-seed oil	–	160	230
Mustard oil	–	175	240
LDPE + Pl mixtures (1:1 by mass):			
PE + mustard oil	104	210	295
PE + Vaseline oil (VO)	92	197	225
CI:			
VTA	98	155	163
Tr	120	–*	–*
ATr	155	157	208
Tet	160	163	233
MTet	149	152	225
ATet	164	165	220
PhTet	205	215	255
KPhTet	352	378	398
NaPhTet	303	325	340

*T_{di} of Tr was not determined because of its high volatility

Table 1.7. CI (PHC) solubility in Pl at normal (20°C) and elevated (60°C) temperatures (Pl to CI ratio 3:1 by mass)

	Pl and Solution Temperature, °C											
	Oleic Acid		Benzyl Alcohol		Glycerin		Morpholine		DMSO		Di(2-ethylhexyl) Di(sebacate)	
PHC	20	60	20	60	20	60	20	60	20	60	20	60
Tr	+	+	−	−	+	−	+	+	+	+	−	+
ATr	−	+	+	+	−	−	+	+	+	+	−	+
Tet	−	−	−	+	−	+	+	+	+	+	−	+
MTet	−	+	+	+	−	−	+	+	+	+	−	+
ATet	−	+	−	+	−	+	+	+	−	+	−	+
KPhTet	−	−	−	+	+	+	−	+	+	+	−	+
PhTet	−	−	−	+	−	−	+	+	+	+	−	−

When the Pl serves as a liquid-phase carrier in the inhibited plastic, it is important to take into account the compatibiity of these components with the polymer base and with each other. Thermodynamic compatibility of CI and Pl depends on their solubility in one another. The solubility parameters of CI of the PHC series in high-boiling-point liquids plasticizing PE are presented in Table 1.7 [50,51].

It is a matter of common experience that PHC do not dissolve in mineral and vegetable oils and esters of aromatic dicarboxylic acids even under elevated temperatures. Only polar Pl like morpholine and DMSO can dissolve such CI. Above 60°C these can be esters of aliphatic dicarboxylic acids (e.g., di (2-ethylhexyl) sebacate), alcohols (glycerin, benzyl alcohol), and high fatty acids (oleic acid). Some of the mentioned Pl able to dissolve PHC (morpholine, oleic acid) exhibit certain CI properties [46].

Along with the empirical approach to forecasting the compatibility of the components of liquid-filled inhibited plastics, quantitative theories of estimating solution capacity of low-molecular-weight liquids are also used with regard to different nature polymers.

The Hildebrandt-Sketchard theory of regular solutions is the most accepted and used [71]. To characterize the dissolving capacity of substances, this theory uses so-called solubility parameters (Δ) calculated as the square root of the ratio between evaporation energy (E_{ev}) and the molar volume of the substance (V_M):

$$\delta = (E_{ev}/V_M)^{1/2} , \tag{1.6}$$

where $E_{ev} = \Delta H_{ev} - RT; \Delta H_{ev}$ is the evaporation heat; and RT is the product of the universal gas constant and the absolute temperature.

The solubility parameter for polymers can be found by the empirical formula [71]:

$$\Delta_{\mathrm{p}}^2 = T\alpha/\beta \ , \tag{1.7}$$

where α is the thermal expansion factor and β is the isothermal compressibility ratio.

The solubility parameter of the compounds, whose structural formula is known, can be calculated by Small's method based on the supposition that cohesive forces of separate atomic groups and radicals entering the composition of a low-molecular-weight substance or an elementary section of a macromolecule exert an additive effect. The respective calculations make use of the so-called molecular attraction constant F_{i}, whose values for different functional groups are cited in special literature [74]. These constants are related with solubility parameter δ

$$\Sigma F_i = \delta M/\rho \ , \tag{1.8}$$

where M and ρ are molecular mass and density of the substance, respectively.

A requisite condition of the forming homogeneous, thermodynamically stable solution during polymer mixing with a solvent is a decrease in free energy (U)

$$U = \Delta H - T\Delta S < 0 \ , \tag{1.9}$$

where ΔH is the enthalpy change (heat effect of mixing); ΔS is the entropy change; T is the absolute temperature.

A specific feature of the polymer-solvent system is positive values of the entropic member ($T\Delta S$) of (1.9) because of the large number of possible conformations of macromolecules in diluted polymer solutions. The enthalpy member value (ΔH) can be below zero if dissolving is reached via strong interactions between the polymer and solvent, e.g. solvation, generation of hydrogen and coordination bonds. Moreover, there is a variant when $\Delta H > 0$, i.e. the energy of interactions between molecules in pure substances and the polymer exceeds that in the polymer-solvent system. In this case, the polymer solution in the solvent or its combination with Pl, liquid CI or other liquids and polymers will take place only if $\Delta H > 0$ and $\Delta H < T\Delta \mathrm{S}$.

When no specific interactions are observed between the components of the polymer-solvent system, the Hildebrand-Sketchard equation is true [2]:

$$\Delta H_{\mathrm{mix}} = (V_{\mathrm{p}}x_{\mathrm{p}} + V_{\mathrm{s}}x_{\mathrm{s}})\varphi_{\mathrm{p}}\varphi_{\mathrm{s}}(\delta_{\mathrm{p}} - \delta_{\mathrm{s}})^2 \ , \tag{1.10}$$

where ΔH_{mix} is the excess internal energy of polymer mixing with the solvent; V_{p} and V_{s}, x_{p} and x_{s} are the molar volumes and shares of the polymer and solvent, respectively; φ_{p} and φ_{s} are their volume share; δ_{p} and δ_{s} are the solubility parameters.

Comparison of (1.9) and (1.10) shows that the most favorable conditions of mixing are attained at $\Delta H_{\mathrm{mix}} = 0$, when $\delta_{\mathrm{p}} = \delta_{\mathrm{s}}$. Based on this, polymer blends with the components having close solubility parameters will show the best compatibility.

Table 1.8. Solubility parameters $(\mathrm{mJ/m^3})^{1/2}$ of polymers and solvents

Polymer	δ_p	Solvent	δ_s
Cellulose acetate	25.9	Butyl alcohol	23.05
PA	20.8	Water	47.80
PVA	23.0	Glycerin	43.78
PVC	22.4	Dimethyl sulfoxide	26.38
Polymethyl methacrylate	23.1	Diethylene glycol	29.78
PS	20.0	Cellulose methyl	24.6
PFC (novolak)	23.1	Morpholine	21.46
PFC (resol)	26.7	Tetralin	19.39
Chlorinated PP	20.8	Cyclohexylamine	18.69

Solubility parameters of some polymers, solvents and liquid CI are presented in Table 1.8 [38, 75].

The excess free energy per solvent molecule of polymer solutions is characterized by a semi-empirical Flory-Huggins parameter, χ, which is a function of temperature for a given polymer-solvent pair. To estimate the compatibility parameter experimentally, it is necessary to define the χ value for each polymer-solvent pair and compare it to its critical value calculated by the equation

$$\chi_{\mathrm{cr}} = 0.5(1 + m^{-0.5})^2 , \qquad (1.11)$$

where m is the degree of polymerization.

The value $m^{-0.5}$ is small enough for polymers and is neglected by taking instead $\chi_{\mathrm{cr}} = 0.5$ for the polymer-solvent pair. In the case that χ exceeds this value, it's impossible to combine the polymer and solvent fully. The procedures of experimental defining χ values have been described elsewhere.

The parameters of PVC interactions with some Pl are given as an example in Table 1.9 [75]. It is quite evident that butyl alcohol and butyl ricinoleate do not fit for PVC as a Pl since they are incompatible with this polymer.

Table 1.9. Interaction parameters of PVC and plasticizers

Pl	$\chi(\mathrm{mJ/m^3})$		Pl	$\chi(\mathrm{mJ/m^3})$	
	53°C	76°C		53°C	76°C
Tributyl phosphate	−0.65	−0.53	Tricresyl phosphate	0.38	0.38
Dibutyl phthalate	−0.04	−0.01	Diethyl phthalate	0.42	0.40
Dioctyl phthalate	0.01	0.03	Butyl ricinoleate	1.22	1.20
Diethyl sebacate	0.17	0.18	Butyl alcohol	1.74	1.58

Some authors believe that more accurate forecast of compatibility values can be obtained on the basis of calculated solubility parameters instead of experimental ones. It is recommended to carry out the calculations in the following order [71]:

1. The polymer solubility parameters are calculated by (1.8).
2. Interaction parameters between components of the polymer-solvent system are estimated on the base of a simplified Hildebrand's equation at 25°C

$$\chi = (\delta_p - \delta_s)^2/6 \tag{1.12}$$

3. The critical value χ_{cr} is calculated by (1.11)
4. Determined χ and χ_{cr} are compared. In the case that $\chi > \chi_{cr}$, the components are incompatible at a given temperature; and vice versa, if $\chi_{ps} < \chi_{cr}$, the polymer and solvent can be combined completely.

Process compatibility of components of polymer CM is evaluated by the following major criteria:

1. physical-chemical properties characterizing processing of the compositions, including melt flow, moldability, mechanical and thermo-mechanical stability of the blend in a melt, presence of phase separation, absence of thermal destruction of components;
2. physical-mechanical characteristics of CM.

Process compatibility of components in a polymer composition at their joint processing may exclude the formation of defects and assist in attaining high physical-mechanical and functional characteristics of the article produced.

One of the reasons of defective plastics with low anticorrosion properties (especially films) is phase instability and thermal destruction of CI and other low-molecular-weight components under elevated temperatures. In this regard, the thermal stability values of LDPE, Pl and their blends with PE and some VCI of the PHC series are also cited in Table 1.6 for comparison [50, 51]. Above considered low-molecular-weight components have T_{di} at or above the LDPE processing temperatures and will not undergo thermal destruction during combination with the polymer binder melt. High thermal stability of VCI of PHC series is one of the merits of inhibited plastics.

Plastic products formed of the blends of compatible components commonly acquire perfect mechanical properties. Nevertheless, sometimes process peculiarities and specifics of their operation conditions as well as economical reasons force the manufacturer to introduce ingredients that are either incompatible or have restricted compatibility with the polymer binder and other CM components. For instance, CI of inhibited plastics are, as a rule, highly polar materials poorly compatible with non-polar polymers, including PE, PP, PS and other thermoplastics widely adopted as binders (see 1.3).

In the case that the components are incompatible, the material strength is impaired due to imicro-inhomogeneity. Mechanical loads applied to such material will be distributed unevenly within the volume and can be higher on some portions than on other. This leads to violation of the material continuity in most loaded sites. Structural defects formed as a result quickly propagate through the whole material leading to its failure. As can be seen, mechanical compatibility of components is critical for plastic materials.

As long as mechanical properties of CM are governed chiefly by the mechanical and physical-chemical interactions of the ingredients at the interfaces, it can be regulated by changing the conditions of these interactions.

Compatibility can be regulated by the following methods:

(1) thorough choice of CM formulation to ensure that the solubility parameters of the polymer match those of the introduced substances,
(2) formation of developed filler surface and its uniform distribution within the polymer binder (via mechanical, US or other methods of dispersion),
(3) creation of porosity and developed surface of the polymer matrix to raise the sorption capacity of the polymer carrier towards the functional components (e.g. polymer foaming and pore filling by CI in liquid- and gas-filled inhibited plastics),
(4) chemical surface modification of the incompatible component to intensify adhesive interactions at the polymer-filler interfaces,
(5) impregnation of auxiliary substances to improve compatibility (these are often Pl possessing certain compatibility with polymers and CI),
(6) stimulation of electric and magnetic interactions between the filler and binder [41].

Proceeding from the above cited information it seems significant to underline that special attention should be paid to the problems of thermodynamic and process compatibility of ingredients when developing compositions of and production technologies for inhibited plastics.

1.5 Manufacturing Methodology of Inhibited Plastics

There are many problems in the regulation of compatibility of the ingredients, when commercial structural polymers and CI are combined in a composite material. The degree of compatibility is regulated by the following criteria:

(1) Inhibitors should be impregnated into the polymer matrix to adhere to it so strongly as to ensure durability of the composite as an integral whole avoiding random factors able to separate the components;
(2) Physical-chemical structural and technological bonding of components should not be too tough as to admit metered (preferably regulated) liberation of inhibitors from the polymer matrix.

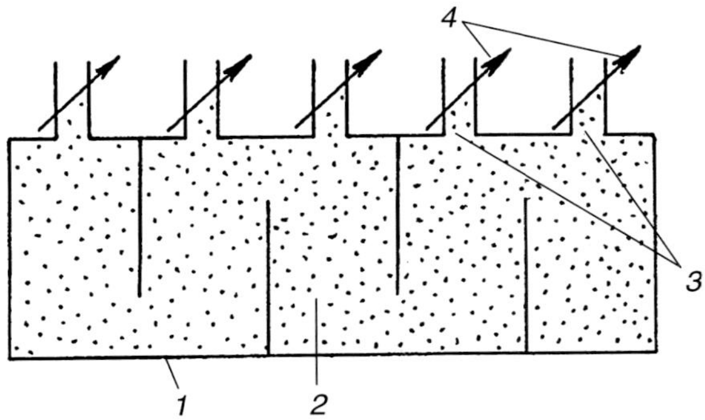

Fig. 1.14. Functional scheme of inhibited plastics: (1) polymer matrix; (2) inhibitor; (3) channels; (4) regulation system of inhibitor liberation

The described system (Fig. 1.14) is implemented in the anticorrosion technology via active, adaptive and smart composites. The polymer matrix adds strength to the material and serves as a container for a corrosion inhibitor. The latter can move within the matrix in the liquid phase and liberate into the ambient medium. The channels through which the inhibitor is liberated match a regulation system, which is actuated either spontaneously or under control factors. The thermodynamically non-equilibrium composite structure starts to approach its equilibrium state as the inhibitor is liberated. This process is accompanied by dissipation of the energy accumulated in the composite during forming.

1.5.1 Mass Transfer of Inhibitors within the Anticorrosion System

Liberation of CI should be in line with the following system parameters: (i) the corrosion system consisting of a metal article and corrosion medium, (ii) the protecting system against corrosion of the article, whose main component is the inhibited plastic element.

In the case of optimum supply of corrosion inhibitor into the protecting system the following phenomena take place:

- the inhibitor is adsorbed on the metal article surface in amounts sufficient to avert corrosion;
- the protecting concentration of the inhibitor is achieved rather soon and is preserved within a given time.

Let us review the schema presented in Fig. 1.15.

The corrosion system presented in Fig. 1.15a, includes metal article 1 and corrosion medium 2. Polymer composite 3 incorporating a volatile corrosion inhibitor of metals is a part of the system protecting against corrosion.

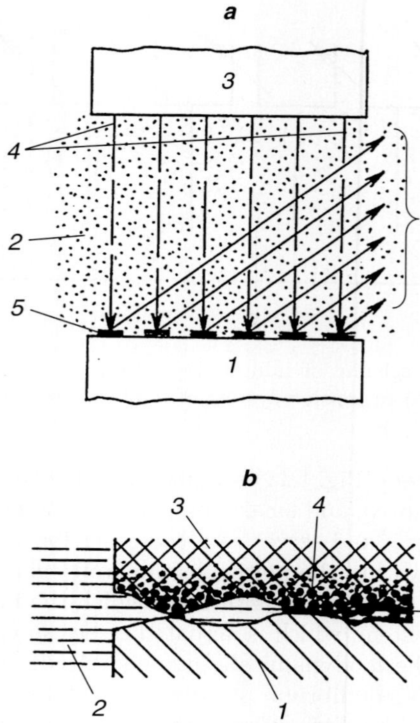

Fig. 1.15. Anticorrosion schema of metal articles based on plastics containing volatile (**a**) and contact (**b**) CI. (1) metal article, (2) corrosive medium, (3) plastics, (4) inhibitor, (5) inhibitor's adsorption layer, (6) vapor flow of the inhibitor desorbed from the article surface

It liberates vapors (4) of the inhibitors whose particles are transported via diffusion to the article and adsorb on its surface resulting in layer 5. Liberation, transportation and adsorption velocities of the inhibitor on the article surface should be reasonable from the point of view of protecting the article against corrosion. This means that formation of adsorption layer 5 is to occur before the onset of the corrosion process or at least before it becomes stationary. Since the protecting system is not absolutely leak proof, losses of the inhibitor from layer 5 may occur. New inhibitor supplied from composite 3 replenishes this unavoidable volatilization of the inhibitor vapors. The adsorption equilibrium in layer 5 is in line with the schema:

$$\begin{matrix}\text{An inhibitor molecule} \\ \text{in the gaseous phase}\end{matrix} + \begin{matrix}\text{A free adsorption center} \\ \text{on the article surface}\end{matrix} \rightleftarrows \begin{matrix}\text{An adsorption complex} \\ \text{(occupied center)}\end{matrix}$$

When a metal product is packed into a polymer film containing a volatile inhibitor, the product will turn to be protected from corrosion if [76]

$$\tau_0 (m_1 - m_3) + \tau_s (m_2 - m_3) \geq \mathcal{Q}_{\min} , \tag{1.13}$$

where m_1 and m_2 are the liberation rates of inhibitor vapors inside the package (m_1 – immediately after packing, m_2 – after time τ_0 expiry); m_3 is the volatilization velocity of vapors from the package at storage; τ_0 is the time in which atmospheric corrosion of the article reaches its stationary regime; $\tau_s > \tau_0$ is the full time of article storage in the package; $\mathcal{Q}_{\min}$ is the minimal amount of inhibitor vapors in the package ensuring protection of the article from corrosion.

It follows from equation (1.10) that after packing a considerable amount of inhibitor vapors are isolated inside the package, i.e. $\tau_0(m_1 - m_3) > \mathcal{Q}_{\min}$. The purpose of this shock dose is to avert corrosion of the packed article that has started already in atmosphere and to shorten time τ_1 of accumulation a critical amount of inhibitor vapors from $\mathcal{Q}_{min}$ until $\tau_1 < \tau_0$. After this, the inhibitor may be released at a steady speed $m_2 \ll m_1$ or during a certain time, e.g. $m_2 < m_3$, providing the condition (1.13) is fulfilled. To guarantee this regime of inhibitor release, it is necessary to: (i) introduce a mixture of quickly and slowly evaporating inhibitors in the polymer matrix or (ii) bind the inhibitor with the polymer at two structural levels to maintain different evaporation rates, or (iii) use other technological means of varying evaporation of the inhibitor from the polymer composite.

Acceptable parameters of vapor leakage from the anticorrosion system are defined on the grounds of the following arguments. A prolonged protecting effect takes place when penetrability of the system as a diffusion barrier for inhibitors is less than half that for water and corrosion activators from the environment. The diffusion factor of water is $D \sim 10^{-8}\,\mathrm{cm^2/s}$ and volatile acids $D = 10^{-6} - 10^{-7}\,\mathrm{cm^2/s}$ in PE, therefore the optimum diffusion factors of CI in a polyethylene packing film are of the order $D = 10^{-9} - 10^{-10}\,\mathrm{cm^2/s}$. It is evident that accomplishment of such conditions is connected with complication of the structure, production technology of the inhibiting film, and addition of special layers functioning as a diffusion barrier that increment material capacity.

Corrosion protection of metal ware by plastics containing contact inhibitors can be exercised according to the schema in Fig. 1.15, *b*. An anticorrosion plastic element 3 contains an inhibitor in the polymer matrix micropores. Element 3 contacts the protected metal surface 1. Corrosive medium 2 penetrates into the contact gaps and wets the article surface. Corrosion inhibitor 4 either spontaneously or under some effect, e.g. of mechanical forces, releases from the micropores of the polymer matrix into the contact gaps and adsorbs on the metal article surface. In case of mutual displacement of the contacting article 1 and element 3, spreading of the adsorption layer over the article surface is accelerated.

The article will be protected from corrosion if the adsorption layer of the inhibitor covers the whole surface. It is apparent that, to meet this condition, the inhibitor has to be a surface-active substance. The main growth mechanism of the adsorption layer is surface diffusion, i.e. motion of inhibitor molecules over the article surface not transferring into the volume phase. The kinetic relation for surface diffusion is:

$$\Theta = \Theta_e \left[1 - \exp\left(-k\tau\right)\right] \ ,$$

where $\Theta = a/a_{\mathrm{m}}$, a and a_{m} are the total number of active centers on the article surface and number of occupied centers; τ is time, $\Theta = \Theta_{\mathrm{e}}$ at $\tau \rightarrow \infty$; $k = k_{\mathrm{a}} + k_{\mathrm{d}}$, k_{a} and k_{d} are the adsorption and desorption velocity constants of the inhibitor. Surface diffusion of the inhibitor can keep to the schema in Fig. 1.15a, providing adsorption of the inhibitor on the surface protected takes place from the gaseous phase. So far, condition (1.13) can be observed even at $m_2 < m_3$.

The specificity of corrosion protection of the inhibited plastic-metal friction pair is liable to tribochemical transformation of the polymer matrix. The products of tribochemical reactions enter into interactions with the elements of the corrosion system and CI liberated from the plastics:

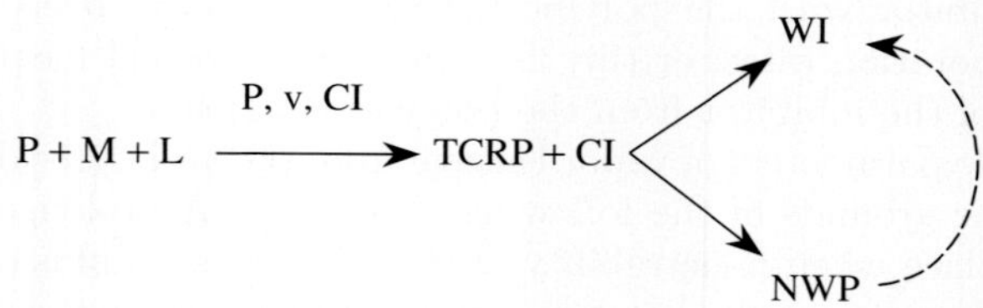

P, M and L are the polymer, metal and lubricating medium; P, ν, CI are the frictional factors, i.e. pressure, velocity and the presence of CI in the friction zone; TCRP are the tribochemical reaction products. The latter can fulfill the function of wear inhibitors (WI) during physical-chemical interactions with the inhibitor on the metal friction surface or form neutral wear products (NWP) affecting neither corrosion nor friction. The task is how to transform these products into useful ones during friction.

It is peculiar for a friction joint as a corrosion system to intensify transfer of the substances participating in tribochemical and electrode reactions. These static processes following either sorption or diffusion mechanisms are speeded up drastically by mutual displacements of the frictional surfaces.

1.5.2 Processes of Combining Polymers and Inhibitors

Thermodynamic compatibility of polymers and CI, and the technology as well, are involved in an integral composite material. Difficulties arising can be overcome by creating quasi-compatible composites. This is the direction in which materials science of inhibited plastics is progressing today. In Fig. 1.16

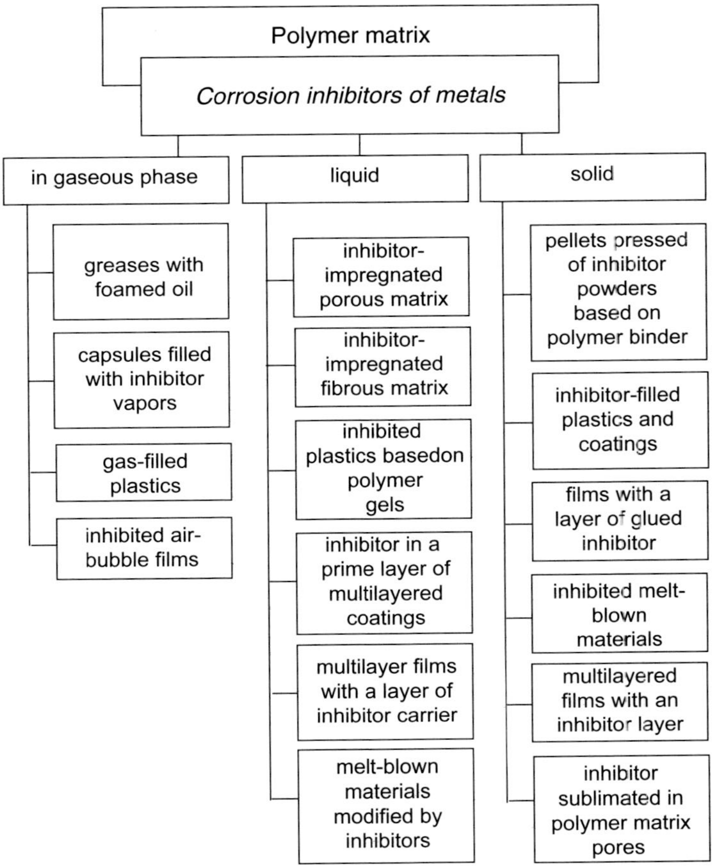

Fig. 1.16. Combinations of polymers and CI of metals in engineering materials

the examples of kinetically equilibrium polymer composites incorporating CI in all phases are illustrated. Let us analyze the technologies presented below.

Plastics containing inhibitors in the gaseous phase are used in engineering comparatively rarely.

This class of materials includes inhibited greases produced by impregnation of a polymer thickener into liquid oils foamed by the inhibitor vapors. The polymer forms a skeleton that, under high loads, imparts the properties of a solid to the lubricant ($p < 0.1 - 2.0\,\mathrm{kPa}$). Thanks to the high viscosity of the foamed oil, it localizes in the polymer skeleton cells so that the system becomes quasi-compatible. It can be roughly referred to as a plastic, although showing the properties of both solid and liquid polymer materials.

Hollow polymer capsules (or microcapsules) filled by inhibitor vapors are intended for fighting corrosion in oil and gas wells. Capsules are pumped into the well with a liquid flow. As soon as the capsule shell ruptures under mechanical loading, it liberates the corrosion inhibitor. The inhibitor molecules adsorb to the well walls and promote local protection against corrosion in places to which the capsules have stuck. Since the capsules appear to be either glued or sintered, they form a composite material. Containers from these materials can be made for storing and transporting metal ware to protect from corrosion, vibration and impacts.

Gas-filled plastics for tribo-engineering purposes are antifrictional materials that liberate vapors of CI in the course of wear and thus protect friction joints from mechano-chemical wearing under operation in hostile media and during stoppage.

An air-bubble packing film consists of a pair of adhering films – the base and the coating. The coating film contains molded cylindrical voids in the form of tablets. The air-bubble films are perfect dampers of impact loads, and bubbles filled with corrosion inhibitor vapors are capable of protecting products packed into them.

Plastics with liquid CI are the largest class of inhibited materials.

The simplest representatives of this class are porous plastics or fibrous polymer materials saturated with liquid (volatile or contact) CI.

Inhibitor-modified polymer gels are also representatives of a broad group of plastics, which are structurized systems formed at solidification of polymer solutes or at a solid polymer swelling. Their structural peculiarities and other properties were reviewed in Chap. 1.4.

Original structural and engineering solutions lie at the base of multilayered polymer coatings containing CI in their prime layer as well as film materials, where an interlayer represents an inhibitor carrier.

Among the described types of materials are those modified by CI melt blowing. Polymer melt-blown materials are produced by a special kind of extrusion when the fibers in a viscous-flow state are squeezed through a spinneret, dispersed by a gaseous flow and then deposited on a substrate [77]. Such polymer materials can be processed jointly with liquid CI, or the fibers can be modified by inhibiting aerosols in the course of dispersion and deposition.

Plastics incorporating solid CI were the first to appear among the inhibited polymer composites.

It is now common experience to use pressed pellets of inhibitor powders based on a polymer binder for packing and preservation of metal ware. They are easy to handle and can ensure prolonged isolation of the inhibitor inside the packing space.

Furthermore, an idea of obtaining inhibited plastics by filling polymer binders with solid inhibitor particles has apparently been realized. A wide range of structural plastics, films and coatings containing powder inhibitors is produced by the major companies of the world at the present time. The

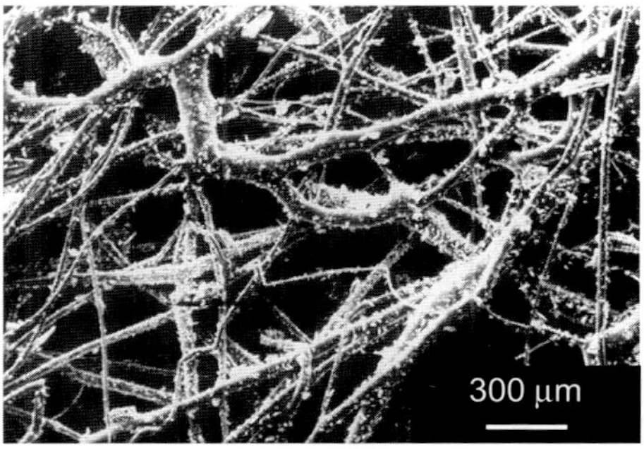

Fig. 1.17. Electron microphotograph of LDPE-based melt-blown material modified by a solid corrosion inhibitor (particles were injected into the gas-polymer flow). (1) polymer fiber, (2) a solid inhibitor particle

structure and properties of the most effective materials of this class will be considered in Chaps. 2–4.

Early inhibited packing films developed in the USSR in the 1970s were polymer films with glued volatile CI on one side [76]. Further elaboration of this idea has lead to melt-blown materials whose fibers carry particles of CI (Fig. 1.17).

This class of inhibited plastics includes multilayered films containing particles of a powder inhibitor between its polymer layers.

Successes in plastics technology have spurred the progress of composites with a microporous polymer matrix having a layer of solid phase inhibitors deposited by sublimation on pore walls. These materials are mainly designed for protection of friction joints against corrosion under both static and dynamic conditions.

The aforementioned plastics will be described in this book. Unusual testing methods are needed to characterize the structure and behavior of these multifunctional active materials used in numerous domains. Most of these materials conform to the concept of smart materials and have a great future.

1.6 Estimation Procedures of Inhibited Plastics Efficiency

The efficiency of the level of protection from corrosion damage of inhibited plastics is characterized by their protective ability. Estimation procedures are specified by certain standards of the International Organization for Standardization (ISO) as well as by the corresponding standardization organs of a number of countries and industrial branches.

In Russia the problem of protection against corrosion has occupied a separate standardization sphere since 1973 when a unified system of protection from corrosion, aging and biological damage was created. A number of standards within this system have been developed by international technical committees. In 2000 the Russian Research Institute of Standardization created a national system of certification of substances, materials and products with respect to their anticorrosion protection, through which the rules and order of certification works were approved. There is a network of testing centers credited with the right to test certain products based on the requirements developed [78].

Corrosion tests of the means of temporary anticorrosion protection to which inhibited plastics belong, are conducted using both laboratory and outdoor test methods [39, 79–83].

A convenient net of specialized corrosion and climatic stations intended for long-term *outdoor testing* was organized. Their locations in industrial, rural, urban and marine areas ensure the necessary temperature, humidity and aerochemical conditions for the tests [39, 79, 81, 83–85].

The metallic rectangular test specimens in the form of flat sheets or manufactured articles or their parts protected in a certain manner by temporary anticorrosion methods (e.g. by application of some inhibited polymer coating or packing in an inhibited film) are placed in the open air or under shelters for outdoor testing.

Open-air exposure involves direct exposure to all atmospheric conditions and contaminants. Sheltered exposure is exposure with protection from atmospheric precipitation and solar radiation either under a cover or in a partly closed space such as shutter sheds. The total period and the season of exposure depend on the type of test specimen and the purpose of the test. The evaluation is carried out by visual examination, metallographic examination and estimation of mass loss, change in mechanical properties or performance characteristics of the test specimens [84].

The main drawbacks of outdoor tests are that they are labor intensive and the prolonged time needed for reliable results. These results are the most reliable of all types of testing [39].

Accelerated testing of the protective ability of the preservation of hardware by inhibited plastics is carried out by standard methods at the branch, state or international level. The real conditions of storage and transportation of preserved articles are simulated during these tests. With this aim, the system is subjected to a cyclic loading with a salt mist in dry or wet conditions with [87–89] and without moisture condensation [86]. Aside from this certain simplified non-standardized procedures take into account the type and purpose of preserved hardware as well as the required conditions, aims and terms of preservation [49–51, 90–92].

The standardized tests use special apparatuses called exposure cabinets for exposure to salt mist, heat and moisture. Normative documents contain special sections specifying the form, size and quantity of metal samples along with their preparation procedure for the tests. Preparation of the samples consists of cleaning of fatty contaminants and polishing to a certain surface roughness. The majority of the methods of evaluating the results of corrosion tests have been standardized [93–95].

Different evaluation criteria can be applied, e.g.: specimen appearance after the test and after removing corrosion products; number and distribution of corrosion defects; the time elapsing before the appearance of the first sign of corrosion; change in mass; alteration revealed by micrographic examination; change in mechanical properties. Computer optical scanning is a new promising control method for corrosion damage [96].

Various conditions of accelerated corrosion testing are summarized in Table 1.10. According to ISO 14993 [85], water condensation on the test specimens should not occur under wet conditions. The methods of testing organic coatings on metallic materials according to ISO 11997-1 under cyclic corrosion conditions include condensation of water on the test surface during the period of humidity [87].

Russian Standard 9.509-88 [12] presupposes tougher test conditions that include the effect of salt mist, elevated temperature and humidity with and without moisture condensation, the effect of cold and endurance under normal conditions (Table 1.10). A single testing cycle corresponds to the real storage conditions or transportation of the samples indoors with corresponding ventilation without regulation similar to the regions with moderate and cold climate for six months.

Kanzaki's cyclic method for estimating the protective ability of inhibited papers and film materials specified by USA (MIL-1-3420) and Japanese standards is based on cyclic cooling-heating of metallic samples placed in a corrosion cabinet with regulated humidity [90]. This method is attractive due to its simplicity and expressiveness.

The cabinet is made of a 1-liter glass jar (Fig. 1.18), into which a 35% aqua solution of glycerin is poured (10 ml) to create an atmosphere of 90% of relative humidity in the free volume. In the waterproof cap a steel sample is fixed along with two strips of the inhibited film (paper) under study. The steel sample is made as a hollow cylinder forming the bottom of the aluminum tube passing through the cap and tightly fastened in it. The aluminum tube serves as a water vessel.

Cold water at 2°C temperature is poured into the aluminum tube, taken away after three hours and the sample is then examined. Three samples are tested at once. In the case that none of them shows any sign of corrosion the testing is continued.

The use of a diurnal cycle with Kanzaki's cabinet has been substantiated [90]. The cycle consists of a 3-hour exposure of the jar in a fridge at 3 ± 2°C followed by a 3-hour endurance in a thermostat at 55 ± 2°C, again for two hours in the fridge and 16 hours in the thermostat (Fig. 1.19). The surface of the metal samples is examined upon each complete cycle, the area of corrosive damage is recorded and an estimate point is given to corrosion resistance

Table 1.10. Accelerated test conditions. Temperature (T,°C), relative air humidity (RH, %), and test duration (τ, h)

Test Methods	Salt Mist		Elevated Temperature			Elevated Temperature and Humidity						Cold		Normal Conditions
						Without Moisture Condensation			With Moisture Condensation					
	T	τ	RH	T	τ	RH	T	τ	RH	T	τ	T	τ	τ
ISO 14993:2001	35 ± 2	2	<30	60 ± 2	4	>95	50 ± 2	2	–	–	–	–	–	–
Russian Standard 9.509-89	25 ± 2	2	50 ± 3	60 ± 2	12	93 ± 3	35 ± 2	66	94 ÷ 100	25 ± 2	8	–60 ± 2	6	2

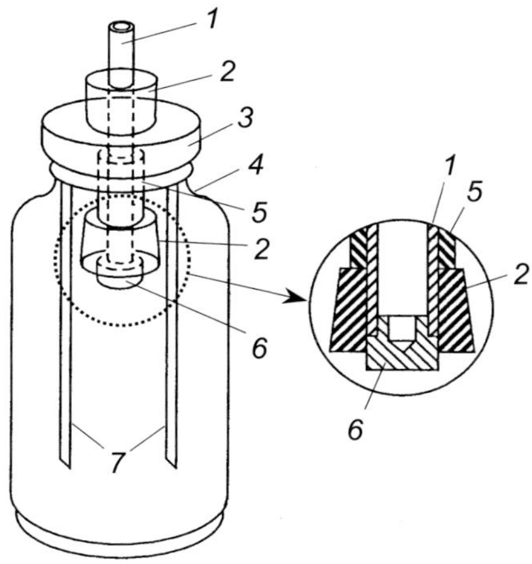

Fig. 1.18. Scheme of corrosion cabinet for testing inhibited film materials by Kanzaki's method: (1) aluminum tube, (2) rubber washer, (3) plastic cap, (4) glass jar, (5) insulating polymer tube, (6) metal sample, (7) strips of studied material

(Table 1.2) based on State Standard 27597 [93]. Kanzaki's method is, for example, employed in South Korean steel mill works for the estimation of the protective ability of packing materials based on inhibited polyethylene-paper laminates.

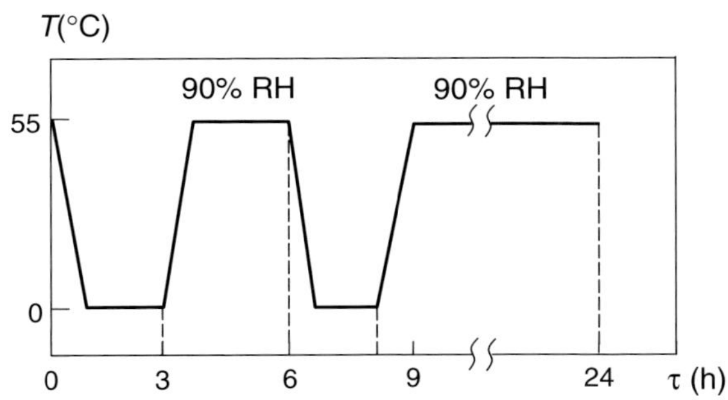

Fig. 1.19. Temperature and humidity conditions of Kanzaki Cycle Test

Independent methods of inhibited PE films of different compositions by the method described have proved that the reproducibility of results depends strongly on the surface roughness of the metallic samples subjected to testing. This method yields, as a rule, comparable results for inhibited plastics modified by highly volatile CI.

Very often the protective ability of inhibited film materials is estimated by laboratory desiccator methods.

Most known procedures of this group of methods are called oxidation and sulfuration tests. In the former case a metal test specimen wrapped in a VCI film material is placed in a desiccator (about 10 l in capacity). The internal atmosphere reaches 100% RH using 20 cm^3 of water. The desiccator is blocked up and is placed in a 50°C constant-temperature tank or in normal-temperature room to promote the growth of rust. In sulfuration test the desiccators about 2.5 l capacity are used. After having adjusted the inside atmosphere to reach 93% RH using 10 cm^3 of a saturated solution of Na_2SO_4, a test metallic strip wrapped in inhibited film material is placed inside. The tests are continued until a corrosive phenomenon is observed.

For a number of years the following desiccator method of corrosive tests was used at MPRI NASB [49–51, 98]. Metallic samples in the form of plates $50 \times 50 \times 2$ in size are wrapped in an inhibited film and placed into the desiccator inside which $98 \pm 2\%$ relative air humidity is maintained by aqua solution of glycerin, and are subjected to cyclic heating-cooling. One cycle consists of the endurance of the samples in the same humidity at a temperature of 55 ± 2°C for 8 h and at a similar humidity and a temperature of 20 ± 2°C for 16 h. The protective ability of the films is estimated by the corrosion resistance of metallic samples after 21 test cycles (Table 1.11).

Table 1.11. Estimation of corrosion damage of metal samples according to Russian Standard 27957

Area of corrosion damage (%)	Estimate point
No damage	10
0–0.2	9
0.2–0.5	8
0.5–1.0	7
1.0–2.5	6
2.5–5.0	5
5.0–10.0	4
10.0–25.0	3
25.0–50.0	2
50.0–100.0	1

The laboratory accelerated corrosion test practiced at Cortec Co. includes exposure of cold-rolled steel plates wrapped in packets of inhibited films and subjected to thermal-humid atmosphere in special cabinets at 54°C temperature and 95% humidity for 21 days (504 h). Upon the termination of the test the corrosion damage of the samples is estimated.

An analogous procedure is described by a Japanese standard JLS K 2246 for evaluation of the protective ability of polymer films designed for preserving precision parts during short-term storage after manufacture at service points. Polished and cleaned steel plates of $1.2 \times 60 \times 80$ mm size tightly packed into inhibited films are exposed to atmosphere at a temperature of 49°C and 98% RH. After 7 and 20 days and one, two and three months the degree of corrosion damage of the samples is examined. For this purpose, the metal plate surface is overlaid with a colorless transparent stencil film having 100 squares 5×5 mm each. Then, the number of squares (n) is determined within which field one or morecorrosion spots are observed. The degree of corrosive damage is subdivided into five categories (K) (Table 1.12).

Table 1.12. Estimation of corrosive damage of metallic specimens according to JIS K 2246 standard

K	A	B	C	D	E
n	0	1–10	11–25	26–50	51–100

The accelerated desiccator estimates of the protective ability of inhibited films intended for dry preservation (without oiling) of bearings and their members are in extensive use in the bearing industry. For instance, the All-Russian Research Institute of the bearings industry recommends the following procedure for corrosion tests. Preliminary degreased parts of bearings (inner and outer rings, rollers, separators, metal and alloy specimens) are tightly packed into the inhibited film and are maintained for 90 days at 18–22°C in desiccators over a saturated KCl solution with glycerin addition (RH 85–90%). After this, the areas damaged by corrosion are estimated according to a ten-point scale (Table 1.11). The samples that withstood the testing are given a one-year warranty period of protection at waterproof packing in the film in field conditions. Kharkov Bearing Plant Co. (Ukraine) suggests a procedure during which the specimens of bearings packed into an inhibited film are exposed in desiccators over a 10% solution of NaCl for 45 days at room temperature. If traces of corrosion are not observed on the specimens, the specimens are given a one-year warranty period of storage and transportation by marine and road transport. The relevance of such conclusions is a result of many years of experience in preserving bearings in inhibited films.

For a qualified express estimate of the protective ability of inhibited polymer coatings, Russian Standard 9.905 specifies a method of measuring polarizing resistance and capacity of metallic samples with a coating exposed to a solution of NaCl. A solution of electrolyte is poured into a glass cylinder for which the coated sample serves simultaneously as the bottom and the working electrode. An auxiliary electrode is immersed into the cylinder with the solution. The protective ability of the coating is estimated by analyzing the dependence of the specific resistance and capacity (per unit area of the sample) versus alternating current frequency. If the capacity is independent of the current frequency and the resistance is inversely proportional to it, the coating is fit for protection [88].

An original and simple method of express estimate of the protective ability of inhibited coatings is described in [83]. The cyclic strength of steel wire samples coated by inhibited polyethylene is estimated upon exposition for two to 12 days in a hostile medium (0.2 n aqua solution of sulfuric acid). The index of cyclic strength is taken as the number of bends in opposite directions through 90 degrees preceding failure of the sample.

In contrast to film materials, express estimates of the protective ability of inhibited structural plastics have been insufficiently elaborated so far. For example, the effect of inhibited plastics on the wear rate of metal samples has been studied by the authors of [100] using a pendulum tribometer containing a plastic-metal friction pair. Corrosion tests of systems such as inhibited structural plastic-metal counterparts are usually conducted in real operation regimes [44].

In addition to the above it should be emphasized that the results of accelerated corrosion tests depend strongly on a number of factors such as test conditions, type of metal specimens (ferrous or non-ferrous metals, metal pairs), preparation methods of the specimens amongst others. The agreement between the results of accelerated laboratory and actual tests is a subject of discussion [90] since universal criteria of estimating the results of actual and accelerated tests have not been as yet substantiated [39]. Reliable conclusions on the protective ability of inhibited plastics should proceed from a comparative analysis of the results obtained by different methods with the benefit of a profound statistical database.

1.7 Designation of Inhibited Plastics

To conclude Chap. 1, we would like to present a classification of inhibited plastics by their properties and application fields (Fig. 1.20). One can find detailed description of different types of inhibited plastics further in this book in corresponding chapters.

An **inhibited polymer coating** represents a layer of a polymer-based composite material bonded with the substrate via an adhesive layer. The composite material may contain a plasticizer, corrosion inhibitor and technological additives in its composition (see Chap. 3). The coatings are subdivided into peelable and unpeelable by their behavior and the intensity of their adhesive interactions with the article being protected.

Peelable inhibited coatings are, in the main, widespread in preservation techniques. Thanks to their sufficient thickness, high physical-chemical and protective properties as well as low adhesion to protected metal surfaces, they are able to function simultaneously as a preserving and packing method. The range of polymer materials constituting the base of peelable inhibited coats is extremely broad; the methods of their formation are numerous and original.

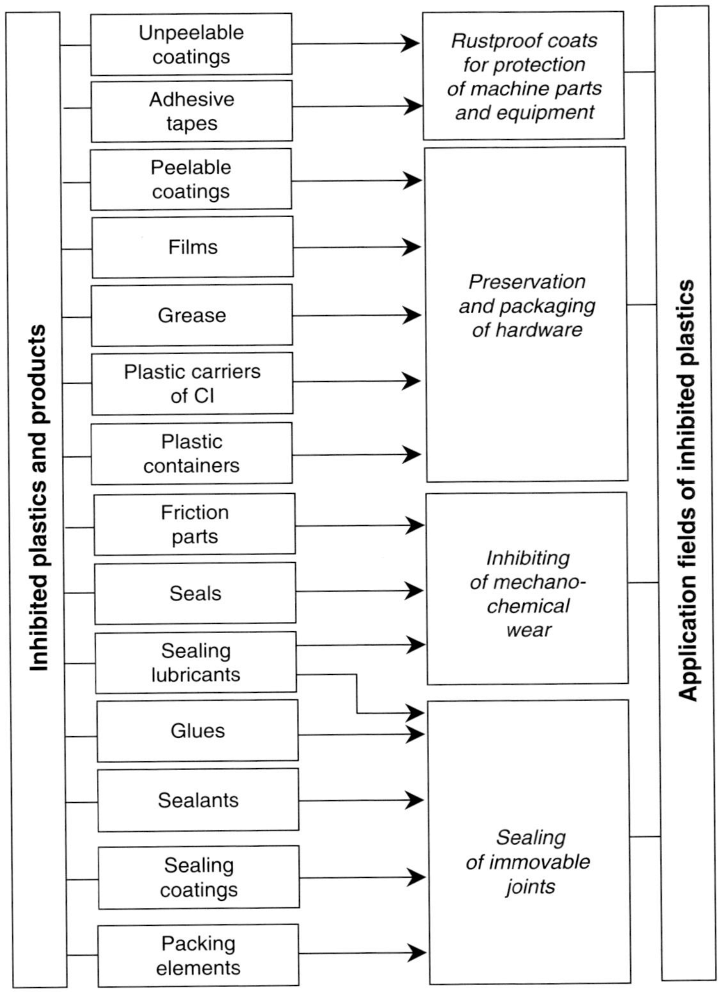

Fig. 1.20. The range of inhibited plastics and main fields of application

The coatings can be melt-laid, applied in suspensions based on polyolefines or their halogen-derivatives, polyesters, acrylates, epoxy resins, cellulose derivatives and numerous copolymers.

Unpeelable inhibited coatings are used largely as primers preceding the paintwork on cars and equipment.

Sealing inhibiting coatings are spreading at present in practically all fields of technology. These coatings serve mostly for imparting anticorrosion properties to bearing members of structures. They can be used as individual sealing elements or as a means of increasing the tightness of contact seals. The sealing joint can have significant toughness and sealing properties by creating a gradient of mechanical properties through the coating thickness. Such coatings are used in threaded joints of hydraulic equipment, oil pipelines and other special equipment.

Inhibited polymer films (IPF) containing contact or volatile CI stand on a par in popularity with preserving coatings (see Chap. 2). The advantages of IPF are high tensile strength, resistance to incision, tearing and multiple bending; they are well fusing and sticking and are indispensable in preserving and packing procedures of hardware. Along with good barrier properties, IPF present another means of preserving hardware with high efficiency of the preservation and depreservation processes, and manufacturability. Moreover, it is possible to combine wrapping with bagging and other packing procedures, including thermal vacuum molding, extrusion etc. The diversity of polymer materials used in films and CI gives grounds for the production of packages fit for service in various climatic conditions. IPF with adhesive layers are used as well in insulating metal pipelines.

Parts of friction joints and seals are often made from inhibited structural plastics (see Chap. 4). Their main role in metal-polymer friction joints is to abate Mechano-chemical wearing of metal counterbodies. Application of inhibited antifrictional materials is considered to be a promising trend for extending the life of friction joints operating in hostile media. The structural polymers or their blends most often contain plasticizers, CI, fillers and modifiers as additional components. The inhibited plastics are employed in plain bearings, sealing elements and other members of oil-extracting equipment and vehicles.

Plastic containers (pallets, boxes, cases, etc.) are commonly made of structural inhibited plastics based on chiefly recycled materials.

A separate group of inhibited plastics includes such engineering materials as glues, sealants, plastic CI carriers, multi-purpose lubricants and others (see Chap. 5). **Inhibited glues** are intended to form tight, nondetachable joints in machine parts operating in hostile media. **Sealants** are able to prevent permeation of the ambient media and create impermeable immovable joints of machine parts.

Plastic CI carriers represent an alternative to inhibited polymer films used for preservation and packaging of metal ware. They are usually made of polymer porous or fibrous materials or may have microcapsules whose voids are filled by volatile CI. Plastic CI carriers like pellets, perforated vessels or fibrous materials are placed inside a tight package to create an inhibited atmosphere that averts corrosion of the packed article.

Inhibited greases were until recently the major and in fact unique material for preserving hardware. Today they tend to be replaced by liquid inhibited lubricants, thin coats and films. **Sealing lubricants** are used to tighten or to alleviate screwing/unscrewing of threaded joints in engineering, machine-tool construction and the oil and gas industry.

Sealing lubricants contain, as a rule, an oily base, polymer thickener, CI and functional additives. The introduction of various fillers, e.g. antifrictional (graphite, molybdenum disulfide), protective (metal powders), dielectric and heat-resistant (talc, resin, asbestos) or others, adds excellent sealing properties to the lubricants. The main fields of assimilation of the inhibited sealing lubricants are sealing of threaded joints, shut-off valves for oil and gas pipelines, manufacture of stuffing-box seals for high-pressure pumps and so on. Inhibited sealing lubricants can also be used as a modifying agent during impregnation of wood-polymer friction joints.

References

1. L.I. Freiman. Corrosion of Metals. *Chemical Encyclopedia,* Moscow, Soviet Encyclopedia, 1990, Vol. 2, pp. 480–482.
2. H.H Uhlig and R.W. Revie. *Corrosion and Corrosion Control. An Introduction to Corrosion Science and Engineering,* 3rd edn. NY, Chichester, Brisbane, Toronto, Singapore, John Wiley & Sons, 1985.
3. M.S. Antelman. *The Encyclopedia of Chemical Electrode Potentials.* NY, Plenum Press, 1982.
4. U.R. Evans. *The Corrosion and Oxidation of Metals.* London, Arnold, 1960.
5. M. Pourbaix. *Atlas of Electrochemical Equilibria in Aqueous Solutions.* NY, Pergamon Press, 1966.
6. Ch.P. Poole and F.J. Owens. *Introduction to Nanotechnology.* Indianapolis, John Willey & Sons, 2003.
7. D.D.L. Chung. *Composite Materials: Functional Materials for Modern Technologies.* NY, Springer, 2003.
8. G.G. Wallace, G.M. Spinks and P.R. Teasdale. *Conductive Electroactive Polymers: Intelligent Materials Systems.* Lancaster, Technomic Publishing Co., 1998.
9. E. Motyl. *Space Charge in Solid Dielectrics Analysis of Measurement Methods and Investigations.* D. Sci. Thesis, Wroclaw, 2000.
10. G.A. Gorokhovsky. *Surface Dispersion of Dynamically Contacting Polymers and Metals.* Kiev, Navukova Dumka, 1972.
11. G. Heinicke. *Tribochemistry.* Berlin, Akademie-Verlag, 1984.
12. A.S. Khanna, M.K. Totlani and S.K. Singh. *Corrosion and its Control* (in 2 Vol.), Elsevier, Amsterdam, 1998.
13. D.A. Jones. *Principles and Prevention of Corrosion,* 2-nd ed. London, Prentice-Hall, Inc., 1996.
14. I.L. Rozenfeld. *Corrosion and Protection of Metals,* Moscow, Metallurgiya, 1969.
15. J.S. Bora. Metal-plastic contact corrosion. *Corrosion Sci.,* 1974, Vol. 14, No. 8, pp 503–506.
16. A.I. Levin, N.I. Shklovskaya, V.F. Lazarev, et al. To the problem of aluminum and its alloys corrosion peculiarities in a closed volume containing polymer materials. *Protection of Metals,* 1974, Vol. 10, No. 3, pp. 314–317.
17. V.A. Goldade. *Investigation of some physico-chemical phenomena at developing and operation of metal-polymer systems.* Ph.D. Thesis, Riga, 1980.

18. I.L. Rozenfeld, F.I. Rubinstein and K.A. Zhigalova. *Protection of Metals from Corrosion by Paint Coatings.* Moscow, Khimiya, 1987.
19. V.A. Goldade, and L.S. Pinchuk. Protective ability of polymer coatings at effect of liquid media. *Transactions of BSSR AS, ser. phys.-techn. sci.,* 1982, No. 2, pp. 50–54.
20. K.R. Trethewey, and J. Chamberlain. *Corrosion for Science and Engineering,* 2nd edn. Longman Scientific & Technical, England, 1995.
21. V.A. Belyi, V.A. Goldade, A.S. Neverov and L.S. Pinchuk. Effect of polymer powders on kinetics of polarization at liquid freezing. *Colloidal Journal,* 1977, Vol. 39, No. 4, pp. 742–744.
22. G.E. Zaikov, A.L. Iordansky and V.S. Markin. *Diffusion of Electrolytes in Polymers.* Moscow, Khimiya, 1984.
23. S.A. Reitlinger. *Permeability of Polymer Materials.* Moscow, Khimiya, 1974.
24. H.K. Lonsdale, U. Merten and R.L. Riley. Transport properties of cellulose acetate osmotic membranes. *J. Appl. Polymer Sci.,* 1965, Vol. 9, No. 4, pp. 1341–1362.
25. H. Yasuda, C.E. Lamaze and L.D. Ikenberry. Permeability of solutes through hydrated polymer membranes. Part 1. Diffusion of sodium chloride. *Makromol. Chem.,* 1968, Vol. 118, No. 2858, pp. 19–35.
26. S. Rosenbaum, H.I. Mahon and O. Cotton. Permeation of water and sodium chloride through cellulose acetate. *J. Appl. Polym. Sci.,* 1967, Vol. 11, No. 10, pp. 2041–2065.
27. C.R. Thomas and R.E. Barker. Moisture and ion sorption in cellulose acetate. *J. Appl. Polymer Sci.,* 1963, Vol. 7, No. 5, pp. 1933–1942.
28. L.S. Pinchuk, V.A. Struk, N.K. Myshkin and A.I. Sviridenok. Ed. V.A. Belyi. *Materials Science and Structural Materials,* Minsk, Vysshaya Shkola, 1989.
29. G.W. Ehrenstein. *Polymer Materials: Structure, Properties, Applications.* Munchen, Carl Hanser Verlag, 2001.
30. V.A. Kabanov, M.S. Akutin, N.F. Bakeev, et al. (Eds.). *Encyclopedia of Polymers* (in 3 Vols.), Moscow, Soviet Encyclopedia, 1972–1977.
31. A. Kelly (Ed.) *Conciese Encyclopedia of Composite Materials.* Oxford, Pergamon, 1994.
32. M. Ash and I. Ash. *Handbook of Plastic and Rubber Additives.* In 2 Vol., NY, Synanse Information Resourses, Inc. 2003.
33. V. Uzdensky. Modification of polymers: improvement of processing conditions and giving special properties to the product. *Plastics. Plastics Processing Industry,* 2002, No. 7, pp. 22–28.
34. H. Zweifel. *Plastic Additives Handbook,* 5th edn Munchen, Carl Hanser Verlag, 2001.
35. M.M. Chanda and S.K. Roy. *Plastics Technology Handbook.* Marcel Decker, 1998.
36. P.A. Schweitzer. *Mechanical and Corrosion Resistant Properties of Plastics and Elastomers.* Marcel Decker, 2000.
37. M.V. Korotky and A.V. Makarevich. Modern trends in anticorrosion polymer materials. *Materials, Technologies and Tools,* 2001, Vol. 6, No. 3, pp. 59–65.
38. V.A. Goldade, A.S. Neverov and L.S. Pinchuk. *Low-Modular Composite Materials based on Thermoplastics.* Minsk, Anoka i Tekhnika, 1984.
39. P.A. Vinogradov. *Preservation of Engineering Goods.* Leningrad, Mashinostroenie, 1986.

40. A.F. Nikolaev. *Technology of Plastic Masses.* Leningrad, Khimiya, 1977.
41. L.S. Pinchuk and A.S. Neverov. *Polymer Films Containing Corrosion Inhibitors.* Moscow, Khimiya, 1993.
42. Yuri. Coached and E.F. Sominskaya. *Protection of Vessels from Corrosion and Fouling.* Moscow, Transport, 1984.
43. I.L. Rozenfeld and F.I. Rubinstein. *Anticorrosion Primers and Inhibited Paint Coatings.* Moscow, Khimiya, 1980.
44. V.A. Goldade, V.A. Struk and S.S. Pesetsky. *Wear Inhibitors of Metal-Polymer Systems.* Moscow, Khimiya, 1993.
45. *Modern Plastics Encyclopedia.* New York, McGraw-Hill, 1996.
46. A.I. Altsybeeva and S.Z. Levin. Ed. L.I. Antropov. *Metal Corrosion Inhibitors.* Leningrad, Khimiya, 1968.
47. I.L. Rosined and V.P. Persiantseva. *Inhibitors of Atmospheric Corrosion.* Moscow, Nauka, 1985.
48. R. Yukhnevich, V. Bogdanovich, E. Valashkovsky and A. Vidukhovsky, *Technika Przeciwkorozyina.* Warszawa, Wydawnictwa Szkolne i Pedagogiczne, 1976.
49. G.V. Rechits. *Creation and investigation of plastics containing corrosion inhibitors for packaging engineering goods.* Ph.D. Thesis, Gomel, 1987.
50. I.V. Tsarenko. *Polymer materials modified by polynitrous heterocycles and the mechanism of their protective effect.* Ph.D. Thesis, Gomel, 1996.
51. A.V. Makarevich. *Physic-chemical and engineering principles of developing active film and fibrous materials based on thermoplastics.* Dr. Sci (Chem.) Thesis, 2000.
52. J. Penninger, K. Wippermann and J.W. Schultze. Molecular structure and efficiency of triazole derivatives and other heterocyclics as corrosion inhibitors for copper. *Werkstoffe und Korrosion,* 1987, Vol. 38, pp. 649–659.
53. K. Wipperman and J.W. Schultze. The inhibition of zinc corrosion by bisaminotriazole and other triazole derivatives. *Corrosion Sci,* 1991, Vol. 32, No. 2, pp. 205–230.
54. P.A. Bradley 1,2,4-Triazoles as a corrosion inhibitor for copper. *Corrosion Sci.,* 1980, Vol. 20, No. 5, pp. 643–649.
55. France Patent 2661690, C 23 F 11/10. Composition protective des metaux, son procede de preparation et procedes de protection des metaux. G. Lapluye and T. Fan, 1991.
56. EP 0462666, C 23 F 11/10. Phenylmercapto-tetrazole/tolyltriazole and/or benzotriazole corrosion inhibiting compositions. C.V. Cha and P.P. Vanderpool, 1991.
57. I.V. Tsarenko, A.V. Makarevich, V.S. Poplavsky and V.A. Ostrovsky. Corrosion inhibiting by five-member polynitrous heterocycles. Part I. Substituted tetrazoles. *Protection of Metals,* 1995, Vol. 31, No. 4, pp. 356–359.
58. Belarus Patent 1813, M C 23 F 11/00. *A method of protecting ferrous and nonferrous metals from corrosion.* A.V. Makarevich, I.V. Tsarenko, L.S. Pinchuk, et al., 1997.
59. I.V. Stabenow, A.V. Makarevich and T.P. Kofman. Corrosion inhibiting by five-member polynitrous heterocycles. Part II. 1,2,4-triazoles. *Protection of Metals,* 1997, T. 33, No. 4, C. 415–417.
60. Belarus Patent 1943, C 08 J 5/18. *A method of manufacturing a polymer film.* L.S. Pinchuk, A.V. Makarevich and A.V. Tsarenko, 1997.

61. I.V. Tsarenko, A.V. Makarevich and T.P. Kofman. Effect of nitro-derivatives of 1,2,4-trizole on corrosion-electrochemical behavior of low-carbon steel in aqua solutions of sodium sulfate. *Electrochemistry,* 1997, Vol. 33, No. 10, pp. 1177–1182.
62. E.R. Booser (Ed.). *CRC Handbook of Lubrication.* Boca Raton, CRC Press, 1992.
63. EP 8802626, C 23 F 11/18, 11/173, F 25 B 47/00. *Inhibiting corrosion in aqua systems.* Dearborn Chemicals, Ltd., 1988.
64. R.S. Barstein, V.I. Kirilovich and Y.I. Nosovsky. *Plasticizers for Polymers.* Moscow, Khimiya, 1982.
65. P.V. Kozlov and S.P. Papkov. *Physic-Chemical Fundamentals of Polymer Plastification.* Moscow, Khimiya, 1982.
66. R. Lambourne (Ed.). *Paint and Surface Coatings: Theory and Practice.* New York-Chichester- Brisbane-Toronto, Ellis Horwood, 1987.
67. V.E. Kuleznev. *Polymer Blends.* Moscow, Khimiya, 1982.
68. S.P. Papkov. *Phase Equilibrium in Polymer-Solvent System.* Moscow, Khimiya, 1981.
69. A.S. Neverov, L.V. Samuseva and D.M. Lugovtsov. Investigation of regularities of phase separation in the polymer – low-molecular liquid systems. *Materials, Technologies, Tools,* 1998, No. 4, pp. 54–59.
70. S.P. Papkov. *Gel-like State of Polymers.* Moscow, Khimiya, 1974.
71. D. Paul and S. Newman (Eds.). *Polymer Blends.* Academic Press, NY, San Francisco-London, 1978.
72. A.A. Tager. Some questions of polymer plastification. *Plastic Masses,* 1990, No. 4, pp. 59–64.
73. D.W. Van-Krevelen. *Properties of Polymers.* 3rd completely revised edn. Elsevier, Amsterdam- London-NY, 1990.
74. S.A. Drinberg and E.F. Itsko. *Solvents for Paint and Varnish Materials.* Leningrad, Khimiya, 1980.
75. B.P. Shtarkman. *Plastification of Polyvinyl Chloride.* Moscow, Khimiya, 1975.
76. L.S. Pinchuk and A.S. Neverov. *Polymer Films Containing Corrosion Inhibitors.* Moscow, Khimiya, 1993.
77. L.S. Pinchuk, V.A. Goldade, A.V. Makarevich and V.N. Kestelman. *Melt Blowing: Equipment, Technology and Polymer Fibrous Materials.* Berlin, Heidelberg, New York, Springer-Verlag, 2002.
78. N.P. Borisov and L.P. Kotova. *Standardization and certification of means and methods of protecting materials and goods from corrosion, aging and biological damage.* 9th Conf. "Destruction and stabilizing of polymers", Moscow, 2001, pp. 27–28.
79. R. Babolan. *Corrosion tests and Standards. Application and Interpretation.* Hitchin, Herts, ASM International, 1995.
80. Yu.N. Mikhailovsky. *Atmospheric Corrosion of Metals and Methods of their Protection.* Moscow, Metallurgiya, 1989.
81. *Corrosion. Understanding the Basics.* Hitchin, ASM International, 2000.
82. R.T. Kenneth and C. John. *Corrosion for science and engineering.* 2nd edn. Essex, Longman Group Limited, 1995.
83. ASM Handbook, Vol. 13. *Corrosion: Fundamentals, Testing, and Prevention.* Hitchin, Herts, ASM International, 1987.
84. ISO 8565:1992. Metals and alloys – Atmospheric Corrosion Testing – General Requirements for Field Tests. Geneva, ISO Copyright Office, 1992.

85. ISO 9223:1992. Corrosion of metals and alloys – Corrosivity of Atmospheres. Classification. Geneva, ISO Copyright Office, 1992.
86. ISO 14993:2001. Corrosion of Metals and Alloys – Accelerated Testing Involving Cyclic Exposure to Salt Mist, "Dry" and "Wet" Conditions. Geneva, ISO Copyright Office, 2001.
87. ISO 11997-1:1998. Paints and varnishes – Determination of resistance to cyclic corrosion conditions – Part 1: Wet (salt fog)/dry/humidity. Geneva, ISO Copyright Office, 1998.
88. Russian State Standard 9.509-89. Means of temporary protection against corrosion. Methods of determining protective ability.
89. ISO 12944-6:1998. Paints and varnishes. – Corrosion protection of steel structures by protective paint systems. – Part 6: Laboratory performance test methods. Geneva, ISO Copyright Office, 1998.
90. L.S. Pinchuk, K.Y. Choi and S.V. Bezrukov. On accelerated corrosion tests of inhibited packaging materials. *Protection of Metals,* 1994, Vol. 30, No. 4, pp. 434–436.
91. Yu.N. Mikhailovsky, V.M. Popova and A.I. Marshakov. Field and accelerated tests of contact and volatile inhibitors of atmospheric corrosion on different metals. *Protection of Metals,* 2000, Vol. 36, No. 5, pp. 546–551.
92. S.G. Tyr, O.V. Moshura and E.V. Tyr. Vistas in adoption of volatile inhibitors. *Protection of Metals,* 2001, Vol. 37, No. 6, pp. 593–597.
93. Russian State Standard 27597-88. Electronic goods. Methods of estimating corrosion resistance.
94. ISO 8407:1999. Corrosion of metals and alloys – Removal of corrosion products from corrosion test specimens. Geneva, ISO Copyright Office, 1999.
95. ISO 11463:1995. Corrosion of metals and alloys – Evaluation of pitting corrosion. Geneva, ISO Copyright Office, 1995.
96. V.A. Kotenev. Use of computer scanners in corrosion control. Thickness and topography of corrosive layers. *Protection of Metals,* 2001, Vol. 37, No. 3, pp. 284–293.
97. Belarus Patent 6309, G 01 N. A device for estimating corrosion damage. A.Ya. Grigoryev, K.Y. Choi, L.S. Pinchuk and N.K. Myshkin, 2004.
98. L.V. Zaborskaya, L.V. Golubova, Ya.M. Zolotovitsky and O.R. Yurkevich. Effect of polymer carrier on efficiency of volatile corrosion inhibitors. *Protection of Metals,* 1985, Vol. 21, No. 2, pp. 301–305.
99. V.A. Belyi, V.A. Goldade, A.S. Neverov and L.S. Pinchuk. Effect of corrosive medium on metal fatigue protected by polymer coatings. *Proc. Belarus AS,* 1982, Vol. 26, No. 6, pp. 517–518.
100. G.V. Rechits, L.S. Pinchuk, A.S. Neverov and V.A. Goldade. On the kinetics of electrochemical processes in friction of polymer composites containing corrosion inhibitors. *Soviet J Friction and Wear,* 1982, Vol. 3, No. 12, pp. 93–96.

2 Films Incorporating Corrosion Inhibitors

A film represents a fine continuous material layer up to 250 μm thick [1, 2] characterized by a large surface-to-volume ratio. In contrast to coatings, films do not adhere to other solid substrates.

Film formation is a specific property of polymers that distinguishes them from low-molecular matter. This is a considerable length and asymmetry of macromolecules capable of forming a strong oriented permolecular structure during stretching of polymer bodies that lies at the base of this property [2]. So far, polymers are considered to utterly suit to a raw material for production of the films.

A highly developed and reactive surface is another distinctive feature of polymer film materials. This feature only furnishes large potential in modification of films using various functional components, including CI and numerous available technological procedures [3–6]. Corrosion inhibitors impregnated into a film may be found in different aggregate states. Despite considerable differences in physical and chemical parameters, films modified by CI are able in most cases to retain their critical characteristics.

Polymer films containing CI have proved to be economical and reliable means of anticorrosion protection of metals. They are referred to as a promising class of preservation and packaging materials in addition to their barrier and corrosion-inhibiting functions. Conservation and packing of metal ware in these films can be united in a single process, and depreservation is essentially simplified. The temporary anticorrosion protection technique of metal ware by inhibited films has a number of advantages over other rust-proofing methods (conservation oils, inhibited coatings, paper, air, etc.), in terms of low material, labor and energy intensity of preservation-depreservation processes, high productivity, safety, automation level and ecological purity [3, 4].

World production of polymer film materials of anticorrosion purposes has been characterized in recent decades by a stable tendency to growth. To date engineering has adopted a vast range of polymer anticorrosion films in various technical domains.

The formulation and production processes as well as equipment for manufacture of inhibited films are the subjects of patent protection since they are, as a rule, know-hows of their manufacturers. The first patents for packaging films containing VCI can be traced back to the early 1950s. Beginning from the 1980s, inhibited polymeric films have come into extensive commercial use. Major scientific centers and industrial companies of the world are engaged in the development and adoption of competitive materials of this type in the international market. Among them are: Northern Technologies International Corporation, Cortec Corporation (USA); Nitto Electric Industrial Co. Ltd., Aicello Chemical Co. Ltd. (Japan), Noncor Korrosionsschutz Technologien und Produkte GmbH, Excor Korrosionsschutz Technologien, SÜD-Chemie AG (Germany); Branopac (Italy); Grofit Plastics (Israel).

Today the application of inhibited films is further subdivided into fields of anticorrosion techniques. This challenges further development of technologies and equipment for the production of this important type of polymer products.

In view of the aggravation of the problem of global contamination of the environment by engineering and domestic waste special attention is paid to the elaboration of non-toxic CI, recycled and biodegradable resins, ecologically friendly methods of salvaging waste and inhibited materials [5–8].

2.1 Classification

Inhibited films can be classified by the following criteria: (1) origin of the chief component (polymer base), (2) film structure, (3) CI type, its aggregate state and distribution pattern in the film; (4) designation of the film material (Fig. 2.1).

There are more than a dozen types of preservative films [3] discriminated by the **polymer base** that governs the physical-mechanical characteristics of the films (Table 2.1) [1,9–11].

Polyolefine films, mostly manufactured from polyethylene (PE), are most extensively used in anticorrosion techniques [8,12–20]. They are characterized by elasticity, low water and vapor permeability, mechanical strength, frost-resistance, inertness, perfect heat-sealing properties and non-toxicity.

The overwhelming majority of films modified by CI are based on *low-density polyethylene* (LDPE). Less frequently *high-density polyethylene* (HDPE), *linear low-density polyethylene* (LLDPE) or *ethylene copolymer with a-olefins* are used with this aim. The chief drawbacks of LDPE are its relatively low softening temperature, insufficient transparency and oil resistance as well as high gas permeability. In contrast to LDPE, HDPE is stronger under tension and heat resistant, less permeable to vapors and volatile matter and resistant enough to oils and fats. HDPE films are, however, more stiff, less transparent and less resistant to tearing and puncture than LDPE films. LLDPE films occupy an intermediate position in terms of their properties between LDPE and HDPE. LLDPE shows narrower molecular-mass distribution as compared to LDPE and surpasses it in chemical resistance and performances both at low and high temperatures. Besides, thanks to high resistance to tearing and puncture, breaking stress at tension and elongation

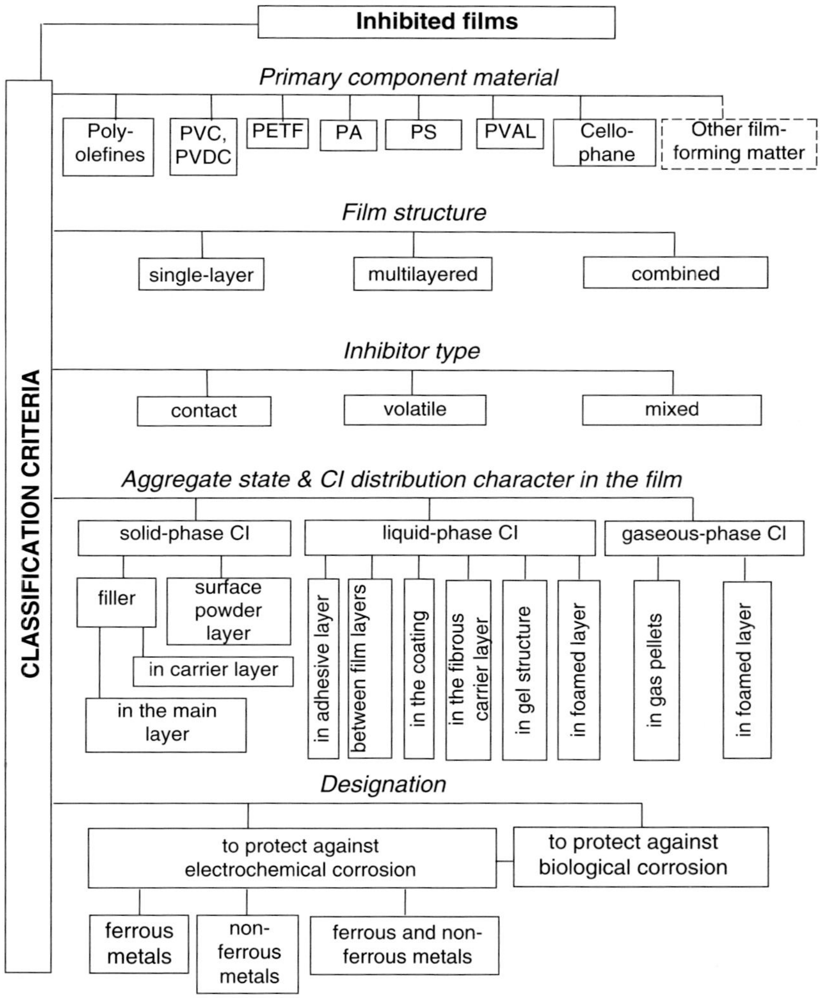

Fig. 2.1. Classification of inhibited films

at rupture (up to 1000%), LLDPE is very useful as a base for the manufacture of single-layer and multilayered extensible films. To improve strength, raise heat resistance, reduce gas and vapor permeability PE-based films are purposefully subjected to radiation modification during which the polymer macromolecules become cross-linked and the films become infusible.

Polypropylene-based (PP) films display low density and are transparent and moisture-proof. They are more rigid, less frost-resistant, but more resistant to heat than PE-based films. A drawback of PP is pliability to oxidation

Table 2.1. Physics-mechanical characteristics of polymer films [9]

Film Base	Thickness, d, μm	Density, g/cm^3	σ_b, MPa	ε_b, %	Elasticity Modulus, 10^{-3}MPa	Heat-Resistance, °C	Frost-Resistance, °C	Moisture Absorption in 24 h, %	Permeability to Oxygen, $10^{-7}m^3/(s \cdot m^2 \cdot Pa)$	Permeability to Vapors (d = 25 μm; 38°C; 90% H_2O), $10^{-4}g/(s \cdot m^2)$
PE	10–300	0.919–0.965	10–50	100–1000	0.1–1.0	70–120	from -40 to -70	0.01	300–1150	0.5–2.0
PVC:										
– rigid	50–250	1.30–1.45	45–120	5–100	1–2	70	–15	0.1–0.5	120–180	3.0–4.0
– soft	–	1.25–1.50	14–35	200–500	0.1–0.5	70–90	-60	0.5–1.0	200–300	1.5–4.0
PP	2–400	0.90–0.91	21–280	40–800	1.0–4.0	130	from -15 to -50	0.005	170–270	0.8–1.4
PET	1–350	1.35–1.40	140–290	60–140	3.4–5.5	130	-60	0.4–0.5	10–15	2.5–3.0
PA	12–200	1.02–1.15	60–350	60–450	0.6–2.8	120–220	from -30 to -60	1–9	5–60	4.0–8.0
PS	4–500	1.05	49–80	3–8	2.7–3.7	70	-50	0.04–0.06	236–665	8.0–17.0
PVDC	40–150	1.59–1.71	56–140	40–100	1.1–1.2	140	from -15 to -35	0.01	1.4–1.9	0.1–0.5
CEVA	15–200	0.915–0.950	10–21	300–700	–	65	-75	0.01	600–1700	5.0–6.0
Cellophane	–	1.45	15–25	15–25	1.4–3.1	150 (carbonizes)	-15	45–115	40–80	0.6–3.0

*PEVS shows the least oxygen permeability – $(0.3\text{–}0.4) \cdot 10^{-7} m^3/(s \cdot m^2 \cdot Pa)$

leading with time to embrittlement of the films. PP-based film materials are characterized by high strength at tension and fair sliding parameters.

The properties of film-forming *copolymers of ethylene and vinyl acetate* (CEVA) approach those of polyolefines. Indices of the properties may change in response to vinyl acetate content in a copolymer molecule. In contrast to LDPE, CEVA melts and can be heat-sealed under lower temperatures, is more elastic, more permeable to gases and vapors and shows higher resistance to puncture and cracking. CEVA films are characterized by stickiness and poor sliding. CEVA can be used to form adhesive inner layers in multilayered wrapping films to ensure that they stick to the packed product surface [20].

Thermoshrinking single- and double-axis-oriented anticorrosion films are produced from polyolefines. Particularly, CEVA-based oriented films show up to 50% shrinkage at 90–100°C temperatures [3].

Ionomers represent a class of film-forming polymers that have only lately found their application as the base for anticorrosion films [21]. Ionomer macromolecules consist of hydrocarbon chains with attached carboxyl groups. The latter are ionized to carboxylated anions and are combined via ion links with metal cations of either alkali or alkaline earth metals. Ionic cross-links appear as a result of interactions of metal cations and carboxyl groups. These links are not as strong as covalent cross-links in traditional cross-linked polymers. Besides, they weaken substantially under elevated temperatures and cannot hamper transformation of the ionomer into the viscous-flow state. Therefore, ionomers can be processed by any methods suitable for thermoplastics. Apart from this, ionic intermolecular links impart specific properties to ionomers, such as fine transparency, flexibility, extremely high stretching values, plasticity at low temperatures, and chemical and wear resistance, rupture and puncture strengths superior to polyolefines [1]. Amongst the ionomers of great practical interest are copolymers of a-olefins with unsaturated mono- and dicarboxylic acids (e.g. ethylene copolymer with acrylic acid) in which carboxylic groups are partially neutralized by Na, K, Mg and Zn hydroxides or salts. A large-scale producer of ionomeric resins of Surlyn trademark is DuPont Company (USA).

Polyvinyl chloride (PVC), which belongs to the family of vinyl polymers, and its *copolymers with vinylidene chloride* (PVDC), *polystyrene* (PS) and its copolymers, *polyvinyl acetate* (PVA), *polyvinyl alcohol* (PVAL), *ethylene copolymers with vinyl alcohol* (CEVA) are also employed in anticorrosion films [3, 18, 20, 22–24].

PVC films surpass polyolefine films in strength, resistance to oils and impermeability to gases and aromatic substances. Stiff PVC films display moldability. Their drawbacks are low thermal-, light- and frost-resistance, poor heat-sealing and propensity to agglutination. Soft PVC films are produced from a plasticized polymer. Introduction of a plasticizer into the polymer base adds elasticity, transparency and glitter, improves frost resistance and resistance to tearing, but reduces heat and chemical resistance, strength and barrier properties.

A magnificent property of PVDC films is their low vapor and gas permeability. Application of even very fine PVDC layers in co-extruded multilayered films imparts them with perfect barrier characteristics. Oriented PVDC films are highly strong, elastic and resist well depression and tearing, although they lose these properties at elevated (60°C) and lowered (−20°C) temperatures.

Oriented PVC and PVDC films show high thermoshrinking properties reaching 60% for PVC and 80% for PVDC, in combination with strong electrostatic sticking, thanks to which they are widely applicable in skin-packaging.

PS is limited in its use and is mostly applied as a film-forming material because of its brittleness, insufficiently high heat resistance and instability in organic solvents. To improve its impact resistance and elasticity PS is often modified by rubbers [20, 22]. Foamed impact-resistant PS is rather seldom used as a layer-vessel incorporating CI in multilayered anticorrosion films.

PVA is not suitable as the base of packaging materials since its physical and mechanical properties are impaired abruptly above the glass-transition temperature (28°C). It is used, nevertheless, along with polyvinyl butyral (PVB), polyvinyl alcohol (PVAL), alcohol-soluble polyamides and polyacrylates as a layer-carrier of CI in multilayered films. The layer is formed via application of emulsions, suspensions or solutions of named polymers containing CI onto the base film [23–26].

Ethylene copolymers with vinyl alcohol (PEVA) are derived by saponification of CEVA. Their critical feature is the unique gas barrier property (Table 2.1), which is combined with elasticity, strength, surface hardness, high wear, oil and atmospheric resistance. Thin, strong barrier layers of inhibited films formed from PEVA are impermeable to both CI and hostile atmospheric vapors and gases [21].

Polyesters are represented in the series of anticorrosion materials first of all by *polyethylene terephthalate* (PET) films [3, 8, 23, 24, 26]. These feature perfect transparency, stiffness, rupture and impact strength, resistance to wear, oils and fats, low water, vapor and gas permeability, serviceability within a wide temperature range (−60 up to 150°C), and good moldability. However, it is in fact impossible to obtain strong sealing seams in PET films since heat-sealing of polyesters is accompanied by shrinkage and embrittlement, besides their high cost limits application of PET films in engineering. To overcome this drawback, the PET film base is either laminated with HDPE or overlaid with a glue layer.

Out of the *polyamides* (PA) mainly kapron (PA-6, the product of caprolactam polymerization) and different grades of nylon (PA-12, the product of dodecalactam polymerization; PA-6.6 and PA-6, 10 synthesized by polycondensation of adipine or sebacate acid with hexamethylene diamine, and others) are of commercial value for processing into films, including anticorrosion ones [18, 26]. PA films are characterized by a wide operating temperature range (oriented films from −60 to 150°C, non-oriented films from –60 to 200°C, perfect elasticity, good mechanical strength, oil and benzene resistance, low gas permeability and are easily heat-sealed. The drawbacks of PA-based films are considerable water absorption and vapor permeability increasing significantly with temperature rise as well as low resistance to oxidants.

Polyurethane (PU) elastomers are used, but rather infrequently, as components of anticorrosion films, particularly polyurethane foam [18,23,24,27].

It is common experience to use *cellulose* for inhibited materials (C), namely *cellophane (cellulose hydrate)* films plasticized with glycerin [3,23,24]. Properties common for cellulose films are perfect transparency, elasticity, strength at squeezing and rupture, resistance to oils, low gas permeability (when dry), besides which they can be easily electrified. Adverse properties of cellophane films are high hygroscopicity and permeability to vapors, loss of strength and impermeability for gases in humid atmosphere and that they cannot be heat-sealed. Cellophane is predominantly used in lacquer films with either single- or two-sided covering, as well as in multilayered and combined materials.

According to their **structural parameters** the inhibited films are subdivided into single-layer and multilayered, when a few layers of one or different polymers are formed, and combined films that contain other materials along with the polymer film layers.

To protect metal ware from external effects and mechanical damages they undergo preservation by packing into inhibited film materials inside which an atmosphere is created that averts corrosion. The requirements imposed on the preservation material depend upon the product to be protected; therefore to create a single-layer film meeting all the requirements is in fact impossible.

It is reasonably efficient to adopt the method of imparting a given combination of properties by forming laminate film structures from different polymers along with combining polymer films with other materials, such as metal foil, paper, fabric network, non-woven and foamed substrates, and so on. Development of named materials began in the early 1960s and has lead to the creation of original designs of films and their specific technologies, as well as the required equipment. As a result, a self-standing commercial branch has originated.

The amount, origin, mutual disposition and thickness of the layers of the laminate and combined films are chosen so as to compensate for the drawbacks of one component through the advantages of the others. Basic properties imparted by separate components to laminate and combined film materials are descriptiond in Table 2.2 [2].

Inhibited laminate and combined films contain, as a rule, layers whose main function is to support mechanical load, to be a barrier to dust, corrosion-

Table 2.2. Dependence of properties of laminated and combined film materials on the origin of components

Component	Properties Imparted to the Material*	Component	Properties Imparted to the Material*
Aluminum foil	St, M, G, HR, Fr, F, Sp, O, Inc	PVC	St, M, G, F, Sp, HS, Inc
		PVDC	St, Fr, F, Sp, HS, Inc
Paper	St, Fr, Sp, O	PETP	St, M, G, HR, Fr, F, Sp
LDPE	St, M, Fr, HS	PA	St, G, Fr, HR, F, Sp, HS
HDPE	St, M, HR, Fr, HS	Cellophane	St, G, F, Sp
PP	St, M, HR		

*St – strength, M – moisture resistance, G – gas impermeability, HR – heat resistance, Fr – frost resistance, F – fat resistance, HS – heat-sealing, Sp – susceptibility to color printing, O – opacity, Inc – incombustibility

active gases, vapors and other atmospheric matter. There are also the layers for formation of welded seams and those for carrying CI [28, 29].

This broad spectrum of properties of *multilayered anticorrosion films* is used either as the base, a CI carrier or adhesive (if required) because of numerous possible combinations of polymer materials.

Polyolefine films are the most applicable, in which the high-density polymer layers fulfil the strengthening and barrier functions and a layer of low-density polyolefine contains a CI [19, 28, 30]. For instance, in a strengthened multilayer Cortec VCI-126 film the outer layer is formed of LDPE with HDPE strips. The inner inhibited layer is made of LDPE.

The most common CI carriers in laminated films are:

- extruded LDPE [28] and CEVA [21] layers; PVAL (for water-soluble CI) [25], and others,
- PVA layers and its blends with PVB, CEVA, cellulose acetyl butyrate, epoxy and acrylate resins (to impart stickiness to the inhibited layer), some PA grades applied on the substrate film from solutions or dispersions [23, 27, 31],
- porous layers based on foamed PU, PS, PE and some other polymers [18, 23, 24].

The mechanical load-carrying gas and vapor impermeable layers are formed of oriented polyolefines, cellophane, polyesters, PVDC, PEVC, ionomers and PU.

Historically the first anticorrosion techniques to appear were *combined films* – paper and cardboard-based materials covered with paraffin, microcrystalline wax, synthetic latex or PE. These materials are accessible, cheap, rather strong, wear resistant, and display elasticity, vapor and light-impermeability, moisture resistance, shape stability, susceptibility to color printing and heat-sealing, and are biodegradable. Besides, they are able to hold considerable amounts of CI.

The production rules for the impregnation of paper coated by PE, PS, paraffin or butadiene-styrene latex with CI are reflected in Russian State Standard 16295 [32]. In particular, such paper has been manufactured since 1970 at Joint Stock Co. Technical Paper, Rybinsk, Russia.

Inhibited Cortec paper, grades VCI-146 and VCI-147, consists of a natural cellulose base coated from one side by a PE film. It is intended for protection of different metals against moisture and hostile gases (SO_2, H_2S) and contains neither nitrates, phosphates and chromates.

A wide assortment of combined paper and cardboard containing layers of inhibited PE is manufactured by Northern Instruments Corporation, including the trademarks Zerust Ferrous and Nonferrous Poly/Paper, Zerust Ferrous Chipboard, Fibreboard and Corrugated Cardboard. Zerust Ferrous and Nonferrous Poly/Paper are produced by an original technique. A coating below 40 μm thick incorporating a CI is formed from the inside of the package. A structure of this type has significant advantages over the standard range of saturated VCI papers. Bulk water is known to have an adverse effect on the inhibiting performance of saturated VCI paper. In Zerust Poly/Paper, CI is already in PE and is not affected by water. The PE coating serves as a vapor barrier and a protective host for its inhibitors. Moreover, Zerust Poly/Paper has a shelf life far longer than standard saturated VCI paper with PE coating and is ecologically friendly. Zerust cartons are fabricated so that the extruded inhibited PE coating becomes the interior surface of the cartons. Cortec Co. produces a packing carton laminated by a thermally activated adhesive film based on ethylene copolymer with vinyl monomers blended with VCI [33].

There is one more method of adding resistance to moisture and anticorrosive properties to carton: by the introduction of CI into the polymer latex coating. The film-forming water or organic solvent-based components are used to form a durable inhibited polymer coating on the paper glued directly on the inner side of the carton.

Metal foil (predominantly aluminum) is introduced into combined film materials to attain vapor, gas and light impermeability, stiffness, thermal and chemical resistance, and shape stability in a package. These materials are often combined with paper. It is common experience to use combined films such as PE-foil, PE-foil-PE-paper-paraffin, PE-foil-PETP, PE-foil-PE-cellophane and many other combinations [8,14,29,32,34–36].

In Fig. 2.2 the structure of UMC brand film materials produced by Japanese Nitto Co. is presented. The materials are used for hydrothermal insulation and anticorrosion protection of low-temperature pipelines transmitting liquefied gas. In these materials the barrier properties of the foil are supplemented by the strength of the polyester layer and the stability of the fluoroplastic layer to atmospheric effects [3].

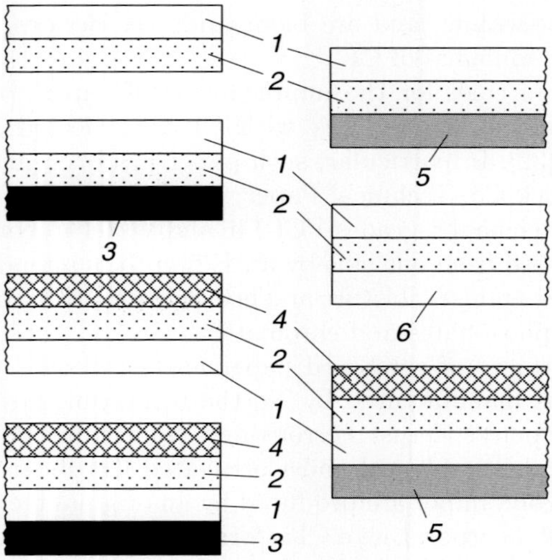

Fig. 2.2. Anticorrosion combined film materials: (1) polyester film; (2) aluminum foil; (3) rubber glue; (4) fluoroplastic film; (5) resin glue; (6) paper

Application fields of materials having thin aluminum layers can be expanded through the use of metallized polymer films [14, 35].

Preservative film materials intended for packing large-size metal ware with sharp edges should of course show elevated resistance to rupture and puncture. With this aim, polymer films are reinforced by glass fibers and kapron by alternating air impermeable and inhibited polymer layers with non-woven fabric [37, 38]. Fibrous materials with developed specific surface and porosity can serve as a container for the evaporating CI [39].

Along with functioning as a CI carrier, the microporous layers of combined materials serve as a damping element of the films [3, 18, 23].

The type and aggregate state of CI form one more classification sign of inhibited films.

Films with contact CI representing a highly viscous liquid (Sect. 1.3) have restricted application in engineering. They are usually single-layered since low-volatile CI does not require sealed packing. The inhibiting liquid diffusing from the film onto the surface of the article being protected is distributed evenly or at a gradient within the polymer matrix [3, 4, 24]. Functions of the contact CI carriers are often fulfilled in the films by mineral or synthetic oils able to reduce viscosity and elevate compatibility of the inhibiting liquid with a polymer base.

Films with volatile CI (in solid, liquid or gaseous phase) are the most used today. The main obstacle to using films incorporating volatile CI is control over dosed isolation of CI from the polymer matrix [3].

Films modified by *CI in the solid phase* most often contain finely dispersed inhibitor particles evenly distributed within the polymer matrix (Fig. 2.3a). Powder CI can be glued to the film surface or embedded within it (Fig. 2.3b).

Films of the first type are obtained by extrusion of thermoplastic blends and powdery CI (chiefly volatile ones) [3, 8, 14, 18, 22–24, 35, 40–43]. To distribute the anticorrosion agent evenly in the polymer matrix and avoid its thermal destruction well compatible low-molecular components are added, which form VCI during interactions [4, 44]. Among the drawbacks of these films is the wasteful loss of volatile CI that are unstable to heating during joint extrusion with the polymer melt. This may result in ecological problems and impairment of the anticorrosion properties of the film. Moreover, CI particles become encapsulated into the polymer matrix and a part of them is wasted. Slight amounts of plasticizers and highly dispersed adsorbents are introduced into the extrudate to lower the processing temperature of the thermoplastic blend and avert its foaming by VCI vapors [4]. Along with this, the plasticizers affect structure formation processes in the polymer matrix thus alleviating diffusive liberation of VCI [5, 6].

From this standpoint, the materials with CI particles fixed in a fibrous [38, 39] or foamed layer [18, 35, 45] or on the film surface are advantageous.

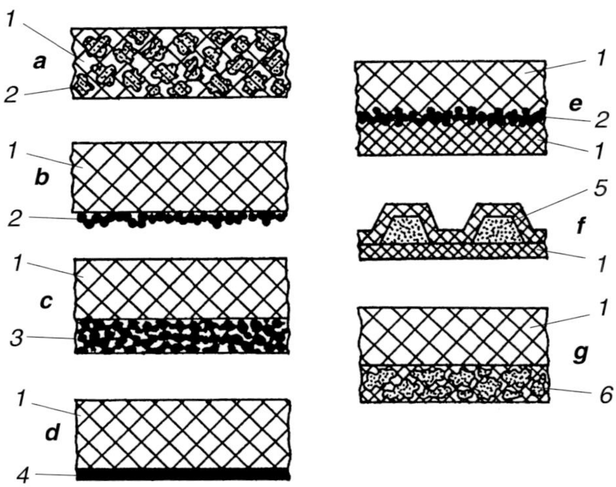

Fig. 2.3. Structure of inhibited polymer films: (**a**) single-layered filled with solid CI particles; (**b**) with a layer of glued particles; (**c**) with a porous layer impregnated with inhibited liquid; (**d**) with a liquid CI in the glued layer; (**e**) with a liquid CI between film layers; (**f**) with CI in the gaseous phase sealed between layers; (**g**) with a layer foamed by an inhibited gas. (1) polymer base; (2) CI particles; (3) porous layer; (4) glued inhibited layer; (5) gaseous pellet of CI vapors; (6) jelly layer with inhibited liquid

A solid, finely dispersed CI is rolled on the surface of a heated thermoplastic film [46] and is either embedded, glued or deposited from the gaseous phase or solid aerosol. It is critical for such films to fix reliably CI on the polymer base surface. A Japanese company, Nitto, manufactures a film with dicyclohexylamine nitrate powder embedded into its surface layer over 1 cm square areas. As a result, the film can be bent and deformed on the areas free from the CI, and the particles will not crumble away [3]. CI can be deposited from the vapor phase on definite sites of the film in a given pattern to occupy about 25% of its area [48].

Liquid CI can be contained in a specific carrier layer (Fig. 2.3, c, d) or in the space between layers of a multilayered film (Fig. 2.3e), as well as in a jelly-like polymer of a single-layered film.

The simplest modifications of named materials are multilayered and combined films in which either solid [41] or porous polymer layers (foamed, fibrous, etc.) are filled with the inhibiting liquid [4, 18, 23, 24, 30, 39]. In addition, film materials where the adhesive layer is filled by a liquid CI are used in packaging techniques [4, 23, 37].

Gel-based polymer films in which the inhibiting liquid is the liquid phase have gained widespread adoption. Gels can be characterized as structured systems consisting of high-molecular substances and a low-molecular liquid occupying a larger portion and displaying propensity to strong reversible deformation with almost no fluidity [49]. They are formed by swelling of solid cross-linked polymers in liquids (gels of the first kind) or by setting of liquid polymer solutions (gels of the second kind). Liquids spontaneously isolate from the polymer matrix of gels as a result of relaxation processes.

The simplest technological process of manufacturing gel-based inhibited films is co-extrusion of a polymer blend (PE) with CI or its solution in a plasticizer, which was realized in the USSR in the 1980s. This, however, worsened properties of the CI due to joint processing with PE melt and irrational consumption since CI isolated from both sides of the film. The latter feature was partially eliminated by application of a lacquer or enamel layer on the hose film surface [50].

More perfect and ecologically safe techniques for manufacturing the jelly layer as a CI carrier on the inner surface of hose single-layer films have since been elaborated [51, 52]. Furthermore, the means of fixing this layer to the barrier base surface have been developed [3].

When introduced into the space between layers or in the middle layer of a multilayered film (Fig. 2.3e) the CI can be isolated into the packing space at a controlled rate to ensure the rustproof environment, on one hand, and sufficient resource of the film material and safety, on the other [3].

The problem of regulated sealing of CI phase inclusions in films has been solved by the method of structural encapsulation at polymer deformation in an inhibiting liquid [4]. Regulation of CI isolation rate is attained via varying the capsule size, stress level in the film, composition of the inhibiting substance and so on. Film materials with a liquid-phase CI (30–50 μm capsule size) microencapsulated in the surface layer are produced by sequential treatment of the film base with two agents, e.g. acidic and base types. As a result of the reaction between the two agents, a target CI is obtained, which is accompanied by heating of the polymer and bulging of the capsules near the film surface [4].

Elastomeric films with CI dispersed in the polymer matrix volume as liquid drops are formed by pouring of emulsions in which the dispersive phase is the inhibitor and the dispersion medium is the elastomer solute [53].

Films with *CI in the gaseous phase* represent either multilayered or combined materials in macro- (Fig. 2.3f) or microcavities (Fig. 2.3g) where the volatile anticorrosion agents are sealed.

Air-bubble films (Fig. 2.3f) consist of a film base with molded cavities in the form of pellets and a flat film coating bonded to the base and tightly covering the cavities. The cavities contain CI vapors diffusing predominantly through the flat regulating polymer layer, while the layer with molded cavities serves as a barrier. Along with anticorrosion properties, air-bubble films protect products packed in them against vibration and impact effects [3].

Apart from this, combined and multilayered films containing a damping layer foamed by CI vapors are known [3, 24].

The problems arising during operation of these materials are mainly connected with difficulties in regulating leakage of CI from the sealed cavities in the films. The best conditions for provision of protection against corrosion are created when sealing of cavities is maintained during operation, e.g. by friction, relaxation processes, aging, etc.

By their designation, inhibited films can be subdivided into those *protecting metals against electrochemical or microbiological corrosion.* As a rule, under service conditions, corrosion processes, whether atmospheric, soil or water, go hand in hand. A number of CI used as modifiers of polymer films combine the properties of inhibitors of both electrochemical and microbiological types of corrosion [7, 54].

What's more, inhibited films protecting ferrous as well as nonferrous metals exist in the range of anticorrosion materials [8]; films for *multi-metal* protection are also known.

The development of universal preserving films able to protect both ferrous and nonferrous metals from electrochemical and microbiological corrosion is common in anticorrosion techniques. In this regard, new film modifiers, including highly efficient CI and their synergetic compositions are lacking. Procedures for their combination with the polymer base remain to be developed and smart inhibiting materials should be created. These inhibiting materials will possess the property of responding to variations in the multimetal object being protected and exert the anticorrosion effect required at a given moment.

As can be seen, inhibiting anticorrosion films represent a vast range of composite materials varying in composition, structure, properties and application field. Their components can be, in fact, any barrier, protecting, inhibiting and auxiliary materials known and available today for anticorrosion techniques.

2.2 Production Methods

Inhibited films of different structures, properties and designation are manufactured by various methods using a wide range of film-forming polymers and CI (see Sect. 1.3 and 2.1).

2.2.1 Principles of Manufacturing Inhibited Films

Film materials incorporate a high-molecular base and a low-molecular CI coexisting in different aggregate states and so having essentially different physical-chemical and processing characteristics. Inhibited films should fulfill the following three main functions: support the mechanical load, form a diffusive barrier to the hostile media and be a CI carrier. Two contradictory problems are to be solved during their production, that is, to reliably combine CI with the polymer matrix so as to guarantee film operation as an integral whole, and to make polymer links with the CI weak enough to allow its free liberation from the film. Solution of the problem lies within the thermodynamic and process compatibility spheres of the ingredients of the inhibited polymer materials (see Sect. 1.4)

The production technique for inhibited films presumes joint processing of their components during which the polymer and CI are brought into contact.

On one hand, inhibited films employ traditional process methods and equipment used for molding film materials. On the other hand, special operations and auxiliaries are needed to modify the polymer base with CI (Fig. 2.4). A number of methods of combining CI with the polymer matrix are unique and patented.

The production technique for inhibited films responds to the requirements imposed in terms of service characteristics, processability of the polymer and CI, and their aggregate state.

The method of combining CI and the polymer is in any case associated with the most suitable procedure of forming the film base whose essence, advantages and drawbacks are described in this chapter.

2.2.2 Basic Production Techniques of Polymer Films

The procedures for obtaining films from melts, including extrusion, calendering, molding, casting of polymer solutions and latexes, planing and rolling of

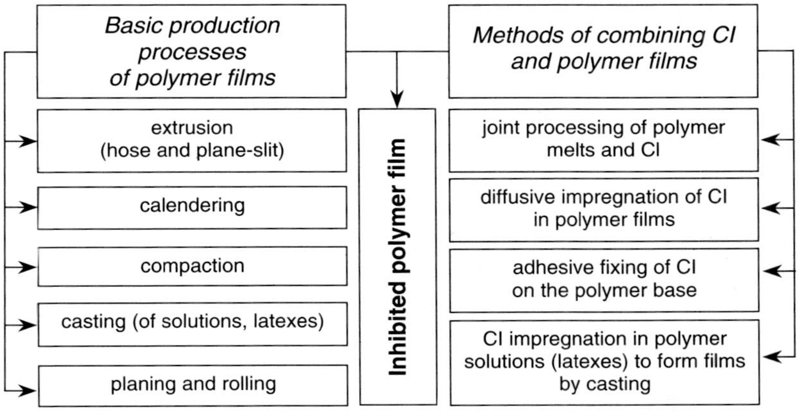

Fig. 2.4. Production techniques for inhibited films

solid polymer materials belong to traditional commercial methods of manufacturing polymer films [1–4, 55].

Extrusion is widely applicable in production of films from thermoplastics. The polymers used in this process should be resistant to thermal-oxidative failure on transition to the viscous-flow state. The heads of the extruders are equipped with either annular or planar slits.

Hose extrusion includes forcing of the polymer melt through the annular slit of the extrusion head, pneumatic blowing of the hose, folding and removal in vertical or horizontal direction. This method produces biaxial films with perfect physical-mechanical characteristics, it is rather simple and economical, and is suitable for the production of laminated film materials by co-extrusion.

During *plane-slit extrusion* the melt leaving the head outlet is rapidly cooled down on cold rollers or in a water bath. Predominantly crystalline polymers are processed by plane-slit extrusion to obtain highly transparent and even thickness films, laminated or combined film materials. Special-purpose devices are needed to attain a uniaxial or biaxial film tension.

Calendering consists of continuous forcing of the thermoplastic polymer through a gap between calender rollers. The plastic mass is fed into the gap between feeding rollers and is shaped into a film blank, which is then passed between the remaining heated rollers of the calender. The thickness of the film depends on the gap size between the last pair of the rollers and surface quality (bright, dull or structured) on the last contact with the roller, whose surface was subjected to a specific treatment. After leaving the calender the film is cooled between the cooling rollers and is reeled onto a bobbin.

Calendering takes place at a transient temperature between the polymer viscoelastic and viscous-flow states. Thermoplastic materials displaying a wide range of the flow temperatures and sufficient melt viscosity (e.g. PVC) are usually processed by this method. Calendering produces homogeneous, smooth and uniform thickness films removed from the calender rollers without damage or stretching.

Films subjected to calendering are better in quality than extruded ones, although the process is labor intensive, energy consuming and requires high temperatures and pressures.

Single-layer and combined films can be produced without heating by ***compaction*** or rolling of powder polymer compositions followed by sintering in an oven. Combined film materials are mainly formed by pressing with heating. Because of low productivity and imperfect quality of films produced this way compaction is infrequently used in industry [3].

Casting of polymer solutions on the film-forming substrate surface is the oldest commercial technique of obtaining polymer films and has not lost its significance today for polymers whose viscous flow temperature is below that of their thermal destruction (cellulose esters, PVC, some of PA and other). The main stages are: preparation of the polymer solution, filtering, pouring onto a smooth surface of the moving continuous tape or rotating drum, casting of the film, drying and cutting. The film is cast during evaporation of a volatile solvent, coagulation and chemical transformation of the film-forming polymer. For instance, a cellophane film is formed as a result of cellulose transformation into a soluble product called cellulose xanthate. Its alkali solution is forced through a plane-slit spinneret and is converted into a glassy state by a chemical treatment of the film, as a result of which cellulose saponifies and forms a cellulose hydrate (cellophane).

Casting of polymer solutions is rather efficient but expensive and ecologically unsafe because of inflammable and toxic organic solvents, whose recuperation presents a separate technological stage. Nevertheless, this method is carried out at moderate enough temperature regimes and the films can be modified by CI with low thermal resistance.

The initial material for casting can be a polymer paste called plastisol. The film is formed as a result of sol coagulation and gel formation followed by attenuation of residual stresses in the gel by thermal treatment [9].

The process of casting polymer latexes is similar to that described above. Latexes consist of colloidal aqua dispersions of polymers with particle size of the dispersed phase within a range of 10 to 10^3 nm in fine-dispersed latexes and to 2–3 μm in coarsely dispersed ones [9]. In contrast to their solutions, latexes are preferable for casting polymer films from the standpoints of ecology and energy saving.

Planing and rolling are rarely used for manufacturing polymer films. Polymers with high molecular mass, heat resistance and melt viscosity (PTFE, celluloid) can be processed into films by these methods. With this aim, a rotated cylindrical polymer blank is cut off by a cutter into a continuous sheet and is rolled out on a rolling mill. Among the advantages of this method is the shortening of the technological cycle through excluding heating and cooling operations and, as a result, elimination of thermal destruction of the polymer [55].

2.2.3 Methods of Combining Corrosion Inhibitors with Polymer Films

Mixing of CI with the polymer base is considered to be the most important operation that governs anticorrosion properties of the film material. Modern technologies allow binding of CI with the polymer by the encapsulation of the inhibitor particles into the polymer binder with free fixing on the film surface.

The existing methods of CI introduction into the polymer films can be subdivided into four groups (Fig. 2.4).

The *first group* of methods presupposes joint processing of polymer melts with CI in either liquid or solid phases. Easily volatile CI may be transferred into the gaseous phase and foam the polymer during processing [3,4,7,14,23, 24].

The *second trend* is based on diffusive saturation of the polymer films with CI in the liquid or gaseous phases [3,4,30,52].

The *third direction* consists in adhesive fixing of CI or a polymer carrier containing the CI either on the film surface or between its layers (in the form of glued particles [47,56,57], on a lacquer surface [31], adhesive [31,58] or fibrous [38] layer).

The *fourth group* of methods is connected with casting of polymer solutions and latexes or monomer compositions polymerizable at film formation and containing CI for the inhibited films [4,41,53].

The choice of method of CI introduction is predetermined by the CI properties, volatility, thermal resistance, toxicity, compatibility with the polymer base or other inhibitor carriers.

Recommendations on the choice of modification methods of PE films by CI of the tetrazole and triazole kinds (polynitrous heterocyclic compounds (PHC)) are presented in Table 2.3 [5,6].

The principles of efficient arrangement of production processes of inhibited films taking into account the specifics of CI mixing with the polymers have been formulated elsewhere [3]. These principles can be briefly set forth as follows:

- CI losses on evaporation, thermal destruction of volatile matter and joint extrusion should be minimized, so preference is to be given to operations presuming a short-term contact of the CI with the melt.
- The methods that are most economical and capable of preserving the original properties of the CI are diffusive saturation of the polymer matrix by inhibiting liquids, their introduction into the polymer solutions (latexes) or application of polymer layers (varnish, glue, etc.) onto the film base carrying the CI.

Table 2.3. Methods of manufacturing PHC-containing PE-based films

CI Mixing Method with Polymer Base	Requirements on CI*	Types of PHC Meeting the Requirements
(1) Joint extrusion of the polymer and solid or liquid CI (dissolved in a plasticizer)	Thermal resistance ($T_{\mathrm{di}} > T_{\mathrm{extr}}$), moderate volatility (losses $\leq$ 5–6 % from CI evaporation at T_{extr}), high efficiency, low toxicity	NTr, ETet, PhTet, KPhTet, NPhTet
(2) Thermal diffusive impregnation of plasticized polymer base of CI in gaseous phase	High volatility ($p_{\mathrm{s}} > 0.1$ Pa), solution in a plasticizer	Tr
(3) Thermal diffusive impregnation of plasticized polymer base of CI in liquid phase (by CI solution in plasticizer)	Thermal resistance ($T_{\mathrm{di}} > T_{\mathrm{l}}$), moderate volatility, solubility in plasticizers	ATr, Tet, MTet, ATet, BTA
(4) CI impregnation into lacquer layer on polymer hose at blowing	Solubility in lacquer	Broad series of PHC
(5) Roll-on of solid CI particles onto polymer film by heated rollers	High dispersion	Broad series of PHC
(6) Combined methods: (1) + (2); (1) + (3); (1) + (2) + (3); (3) + (4) and other	Depending on the combination of methods	Depending on the combination of methods

*T_{di}– temperature of CI destruction initiation; T_{extr}– extrusion temperature; T_l – temperature of the inhibiting liquid fed into the extrusion head die [31].

- The method of roll-on or pasting solid CI particles onto the polymer film with the help of heated rollers is rather economical, although inefficient since CI particles inevitably crumble off during operation.
- Manufacture of foamed and gas-filled films using CI does not impose any rigid temperature requirements but complicates liberation of the gaseous CI from cavities in the film.

The examples of different methods of combining CI with polymers realized in practice are discussed further in Sect. 2.2.4–2.2.6.

2.2.4 Manufacture of Inhibited Films by Hose Extrusion

Extrusion by polymer hose blowing may produce both single-layer and multilayered inhibited films. It is highly productive, quiet simple and facilitates regulation of dimensions, structure, physical-mechanical and other characteristics of the film materials within a wide range.

Modifications of the method for the *single-layer inhibited films* are illustrated in Fig. 2.5 [3, 59].

Extrusion of polymer blends containing CI is the simplest method of inhibited film production [60, 61] (Fig. 2.5a). The polymers, CI and technological additives are mixed during one or several stages and extruded with hose blowing. A film with an even distribution of the solid, liquid or gaseous CI in the polymer matrix is thus produced. In the latter case the conditions for CI evaporation or sublimation in contact with the polymer melt are created.

One of the difficulties in realizing this technique is how to preserve the initial properties of the CI as it is processed jointly with the polymer melt. If one compares the extrusion temperatures of PE (120–250°C) and T_{di} of VCI (Table 2.4), it becomes evident that their joint extrusion may result in the loss of the volatile fractions and impairment of the inhibiting property of CI. Therefore, the problem of finding heat-resistant VCI comes to the forefront. These properties are intrinsic to novel CI of the PHC series [5, 6, 62]. Besides, this problem can be solved by impregnation of matter that is relatively resistant to temperature and has a low volatility, such as monoamines and organic or inorganic acids that form VCI during interactions at extrusion [44].

A serious technological problem is the extrusion of blends with a high content of inhibitor liquid. The low bulk friction of the blend affects the extrusion output. Alongside, there arises the problem of keeping and evenly spreading the inhibiting liquid inside the extruded blend. To this end, an original procedure of mixing components is used [63], during which a polymer powder is impregnated with a liquid CI, plasticizer and this blend is then mixed with granules of the same polymer. The polymer powder represents here a carrier of the liquid-phase components, while granules add the friability necessary for transporting and charging.

The anticorrosion characteristics of inhibited films depend strongly on CI liberation kinetics. During joint extrusion the CI become encapsulated in

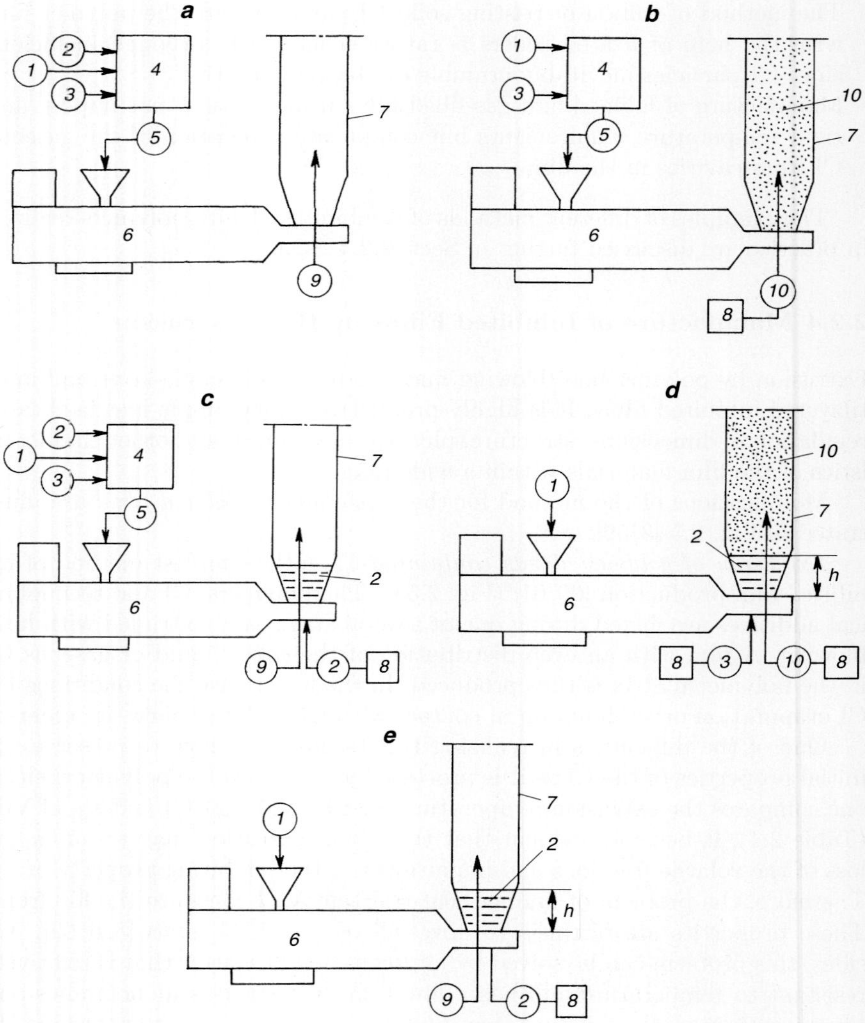

Fig. 2.5. Production diagrams of single-layer inhibited polymer films: (**a**) joint extrusion of polymers and CI; (**b**) film saturation with CI in a gaseous phase; (**c**) dissolving of the CI in plasticizer impregnated into the whole film bulk; (**d**) dissolving of CI vapors in the plasticized film layer; (**e**) modification of non-plasticized film by an inhibitor liquid; (1) polymer; (2) CI; (3) plasticizer; (4) mixer; (5) mixture of components; (6) hose-film setup; (7) hose; (8) device for CI supply into the hose; (9) air; (10) mixture of air and CI vapors

Table 2.4. Temperature of destruction initiation of CI (T_{di}) recorded by DTA [5,6,24]

Russian Trade Name	T_{di}, °C	Russian Trade Name	T_{di}, °C
IFHAN-1	80	VNHL-20	170
IFHANGAZ	120	VNH-5	240
NDA	120	Benzotriazole	155
VNHL-49	150	5-Phenyl tetrazole	215
MSDA	160	Guanazole	205

the polymer matrix and the rate of VCI vapor liberation from the polymer film decreases by a few orders compared to their evaporation rate in a free state. Introduction of a Pl into the mixture intensifies CI release from the film [64–66], which is attributed to the formation of a gel-like structure of the film material. As a result of syneresis, CI migrate together with the Pl to the film surface and ease further CI transfer to the surface of the protected product or transformation of VCI into the vapor phase (see Sect. 2.3).

A disadvantage of this method is equal probability of the CI liberation on both sides of the film. Such films are, as a rule, more permeable than those of the initial polymer. One of the ways of eliminating this drawback is to apply a polymer coating on the outer surface of the extruded hose, which may also serve as a diffusion barrier [50]. The coating material filling the polymer matrix pores acquires high mechanical adhesion to the film (Fig. 2.6a).

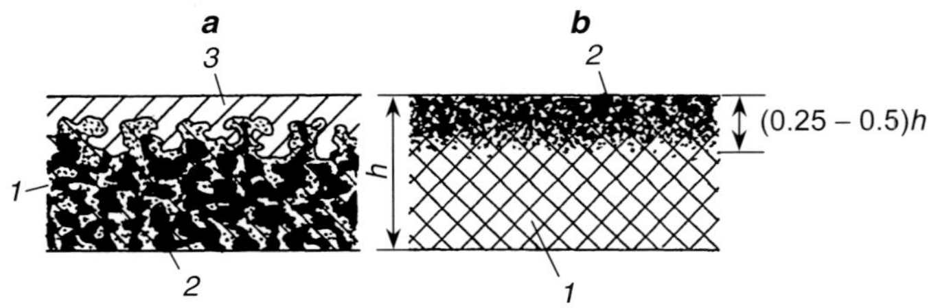

Fig. 2.6. Structure of the inhibited polymer films: (**a**) film with a coating; (**b**) film with a plasticized layer; (1) polymer matrix; (2) dispersed inhibitor liquid; (3) coating

The walls of the pores in the polymer matrix gain an oriented structure and higher strength as compared to the isotropic polymer material, so the system of the pores creates favorable conditions for strengthening of the film. An extruded sample of the gelled film, whose thickness exceeds by an order that of a completed film, undergoes rolling. The material strengthens significantly as a result of the pore walls collapse during rolling and stretching [67].

Diffusive saturation of the films with CI from the gaseous phase is based on the use of polymer Pl as carriers of the inhibitors [68–71]. They are absorbed by the plasticizing liquids within a shorter time and in greater amounts than by the polymer materials.

Information on CI vapor sorption by PE-based films plasticized by the mineral oil is presented in Table 2.5 [24]. Films with oil content below 10 wt% turn out to be poor sorbents of the gas-phase CI. The process is intensified at Pl concentration reaching 15–20 wt% due to the formation of a polymer matrix with a system of communicating pores filled with the liquid [72].

Table 2.5. Sorption (g/m^2) of CI vapors by PE-based films in 2 min at 95°C

Film Components, wt %		Russian Trade Names of VCI				
PE	Mineral Oil	IFHAN-1	IFHANGAZ	MSDA	VNHL-20	NDA
100	–	0.42	0.28	0.14	–	–
95	5	0.71	0.57	0.28	0.42	1.14
90	10	0.99	0.99	0.71	0.57	0.38
85	15	1.84	1.42	1.13	0.85	0.85
80	20	2.55	1.70	1.27	1.27	0.99
75	25	2.83	2.55	1.98	1.69	1.27
60	40	3.40	3.12	2.83	2.27	1.28

The saturation by VCI is quicker when the temperature, pressure of the saturated inhibitor vapors and their solubility in Pl are higher (Fig. 2.7).

These processes are illustrated in Fig. 2.5b. A mixture of the polymer and Pl is prepared and charged into the extruder. The CI vapors are pumped into the polymer hose using a special device and dissolve in the Pl. In spite of the fact that the Pl is evenly distributed in the film, the CI show a concentration gradient in the direction of the inner surface of the hose and is predominantly isolated from this side of the film.

The described technique is used for films containing deficient, costly though highly efficient CI. Such CI should be under above 0.1–1 Pa pressure of saturated vapors to attain the required antirust effect at low concentration of the inhibitor in the film.

The technique helps to produce films with CI having low thermal stability. This minimizes losses of CI due to escape and thermal destruction during extrusion.

Solution of liquid CI in a plasticizer impregnated into the film bulk is used when the inhibitor is poorly compatible with the polymer [69].

The hose is extruded from the polymer and plasticizer mixture (Fig. 2.5c). Liquid CI are fed to the extruder mandrel, where the hose serves as a cavity for the inhibiting liquid using a special device. Upon cooling the hose is transferred into a gel-like state in which CI together with Pl are encapsulated into

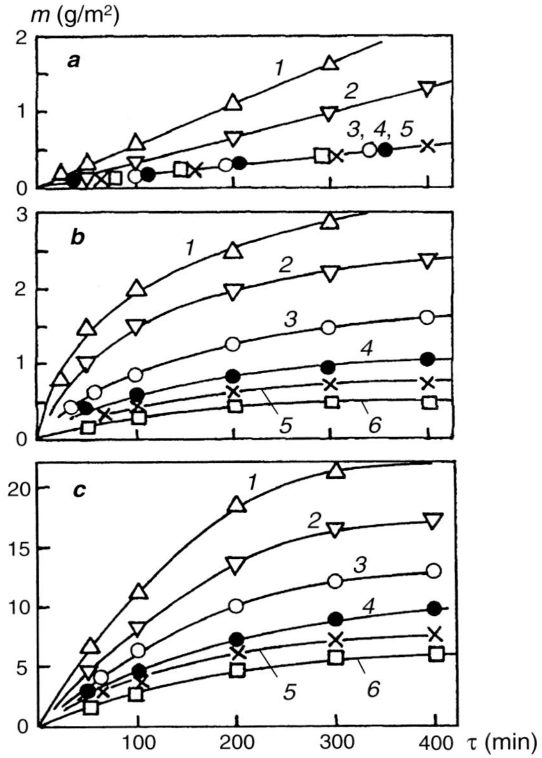

Fig. 2.7. Sorption kinetics of CI vapors by films with the composition 85 wt% of PE + 15 wt% of mineral oil depending on temperature: (**a**) 18°C; (**b**) 55°C; (**c**) 100°C; (1) IFHAN-1; (2) IFHANGAZ; (3) MSDA; (4) M–1; (5) VNHL-20; (6) NDA

the polymer matrix pores close to the inner surface of the hose. Additional cooling of the hose with a liquid CI may hamper its blowing, so the temperature of the CI and air flows used for blowing and fanning is monitored.

It is possible to increase the amount of sorbed CI by raising the plastification degree of the polymer (Table 2.6).

As CI concentration in the film exceeds 10w%, it acquires the structure of an open-pore polymer matrix filled with the Pl. The sorption mechanism of the liquid CI by such film is, evidently, similar to that of mass exchange. The inhibitor partly forces Pl out of the pores of the polymer matrix and, if its density is lower than that of the Pl, this sorption will be accompanied by a mass drop of the film (Table 2.6).

Elevated permeability of the film for corrosion-active media due the uniform distribution of Pl across the polymer bulk is among the disadvantages of this method.

Table 2.6. CI sorption values (IFHANGAZ) by PE films plasticized by mineral oil (95°C, 10 min)

Parameter	Sorption Value (g/m^2) at Mineral Oil Concentrations (wt%)					
	0	5	10	15	20	25
Sorption according to titrimetric data	4.3	6.5	9.0	12.1	14.5	16.3
Change of film mass as a result of CI sorption*	6.6	6.5	2.4	0.3	−0.2	−0.4

*Densities of CI (IFHANGAZ) and mineral oil are 0.85 and 0.90 g/cm^3, correspondingly.

Such *solution of CI vapors in the plasticized layer* of the film is used if the pressure of saturated vapors is about 0.1–1 Pa and the inhibitor is not compatible with the polymer but dissolves perfectly in Pl.

The plasticizer is fed to the mandrel and as soon as it contacts the hose in the viscous-flow state, it forms a layer of colloidal solution, becoming an integral part of the film thickness (Fig. 2.5d). When the colloidal solution separates into phases this layer becomes gel-like. As a result, a film is formed with a structure that is inhomogeneous across its thickness and that has a gradual change from the pore-free polymer layer to that possessing a system of communicating pores filled by a Pl (Fig. 2.6b). CI vapors supplied into the hose dissolve in the plasticizer and the inhibitor thus fills the film pores.

The inner surface area of the modified film hose broadens thanks to the relief structure, which intensifies CI absorption from the gaseous phase during film molding.

Modification of a non-plasticized hose by CI is exercised when the CI serves as a polymer Pl or the inhibitor solution in the Pl is used as a modifying liquid [73, 74].

The hose is formed from a non-plasticized polymer and is brought into contact with the inhibitor liquid (Fig. 2.5e). Such films have a structure similar to that shown in Fig. 2.6b. The pore-free layer functions as a barrier, while the porous one is the CI carrier. The porous layer comprises about 25–50% of the film thickness, which is in part dependent on the contact time of the hose and the liquid. According to field tests, the degree of thermal-diffusive filling of the polymer with a 50% solution of CI (M-1) in dioctyl phthalate Pl increments in the course of blowing by 3 to 5.5% as the solution degree over the mandrel rises from 1 to 3.5 cm.

CI liberation from only one side of the PE film is characterized by minimum production and operation expenditures. The method and its equipment have been patented [52, 73–76].

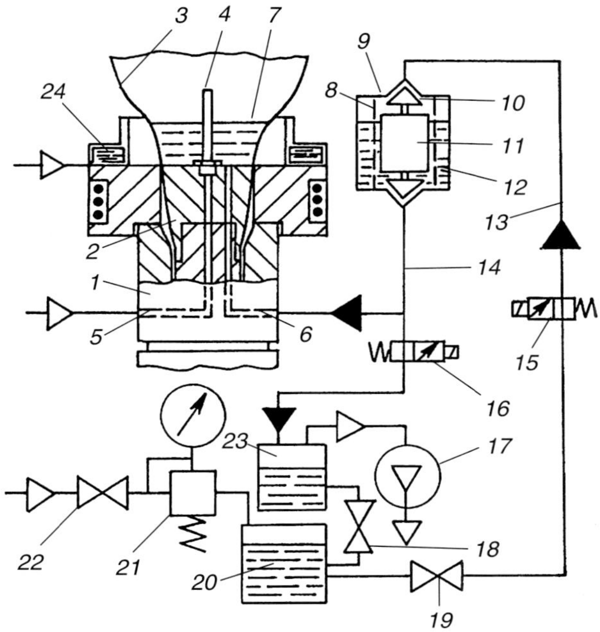

Fig. 2.8. A device for feeding a modifier to the extrusion head mandrel of the hose-film aggregate: (1) extrusion head; (2) mandrel; (3) polymer hose; (4, 5) and (24) tube and channels for compressed air; (6) channel for inhibitor liquid; (7) and (8) liquid levels over extrusion head mandrel and in the feeding chamber, respectively; (9) CI feeding chamber; (10) and (12) valves; (11) float; (13) and (14) mains; (15) and (16) magnetic distributors; (17) pump; (18), (19) and (22) taps; (20) and (23) accumulator tanks; (21) pressure valve

Furthermore, a device that is part of a hose film aggregate for thermal-diffusive modification of polymer films by liquid-phase CI has been developed (Fig. 2.8) [74,75]. Film hose 3 is formed by forcing a polymer melt through an annular orifice of the extrusion head 1. The CI is kept in a pressurized tank 20 into which compressed air is fed through tap 22 and regulated by pressure valve 21. As tap 19 is turned on, the CI flows from the tank to chamber 9 making float 11 rise to the surface and open the duct blocked by valve 12. The CI flows from main 14 and channel 6 into the hose hollow 3 and spreads over the surfaces of mandrel 2. The hose continuously moves up, absorbing CI, as a result of which its level in the hose hollow 7 and 8, and chamber 9 lowers. The CI level is replenished automatically as float 11 goes down and the gap over valve 10 expands. The relationship of CI levels 7 and 8 in the hose hollow and in the chamber depends on the air pressure in the hose and tank 20, correspondingly. The level of CI in the hose hollow can be monitored by moving chamber 9 along the vertical line.

In emergency situations, e.g. rupture of the hose in the chamber, magnetic distributors 15 and 16 shut off main 13 and open main 14 over which the CI drains into tank 23. Simultaneously, vacuum pump 17 is turned on and the CI is evacuated fully from the hose hollow and from chamber 9 until float 11 overlaps by valve 12 the orifice connecting the chamber with the main 14. The device minimizes infiltration of CI vapors into the environment and guarantees safe working conditions for operators.

CI sliding over a heated mandrel of the extrusion head was unfavorable in many respects. In this regard, it was proposed to contract the contact areas of the inhibitor liquid and the mandrel (Fig. 2.9) [77]. With this aim, bushing 4 is installed on mandrel 1 coaxially to channel 2, from which hose 3 is blown out and fanned by air 7. The cavity between the hose and the bushing presents a vessel for inhibiting liquid 5. The outer diameter of the bushing d is related to the mandrel diameter D by $0.83 \leq d/D \leq 0.95$. An opening of channel 6 for CI supply is made in the vicinity of the bushing base.

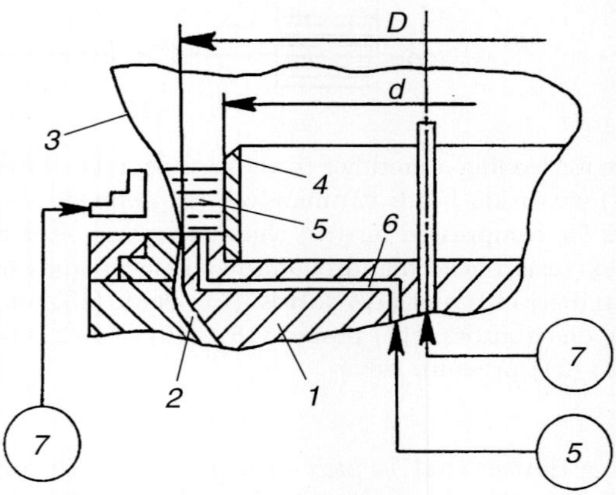

Fig. 2.9. Extrusion head for manufacture of liquid-filled inhibiting film (see explanations in the text)

It is possible to avoid the contact of CI with the heated surface of the mandrel if it, or its solution in Pl, is poured off on the inner surface of the hose at a velocity similar to the extrusion rate from the height of the polymer setting line [78]. It appears that the hydrostatic pressure of the draining liquid is much less than the hose strength at any height of its feeding, so the liquid will not break the hose. The CI flow velocity correlated with the extrusion depends on its viscosity, density and adhesion to the hose in the viscous-flow state.

It is convenient to apply the CI liquid on the hose surface using an elastic tube coiled up in the form of a conical spring. Its upper part is bow-shaped and is permeable to the CI. The upper end of the tube is plugged; the lower end communicates with the CI feeding channel. As the CI is forced into the tube, the spring diameter enlarges and its bow-shaped part approaches the hose. The CI pours out on the hose through the permeable wall.

It is expedient in most cases not to impregnate the whole surface of the film but only the parts that will contact products packed within it, leaving intact the portions to be fused. Using the described technique, it is possible to create an antirust texture on the film surface and thus improve efficiency of CI consumption.

Solid particles of CI can be bonded to *the inner surface* of the hose film during forcing of air with a fine-dispersed inhibitor powder into the hose cavity. CI particles adhere to the hose surface in the viscous-flow state and fix on solidification [47].

The technological potential of manufacturing inhibiting films as well as their range and application fields can be significantly expanded by combining the described methods.

For example, a heat-resistant CI (alone or together with Pl) can be impregnated directly into the polymer mixture, which is then extruded and melt-blown. The hose is then saturated by a volatile CI that is nonresistant to temperature in the vapor or liquid phase. A contact CI can be also employed as a Pl [3].

Furthermore, a procedure is known [79] in which activated lignin (hydrolyzed lignin treated at 200–230°C) is used as a contact CI impregnated into a PE-based polymer blend with benzoic acid. The resultant hose is brought into contact with a dicyclohexilamine melt. Film extruded from this blend filled by the activated lignin becomes a VCI carrier as a result of interactions between benzoic acid and cyclohexilamine.

Multilayered and combined inhibited hose films are mainly produced by co-extrusion methods. Apart from these, original procedures for obtaining adherence to the hose base functional layers (coats) are employed. In one of these [50] a barrier coating is formed on the outer surface of the hose film filled by CI using one of aforementioned methods.

Multilayered inhibited films with CI desorption on one of the sides can be produced by a process illustrated in Fig. 2.5e, in which the liquid forced into the hose cavity is a film-forming inhibiting composition. To manufacture adhesive films on a polyolefine base, a film-forming composition has been developed that contains 2–10 wt% of mineral oil, 4–12 wt% of oil-soluble CI, 3–10 wt% of rubber, 3–10 wt% of rosin, 5–16 wt% of tall pitch in an organic solvent. The composition is brought into contact with the extruded hose at a temperature of T = (0.6–0.9) T_m of polyolefine. As a result, an adhesive coating is formed on the hose surface, which is the source of CI and shows high adhesion to metals (about $10\,N/cm^2$) and to the base film [80].

A method of manufacturing films, during which thermodiffusive modification of the hose blank by a liquid-phase CI runs simultaneously with the

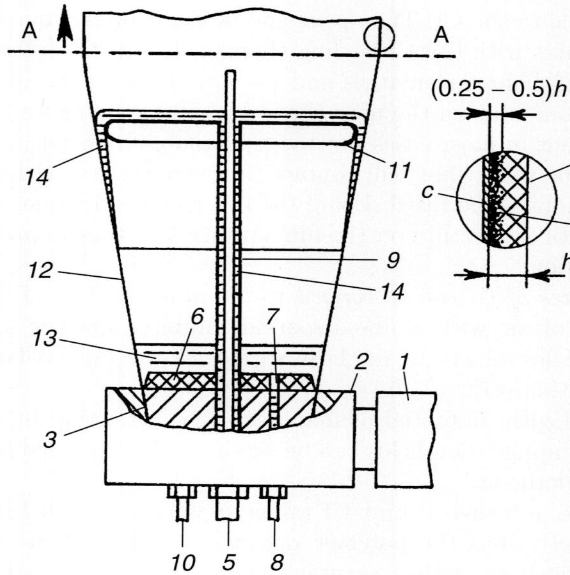

Fig. 2.10. Process layout of hose films with inhibiting varnish layer and their structure: (**a**) polymer base, (**b**) jelly-like layer, (**c**) varnish layer. (1) extruder, (2) head, (3) annular slot, (4) and (5) compressed air main and channel, (6) heat-insulating pad, (7) and (8) pipeline and duct for inhibiting liquid, (9) and (10) pipeline and duct for inhibited varnish, (11) adjustable diaphragm table, (12) hose, (13) inhibiting liquid layer, (14) annular flow of varnish. (**A–A**) solidification line of polymer melt

formation of an inhibited varnish layer on the hose inner surface has been patented [31] (Fig. 2.10).

The polyolefin melt is forced through an annular slot 3 of extrusion head 2 and is blown out into a hose. A solution of CI that is easily volatile in Pl is fed onto the site over the mandrel. The CI solution dissolved in the varnish based on PVB, CEVA or cellulose acetobutyrate is forced to the surface of a diaphragm Table 2.11. The varnish is poured over the inner surface of the rising hose so as to avoid contact between the lower edge of the annular flow of varnish and the CI layer. The varnish contacts the colloidal solution of polyolefine with CI formed in the hose surface layer. Above the solidification line A-A the colloidal solution decomposes into phases and a jelly-like layer is formed. Just in this layer the inhibiting liquid is enclosed in the polymer matrix pores that are thermodynamically compatible with the varnish. The varnish diffuses into the pores and, on setting, forms on the inner hose surface an inhibited varnish coat embedded in the porous layer.

It may occur that the jelly and varnish layers contain CI evaporating at different rates. This fact can be utilized to make a protective atmosphere

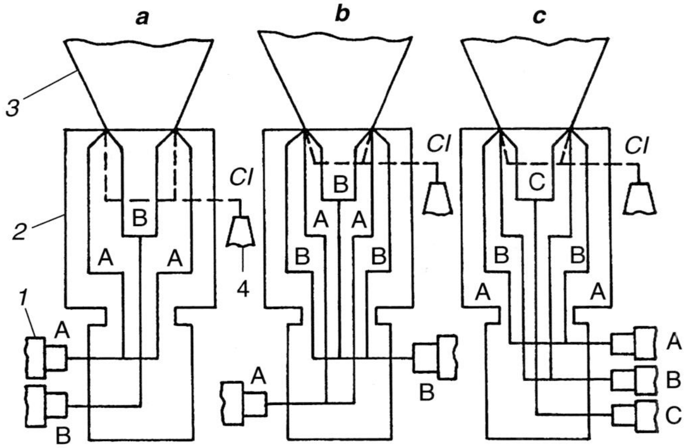

Fig. 2.11. Process schemes for inhibited laminated hose films: (**a**) two-layered and (**b**) three-layered films of two polymer materials A and B, (**c**) three-ply film of three materials A, B and C. (1) extruder, (2) extrusion head, (3) hose, (4) pumping unit

inside the package thanks to the intensive evaporation of the readily evaporating CI and to maintain it for a prolonged time via the gradual evaporation of the moderately volatile CI.

In addition, process schemes for manufacturing laminated inhibited films by co-extrusion are presented in Fig. 2.11. Working volumes of extruders 1 for processing polymer materials (A, B, C) are connected with extrusion head body 2. The head is fitted with several annular slots that form, on blowing, a multilayered hose 3. Two-ply films can be formed from the same [7] or two different (Fig. 2.11a) polymers. Three-layered films can, correspondingly, incorporate two (Fig. 2.11b) or three different (Fig. 2.11c) polymer materials.

CI can be impregnated in one of the layers during joint co-extrusion of the polymer and inhibitor. There exist various methods for feeding CI into the space between the layers prior to pasting them together into a hose (Fig. 2.11) or forcing CI into the forming die of the carrying layer (see Sect. 2.2.6, Fig. 2.24).

The structure of inhibiting laminated films (Fig. 2.12) is distinguished by the solid state of their outer (barrier) layers. They are commonly produced from polymers that are incompatible with CI (Fig. 2.12, layer A). Polymer layers facing the protected metal surface posses a restricted compatibility with CI and contain structural areas incorporating the CI in one of the states (Fig. 2.12, layers B and C).

An original design for the extrusion head has been developed for manufacture of multilayered inhibited film materials with a fibrous melt-blown polymer layer (B) bonded to the inner or outer surface of the film base (F), and the materials of the F-B-F-B-F structure [81]. A peculiarity of this

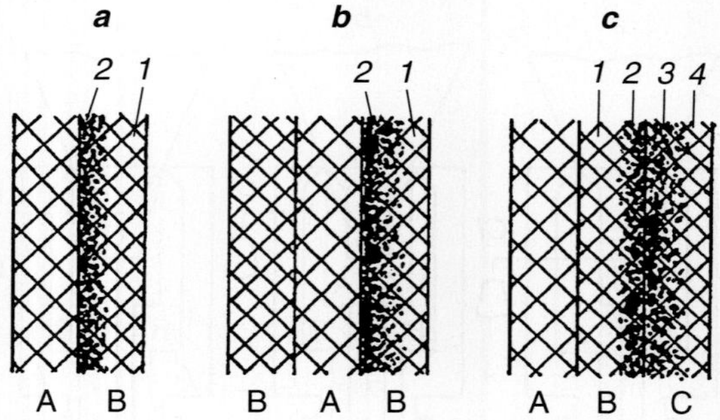

Fig. 2.12. Structure of multilayered inhibiting films: (**a**) two-layered and (**b**) three-layered film composed of two polymer materials A and B, (**c**) three-layered film made of three materials A, B and C; (1) and (4) solid layers, 2 and (3) CI-filled areas of layers B and C

extrusion head design is the presence of new structural elements. Special annular channels are made in the spinneret of the head body either from the inner or outer side of the spinneret orifices at an angle to their axes.

A fibrous mass formed during blowing of the hose is bonded to the film layers. The design can additionally be fitted with a device for the impregnation of CI into the film, into the melt-blown layers or between the layers. The described film bears heat insulating and damping as well as anticorrosion functions.

2.2.5 Some Other Methods of Manufacturing Inhibited Films

Slot extrusion is used to form flat inhibited films and laminated films. To this end, flat co-extrusion, extrusion pasting and lamination methods are used [7, 14, 18, 28, 34, 40, 82]. Commonly, preliminary granulated mixtures of film-forming polymers and CI participate in extrusion. To avert foaming of the melt by VCI vapors, adsorbents like silica gel [7] are introduced into the polymer composition.

A scheme of applying an inhibited coating in a polymer melt onto the film base by extrusion lamination [34] is illustrated in Fig. 2.13. A similar procedure is employed at Cortec and Northern Instruments Co. for paper production with a laminated layer of inhibited PE.

A special setup is used for manufacture of a three-layered inhibited film whose middle layer is embossed and the cells are filled with a non-drying inhibiting composition (Fig. 2.14). The setup consists of extruders 1 with slot heads 3, vacuum-forming drum 4 and CI batcher 2 [4, 83]. The film blank is directed by guide rollers 5 and 6 to the assembly unit of the laminated material equipped with a heat chamber 7.

Polymer films produced by slot extrusion are homogeneous and have even thickness, which is an important factor for modification by a structural

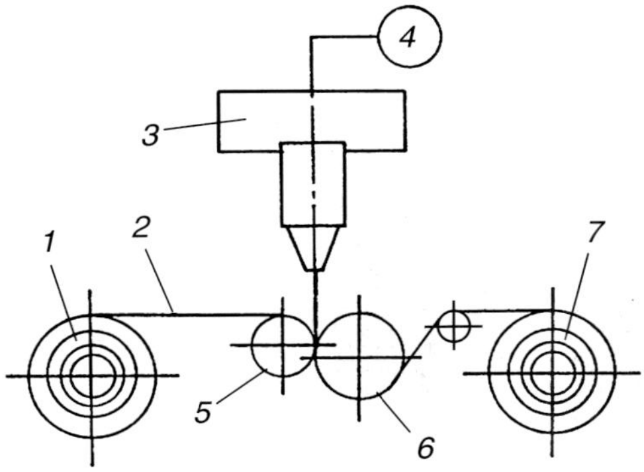

Fig. 2.13. Extrusion-lamination scheme for production of two-layered inhibited polymer film materials: (1) unwinding machine; (2) film base, (3) extruder with a slot head, (4) inhibited polymer composition, (5) pinch roller, (6) cooling shaft, (7) coiler

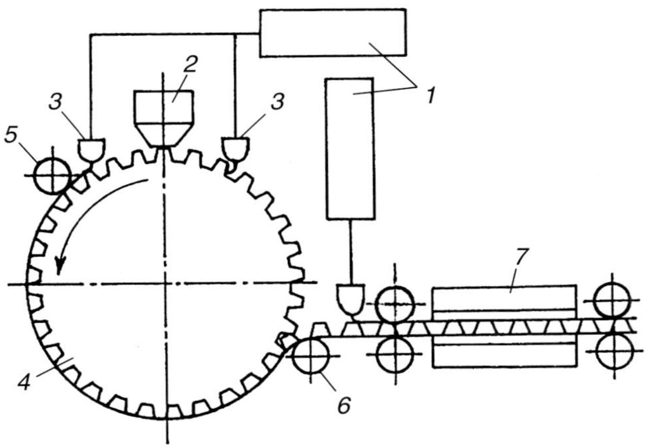

Fig. 2.14. Production scheme of a three-layered inhibited film with a cellular middle layer filled with CI

encapsulation method [4]. This technological process consists of three stages: (i) formation of a flat isotropic film blank, (ii) its stretching in encapsulating inhibiting liquid, and (iii) isothermal treatment of the stretched film.

The method of structural encapsulation eliminates thermal effects for inhibitors known for their low thermal stability. Its drawback is CI loss due to uneven distribution within the polymer matrix and diffusive isolation from both sides of the film. The described films are good as the inner layer of multilayered packages.

Calendering is used for manufacture of laminated and combined inhibited films by pasting, i.e. by bonding film blanks using glues or polymer melts [34].

A diagram of a backing machine VEMA produced by Swiss Interplastica S.A. Co. is presented in Fig. 2.15 [3]. Film materials A and C fed from the unwinding device are bonded in the calendering unit via the adhesive layer B formed of a melt or solution in unit 9 and heated rollers 8.

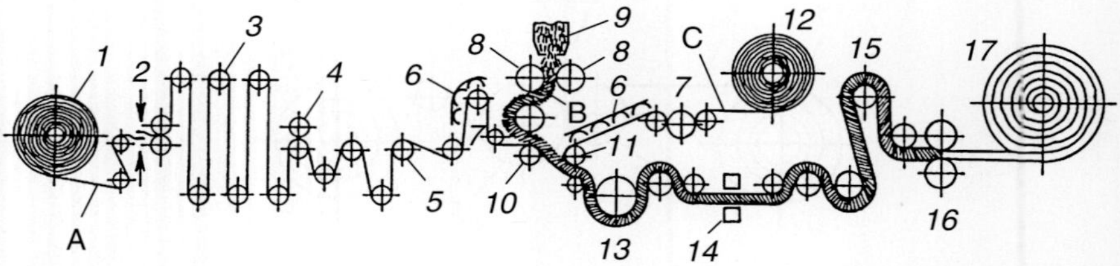

Fig. 2.15. Scheme of a backing machine for laminated and combined film materials: (1) and (12) unwinding device; (2) splicing block; (3) compensator; (4) strainer control; (5) centering unit; (6) IR heater; (7) preheating rollers; (8) shaping rollers; (9) batcher or extruder; (10) rubberized press roller; (11) embossed roller; (13) cooling roller; (14) thickness gauge; (15) cooling unit; (16) edge cutter; (17) coiler; A, B and C – layers of the material

Calendering is a basic procedure of the Italian production line Colines for air-bubble films (Fig. 2.16) [3]. The base film is formed from melt A fed from extrusion aggregate 1. Hollows in the film are formed by vacuum molding in unit 2. Simultaneously, a film coating formed of melt B is glued to the base. When passing between rollers 3 and 4, the air-bubble film can be reinforced by coiled materials C and D fed from unwinding blocks 5 and 6. Then, the film is directed into the cooling device 7, centered in device 8 and wound into a roll 9. The line can be equipped with a mechanism for pumping the inhibiting gas into the hollows between layers A and B.

The method of casting solutions and polymer latexes, as well as application of polymer plastisols containing CI on the forming substrate is used for the manufacture of both single-layer and multilayered inhibited films [3,4,23,25,41,43,53,84] as well as films with an adhesive layer [57,58].

A typical scheme for the production process for single-layer films by casting a solution of a film-forming polymer is illustrated in Fig. 2.17.

A polymer solution containing dissolved or dispersed CI particles is fed from the slot head onto a continuous band 2 whose surface adheres poorly to the polymer and which moves at a constant speed. Film formed on the band is dried by IR heaters 3. It is set and dried completely on contact with the cooling drums 4. The film sheet is then centered, its edges are trimmed by device 5 and it is wound onto roll 6.

Hollow metal cylinders are commonly used as the substrates or moving polymer films when manufacturing multilayered or combined materials by casting. Depending on the volatile properties of the solvent and viscosity of

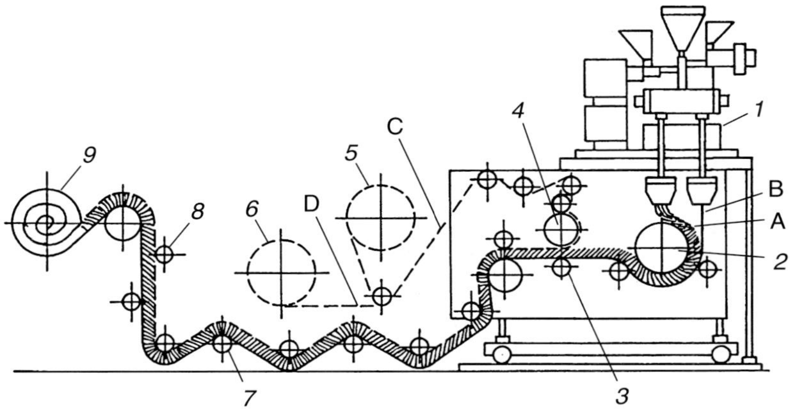

Fig. 2.16. Production line for air-bubble films

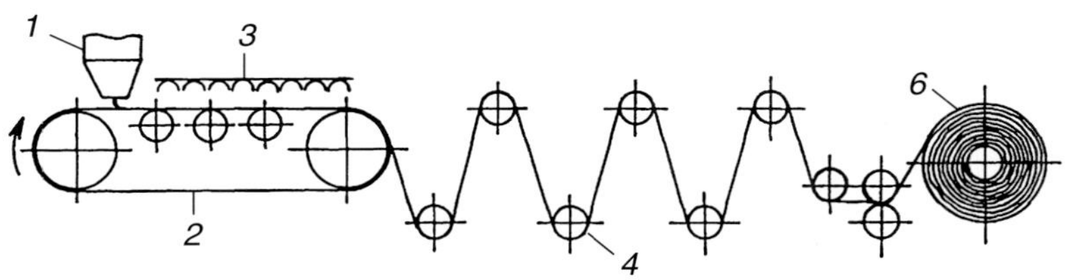

Fig. 2.17. A casting machine for antirust films

the polymer solutions or latexes, spinnerets of spreading type, slot heads or spinnerets with a roller are used for forming such films [4].

The method under discussion creates a moderate temperature regime when molding inhibited films thus eliminating their thermal destruction and VCI losses on evaporation.

2.2.6 Monitoring of Corrosion Inhibitors Absorption by Films Using Physical Fields

Physical fields are used in production techniques of inhibited films firstly to raise process efficiency. Films formed in these conditions undergo structural changes that impart sometimes unforeseeable properties to the film materials and furnish new potential in their application.

Below a few examples are given of production techniques for inhibited films involving electrical and magnetic fields as well as unconventional heat and mechanical effects.

Electrical fields are traditionally used in the manufacture of polymer films with the aim of facilitating printing, dying, pasting and other operations to change the surface layer structure of the films and to polarize the polymer material.

During *treatment in a glowing or corona discharge*, the films are subjected to bombardment by charged particles (electrons and ions) and the effect of active chemical compounds formed in the discharge area. Structural transformations observed in the surface layers are a result of broken macromolecular chains and oxidation-induced conversions of macroradicals. Functional groups of oxidized macromolecules are participate actively in the formation of adhesive bonds with other polymers. Moreover, electrical treatment reduces the wetting angle of the films by polar liquids [34, 85]. These effects are also the reason for improved cohesion of film surfaces to printing ink, lacquer, glue and other engineering materials.

Electrical polarization is a fundamental property of polymer materials transformed into electrets in electrical fields [86–88]. Electrical polarization is used to promote adhesion between layers of antirust film materials, limit permeability of the layers, impede wetting, reduce waterproof properties, raise the quality of printing and intensify CI saturation [68, 70, 88].

Electrical polarization of antirust films is most often based on the production techniques of thermal and corona electrets.

In the former case, films with a temperature close to that of the polymer softening are endured and cooled down in an electric field of $E < E_{\mathrm{br}}$, where E_{br} is the field leading to breakdown of the film. The field is applied by passing the film in the gap between electrodes, one of which is immovable and the other is formed by rotating roller. The intensity of the electric treatment is regulated by varying the electrode voltage, the gap between them and the sliding velocity of the film [3, 88].

To produce combined films without any glue, the film blanks are treated by a corona discharge at room temperature while passing between rotating cylindrical electrodes. Tens to thousands of kilovolts of voltage and at a frequency of hundreds of kilohertz frequency is fed onto the electrodes [88]. Such electric treatment is accompanied by the formation of ozone in the discharge area, which oxidizes the film surface. The treatment efficiency depends heavily on the gaseous medium composition where the discharge occurs. Notice that adhesion of PP film to a PP sublayer rises 100 times after treatment in a corona-forming gaseous medium containing 5% carbon oxide [2].

Along with the above-considered methods it is worth mentioning swelling of film blanks in a solvent followed by its evaporation in a constant electric field, treatment with ionizing radiation, treatment in magnetic fields, and etc. [87].

Electret films can be produced without application of any outer source of electric or magnetic fields. The extruded polymer film blank in a viscous flow is subjected to stretching and brought into contact with shaping parts made of short-circuited unlike metals. After cooling to a viscoelastic state, the film is removed [89].

Modifications of the above method are illustrated in Fig. 2.18. Polymer-melt 1 flowing from extruder head 2 contacts two continuous bands 3 and 4 made of unlike metals (Fig. 2.18a). The bands are close to a bridge 5 fitted with a roller drive 6 whose rotation velocity defines the degree of stretching of the film and its cooling rate. The polymer melt can be extruded into a gap between two foils of different metals unwound from reels 3 and 4 (Fig. 2.18b) or between a continuous band 4 and roller 3 having a metal coating (Fig. 2.18c).

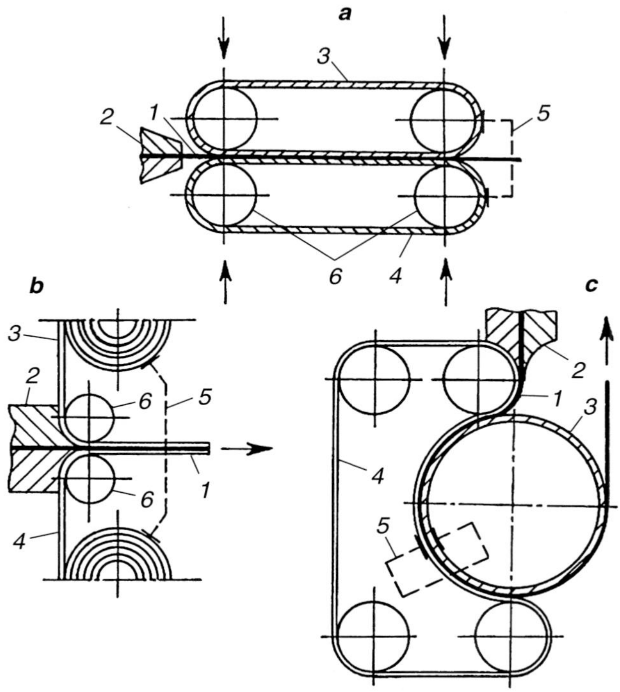

Fig. 2.18. Schemes of manufacturing electret films in contact with unlike metals

Electrical polarization of films in contact with unlike metals is employed during the production of combined materials by pressing [90]. The films can be also polarized by injection of charged fillers into the polymer matrix [91].

Magnetic fields are commonly used to manufacture conducting polymer materials and to process ferroplasts, i.e. polymer composites containing ferromagnetic fillers.

It has been established that the introduction of small amounts (0.5–2 wt%) of a ferromagnetic matter (barium ferrite or strontium ferrite powder) significantly intensifies the saturation of CI in the polymer matrix. With this aim, extruded hose filled with the ferrite is treated in the contact zone

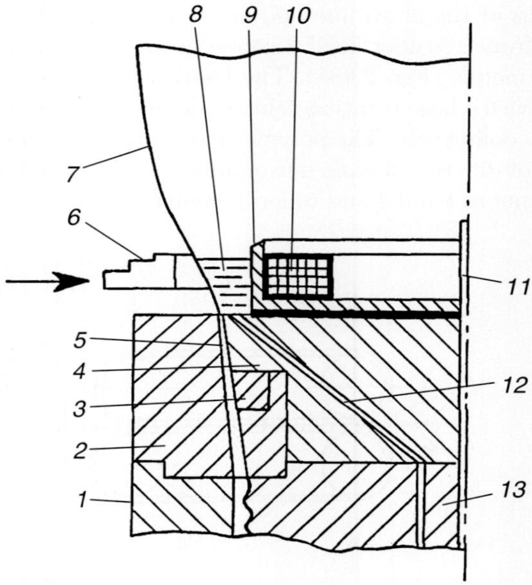

Fig. 2.19. An extrusion head for CI impregnation in the polymer film using magnetic field

with the CI by an alternating magnetic field at a frequency of 100 Hz and an intensity of 240 kA/m. The magnetic particles form pores in the film material during oscillations under the effect of the magnetic field, and CI penetrates into these pores [92].

This procedure employs an extrusion head [93] (Fig. 2.19) that consists of a body 1, nozzle 2 and collapsible mandrel 13 with ducts 12 for CI and compressed air 11. The nozzle and mandrel form in the upper part an annular forming die 5 communicating with the extruder. The collapsible mandrel contains a permanent magnet 3 made of a samarium-cobalt alloy with a plate 4 of diamagnetic chrome-nickel steel placed over it. Bushing 9 is mounted on the mandrel coaxially to restrict the amount of CI fed to the hose 7. Inside the bushing there is a solenoid 10, which is the source of alternating magnetic field. Annular collector 6 is intended for cooling the outside surface of the polymer hose by a flow of compressed air.

The polymer melt filled with ferromagnetic particles is forced through the die 5 and is blown into hose 7 by the compressed air supplied over duct 11. The CI is fed between the hose and bushing 9 over duct 12. The field of the permanent magnet 3 effects the movement of the extrudate along die 5 and shifts ferromagnetic particles spread in it to the inner surface of the hose. The diamagnetic part 4 prevents extraction of the particles amassed near the magnet from the melt. The resultant discontinuities in the extrudate are healed under the action of the stresses created in the melt by the extrusion pressure. As solenoid 10 is actuated, particles concentrated near the inner hose surface start to oscillate thus allowing permeation of CI into the gaps between the particles and binder. The amount of CI absorbed by the film can be regulated by varying the frequency and amplitude of the magnetic field intensity or by changing the height of the solenoid over the mandrel. Note that the CI, similarly to the magnetic filler, is concentrated near the inner surface of the hose and isolates from this side of the film only.

Heat effects on the film during impregnation with CI are chiefly aimed at optimizing the temperature regimes of the film material molding, and regulating heat exchange and physical-chemical transformations in the hose-inhibitor contact.

The specifics of the hose film technique facilitate the use of heat flows generated by eddy motion [94, 95]. The eddy or Ranks' effect appears in the swirl flow of a viscous compressible medium and can be realized in a simple device called a vortical tube [96]. This looks like a smooth cylindrical tube 1 (Fig. 2.20) fitted with a tangential nozzle 2, scroll 3, diaphragm with axial socket 4 and throttle 5. The influent through the nozzle in tube 1 gas forms a swirl flow whose paraxial layers are cooled down noticeably (to negative temperatures) while the peripheral layers, vice versa, are heated (to 200°C and above) and issue through the throttle as a hot flow. As the throttle is turned off, pressure in the vortex tube intensifies and the cold flow discharge through socket 4 increments, while that of the hot flow diminishes. Their temperatures change too.

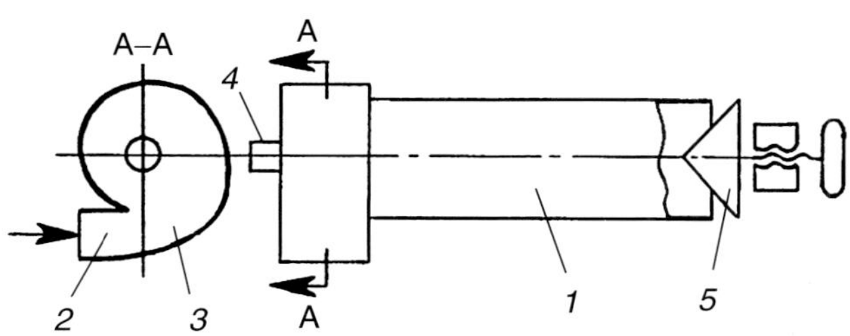

Fig. 2.20. A vortical tube

The extrusion head [94] shown in Fig. 2.21 contains a unit for preparation of the inhibiting mixture that is made as a vortical tube. The polymer melt is fed from extruder 7 into annular slot 6 formed by mandrel 9 and nozzle 5, and is squeezed as a hose 3. The compressed air is fed into nozzle 10 of scroll 11, and into nozzle 15, and CI is fed from tank 14 in a liquid or gaseous state. The compressed air and CI are accelerated in the scroll over the spiral and form a swirl flow of a homogeneous mixture in duct 8. The peripheral layers of the flow become heated and the CI (if it is a liquid) evaporates, whereas

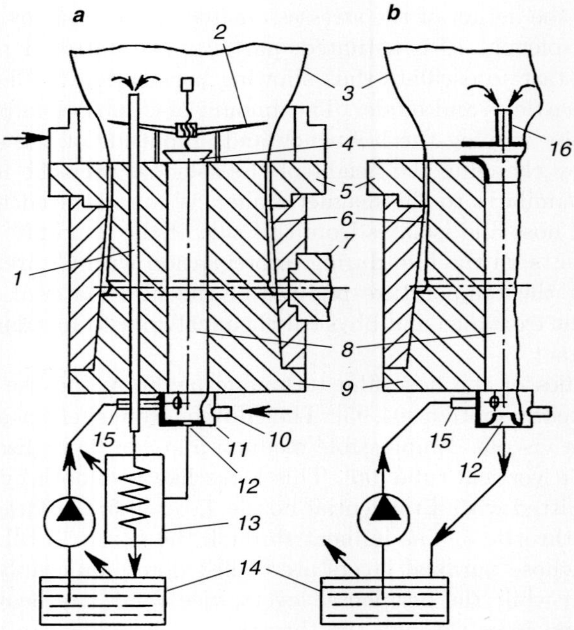

Fig. 2.21. An extrusion head with a vortical unit for CI feeding to the mandrel

paraxial flows are cooled down. The mixture of air and CI vapors issues from throttle 4 into the hose cavity, blows out the hose directed by screen 2 and saturates it with the inhibitor. Exhausted mixture is passed through channel 1 to cooler 13, being cooled by the flow from opening 12. CI vapors condense and the condensate is fed to tank 14. The spent air is ejected to atmosphere.

A diffuser modification at the outlet can be of a slot shape (Fig. 2.21b). The above-described design readily reduces the pressure and assists self-evacuation in hollow 8. In this case, used up CI is removed from the hose through duct 16 in the diffuser face wall. When passing the axial area of the tube, CI vapors are cooled down and condense. The operating regimes of the head are regulated by varying the height of the slot in the diffuser.

The described specificity of plasticized inhibited films promotes monitored isolation of CI from the films in response to heat effects. For example, by cooling PE films containing CI in mineral oil down to 0–40°C it is possible to speed up isolation of the liquid-phase components during operation [97]. This is a simple solution to avert metal ware corrosion in the first hours after packing when the probability of corrosion initiation is conditioned by the rate of CI liberation into the package volume.

Mechanical effects in the film-CI system are able to intensify absorption of the inhibitor by the polymer material.

Several methods using mechanical effects of inhibiting liquids on the polymer material during blow molding have been considered elsewhere [3].

It is not always possible to elevate the liquid level fed to the extrusion head mandrel and escape high hydrostatic pressure (see Fig. 2.5e) lest the hose should rupture. To avoid this, the CI is applied on the hose wall during rotation [98]. The liquid vortex formed this way promotes the required level of the liquid on the mandrel near the hose surface, making use of a minimum amount of CI and changing the rotation velocity of the hose. The dynamic interaction of the hose and the CI intensifies its diffusion into the polymer material, breaks the vapor jacket at the inhibitor-hose interface and facilitates the uniform distribution of CI pressure, minimizing the probability of hose rupture.

The CI angular rotation velocity to extrusion rate relationship is monitored so as to allow a sufficient CI content in the film and escape the danger of hose rupture under the action of the centrifugal force.

An extrusion head (Fig. 2.22a) for the procedure described above is fitted with an impeller 5 installed coaxially to die 6 and shaft 2 mounted on duct 7 for compressed air. The latter, when issuing from tube 1, blows out hose 3 into whose hollow a CI is fed over channel 8. Rotation of the impeller creates a liquid vortex and the CI climbs up the hose wall owing to centrifugal forces.

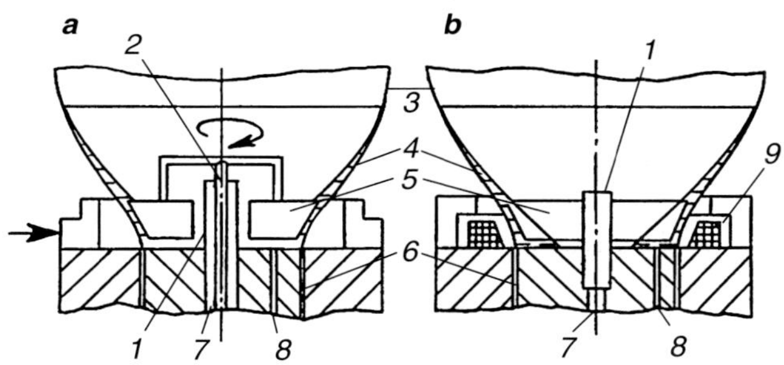

Fig. 2.22. An extrusion head with CI vortex generator: (**a**) mechanical and (**b**) vortex generators

The head design shown in Fig. 2.22b is fit with an impeller 5 made of a ferromagnetic material whose drive is a source of the rotary magnetic field 9, which broadens the range of rotation velocities of the impeller.

To improve weldability of the film and cut its cost via lowering CI consumption, the CI can be applied only on separate areas of the film material. The extrusion head [99] producing a given pattern of CI spots on the hose surface is equipped with a device for airless spraying of CI (Fig. 2.23). The device is installed in mandrel 1 coaxially to die 2 for shaping hose 3 on coaxial tubes 7 and 9 forming channels 8 and 10 for CI and compressed air. Radial openings 6 are made on the projecting part of tube 7 conjugated to bush 12

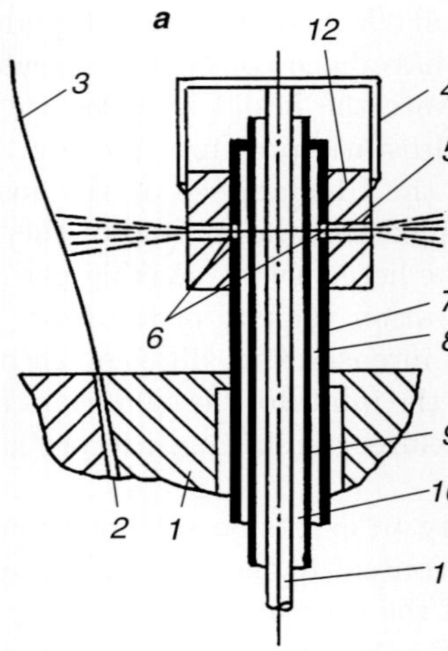

Fig. 2.23. A device for spraying CI on the inside hose surface: (**a**) working and (**b**) idle positions

via lathe dogs 4 fixed on shaft 11. The bush is fitted with radial jet orifices 5 for CI spraying. Shaft 11 is actuated by a drive imparting reciprocating or revolving motion to bush 12.

Upon actuation of the hose film aggregate and stabilizing extrusion and blowing regimes, the CI is forced over channel 10 under a pressure of 0.3-0.4 MPa and the drive of bush 12 is switched on. Position (Fig. 2.23a) illustrates when openings 6 in tube 7 match up nozzles 5 and CI is sprayed on the hose. The spraying is stopped (Fig. 2.23b) as soon as bush 12 is moved above its working position and openings 6 overlap with its wall. The number of openings and nozzles as well as the frequency of bush travels is adjusted so as to make the hose absorb the sprayed CI fully without leaking.

The arrangement of the spots on the film surface depends on the spraying regime and bush 12 travel pattern. Reciprocation of the bush creates vertical stripes, its rotary motion results in spiral curls (combining radial shifts of the CI jets with the vertical travel of the film). Pulse spraying at variable velocity of bush 12 produces intricate drawings on the hose surface.

Forcing of CI into the head die is a multifunctional operation leading to reduced friction of the melt against the die walls, in addition to improved orientation stretching and reinforcement of the film. The distinguishing feature of the employed extrusion head [100] (Fig. 2.24) is the presence of an annular chamber 6 in mandrel 4. The chamber communicates with CI channel 8 and is shut off by a porous wall 5 from the side of channel 7 over which the polymer melt is fed from the extruder into the head. The outside surface (a) of the porous wall 5 and inner surface (b) of matrix 1 are made in the form of a cone whose apex faces the side opposite to the inlet to channel 7.

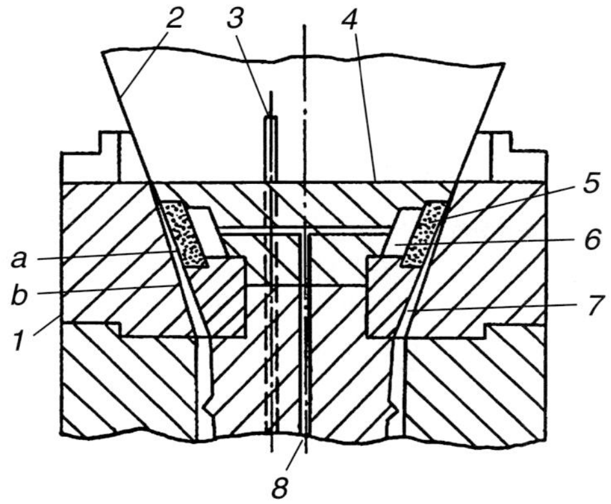

Fig. 2.24. An extrusion head in which CI is forced into the annular forming channel

Once the polymer hose 2 is shaped and blown out by the air coming from socket 3, the CI or its solution in a plasticizer is forced into channel 8 under pressure surpassing that of the melt in channel 7. It is filtered through the porous wall 5 and enters channel 7, forming a hydrodynamic wedge between the melt and surface (a) of the wall. The pressure of the liquid entrained into a constricting gap in channel 7 augments and the melt presses against the matrix wall (b), while the liquid is brought out as a film on its inner surface by the hose. The contact time of the hose and the liquid is sufficient for saturation of the film with CI. Films produced this way acquire 1.5 times higher tensile strength in contrast to those co-extruded from the polymer and CI.

According to the information presented above, the spectrum of available techniques for manufacturing inhibited films is extremely broad. To choose a required procedure, one should refer to the required properties of the components (CI in the first instance), the demands imposed on the film material and the final operating conditions.

2.3 Structure, Properties and Protection Mechanism

Properties of anticorrosion inhibited films are restricted by tough requirements. Along with their main functional property of protecting against corrosion damage, they should display rather demanding deformation, strength and barrier characteristics, resistance to atmospheric effects and good weldability.

The structure and properties of inhibited films may alter within a wide range in response to their composition and production techniques. Their anticorrosion effect depends largely on the mechanism of CI transportation from the bulk film to the article surface and the kinetics of electrochemical processes in the metal-inhibited film-hostile-medium system.

2.3.1 Structural Peculiarities of Inhibited Film Materials

The most commonly used type of inhibited films is produced by joint processing of thermoplastic polymers and highly dispersed CI powders. When the melting points, as well as the sublimation and thermal destruction temperatures, of the CI exceed the viscous flow temperature of the thermoplast, the film acquires the structure of a polymer composite filled with solid inhibitor particles [101]. The particles may partially dissolve in the polymer melt and exert a plasticizing effect on the polymer. Introduction of 1–2% of NDA into PE elevates melt flow by 1.5–1.8 times [10]. Aggregation of solid CI particles may present essential difficulties for their joint processing with polymer melts and result in structural defects or impair the quality of the film as a whole [4].

Homogeneous films with a structure typical of plasticized polymer materials are obtained from thermoplastic melts containing slight amounts of thermodynamically compatible liquid CI [103].

Some processes for the manufacture of inhibited films employ a Pl as a CI carrier. The Pl is impregnated into the polymer matrix in concentrations surpassing the limit of thermodynamic compatibility with the polymer (see Sect. 2.2.4). A mixture of components of Pl and CI transforms on heating until the polymer flows viscously into a homogeneous solution. Further cooling brings about separation of the solution into two amorphous phases. Because of the high viscosity of the system, the stratification localizes on microareas commensurate with permolecular formations of the polymer and leads to formation of a gel, consisting of the polymer matrix with a system of communicating pores filled by the inhibiting liquid (Fig. 2.25). The pore size depends on the composite formulation and temperature regime of the phase stratification [104, 105]. Free diffusion of CI in the liquid phase exercises timely transportation of the inhibitor to the surface of the article protected.

Microstructures of respectively outer and inner surfaces of the extruded hose film and the film produced by thermodiffusive filling of the polymer base

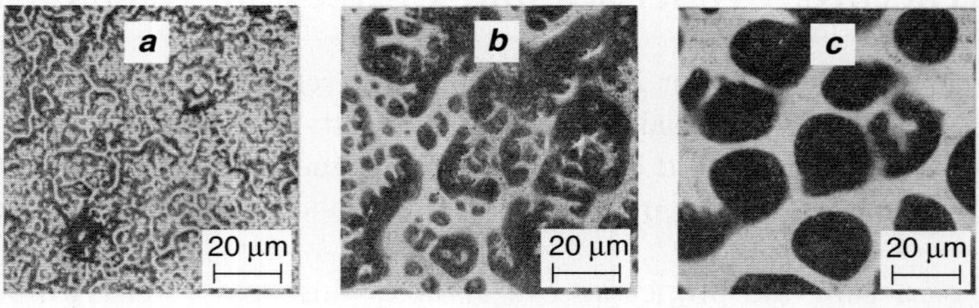

Fig. 2.25. Structure of polymer gels based on PE and mineral oil: oil concentration (**a**) 40 wt%; (**b**) 50 wt%; (**c**) 70 wt%

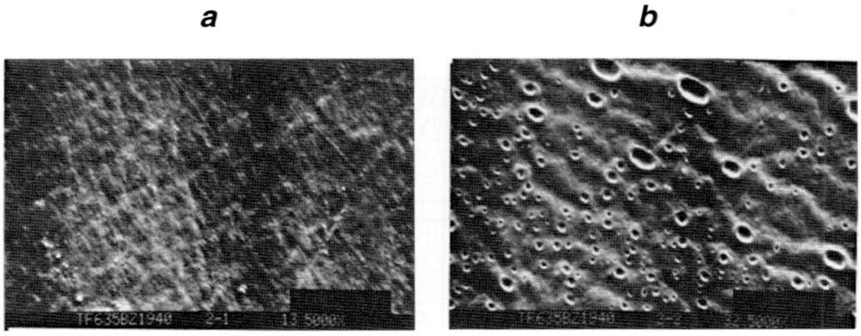

Fig. 2.26. Electron microphotographs of the surface of inhibited PE-based hose film: (**a**) outer surface of the hose; (**b**) inner surface from the treated by CI (PhTet) in Pl (dioctyl phthalate) suspension

with CI suspension in Pl during hose blowing are illustrated in Figs. 2.26a and b.

The outer surface of the hose has a structure characteristic of a solid PE film. This solid layer gradually transforms across the thickness into an inhibited jelly-like layer (see Sect. 2.2.4 and Fig. 2.6b). The inner surface of the hose has a relief with pore openings over which the CI can be transported from the jelly-like layer volume into the film surface. Accelerated exudation of the liquid phase is attributed to structural and relaxation processes occurring in the gel [49].

Inhibited films are sometimes produced by impregnation of CI into a loosened surface layer of a polyolefine film base [30]. The layer is formed by fixing a monolayer of ground granules of the same polymer on the extruded base. After fixing CI in the surface layer, the film undergoes thermal and pressure treatment in a press or roller mechanism. A schematic diagram of transformations taking place in the film during this procedure is presented in Fig. 2.27 [4].

Thermoplastic particles applied on the extruded film base (a) are fused until the shape is close to spherical. The particles are connected with the film via a sufficiently strong cohesive bridge whose diameter is less than that of a particle. CI fills the gaps between particles and concentrates mainly around the particle foot when contacting the film (b). Further temperature and pressure treatment of the film material (c) leads to a total or partial encapsulation of the inhibiting liquid in the film surface layer. Depending on the concentration and shape of the polymer particles bonded in the monolayer, the projections appearing on thermal treatment either merge into a solid layer tightly insulating the liquid microdrops in the film or form a porous diffusive barrier able to isolate CI from the film at a given rate.

Structural peculiarities of inhibited polymer films formed by extrusion encapsulation, i.e. stretching of films in a CI followed by the thermal treatment, have been described elsewhere [4]. The capsules look like cylindrical voids in the polymer bulk filled by a liquid CI and oriented along the stretching axis.

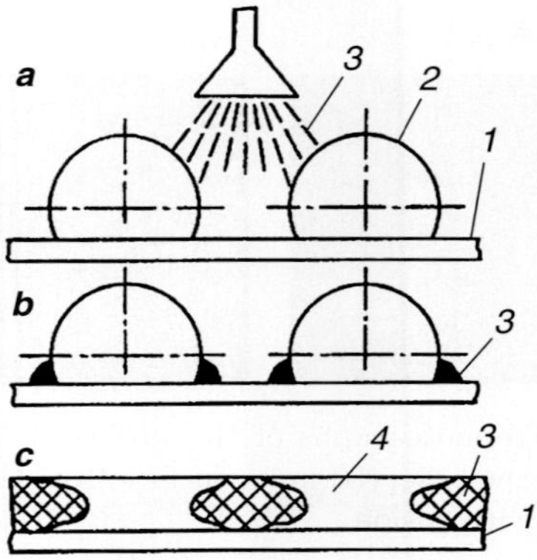

Fig. 2.27. Structural changes in film material modified by CI during production: (**a**), (**b**) and (**c**) – process stages. (1) film base, (2) a polymer particle, (3) CI, (4) layer with particles pressed into the surface

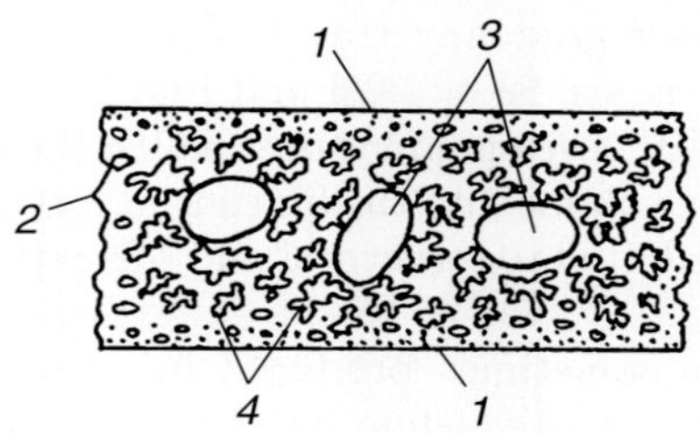

Fig. 2.28. Cross-section of the film area with structural capsules filled by CI: (1) film surface, (2) polymer matrix, (3) structural capsules filled by CI, (4) micropores

The concentration gradient of the encapsulated liquid is directed from the periphery to the center of the film cross-section (Fig. 2.28).

The structural capsules start to be formed in films subjected to deformation in liquids until some tension threshold. Microcracks and microvoids appear and are filled with the inhibiting liquid under tensile stresses exceeding the polymer flow limit. Capillary channels connecting these voids with the process liquid and with each other start to merge or open in the course of structural transformations but do not disappear fully. The liquid may move over the network of the formed channels beyond the polymer matrix limits or concentrate in some voids able under certain conditions to enlarge the manifold. Thermal treatment of the deformed film intensifies the relaxation processes in the polymer matrix, the film shrinks in the tension direction and the capillaries between voids link up densely, thus insulating liquid particles from each other. If the film is treated in the extended state, a more complex mechanism of microcapsule formation is realized [4]. CI liberation from microcapsules is related to their ability to break spontaneously under residual stresses appeared during production or operation. For instance, isolation of CI can be intensified by external mechanical or heat effects on the film.

Certain structural schemes of multilayered inhibited films in which CI carrier is e.g. a varnish coat (Fig. 2.10) or individual layers (Fig. 2.12) have been considered earlier in Sect. 2.2 and 2.1 (Fig. 2.3).

2.3.2 The Main Functional Properties of Inhibited Films

The most important functions that are critical for reliable serviceability of inhibited films as anticorrosion means are the following

- form a vessel for housing CI;
- be a barrier for hostile media permeation to the surface of protected metal ware;
- serve as a skeleton bearing the mechanical loads.

Prolonged liberation of immobilized CI controlled by evaporation and diffusion governs the protective characteristics of the films. Polymer films with contact CI have gained limited application in mainly skin packaging of metal ware [104]. Their anticorrosion properties depend much on the syneresis intensity of the liquid CI during which the inhibitor is transported from the polymer matrix onto the film surface (the physical-chemical bases of this phenomenon have been considered in Sect. 1.4).

A requisite of high exudation rate of viscous liquid CI, to which contact inhibitors belong, is the formation of a jellygel-like structure in the polymer matrix with a developed system of pores. With this aim, polymers are impregnated with Pl that also function as transporting agents for contact CI [3, 104, 106, 107].

For instance, the introduction of mineral oil into PE films (Fig. 2.29) substantially accelerates isolation of highly viscous MSDA inhibitor. As the concentration of the liquid phase rises above 30–40 wt%, syneresis intensifies as a result of, evidently, formation of the open-pore structure in the polymer matrix. Some contact CI (e.g. Vital) are themselves polymer plasticizers and do not necessitate additional introduction of any structural agents or inhibitors [3, 104].

It has been proven in [107] that the exudation kinetics of the inhibiting liquid from film samples produced of a mixture of PE + mineral oil + contact CI (Vital, GRM) by pressing under the polymer viscous flow temperature is described by a function

$$y = a\tau^b , \tag{2.1}$$

where y is sample mass loss (wt %) via exudation; τ is exudation duration (hours); and a and b are coefficients varying within the range 0 to 1 depending on the sample composition.

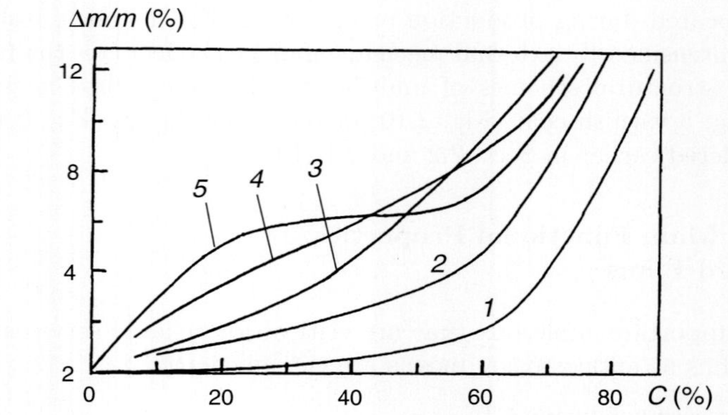

Fig. 2.29. Relative decreasing of PE-based inhibited film mass (during 100 days at 20°C) depending on the initial liquid phase content: (1) MSDA, (2) mineral oil + MSDA, (3) mineral oil + Vital, (4) mineral oil, (5) Vital

It is thus possible to calculate the time needed to attain a given exudation value. In Figs. 2.30a and b one can see the lines of equal time values required for isolation of 2 wt% of the inhibiting liquid. The isolation rate at the initial stage of exudation is seen to have a strong dependence on the film-filling degree of the CI. It is also a parameter that is sensitive to variations in the Pl:CI ratio, which is specifically evident at their low thermodynamic compatibility (Fig. 2.30b).

Polymer films containing VCI are said to be most promising means in inhibiting atmospheric corrosion of hardware, when evaporating VCI create a protecting atmosphere within the packing space. When impregnated into a

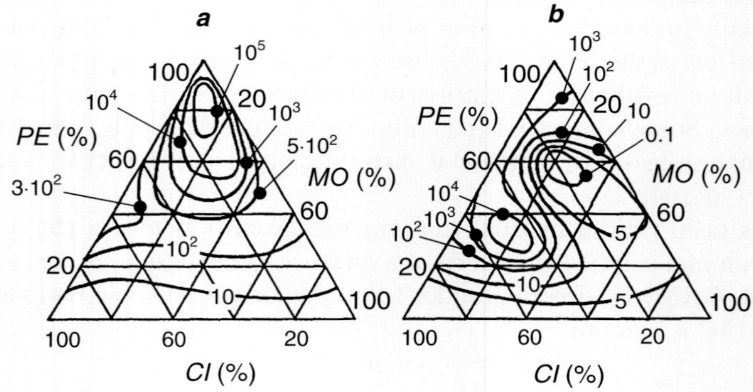

Fig. 2.30. Time (hours) of 2 wt% of inhibited liquid CI exudates from the film samples based on PE + mineral oilMO + CI: (a) Vital, (b) GRM. Exudation temperature 20°C

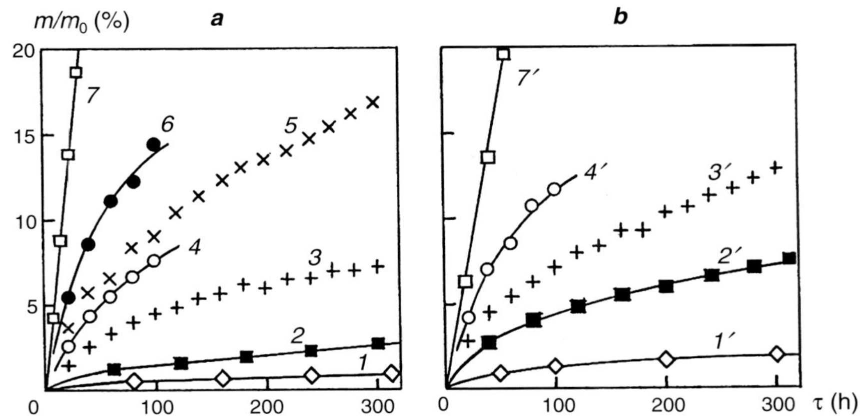

Fig. 2.31. Kinetic curves of sublimation at different temperatures of VCI (G-2, Russia): curves (7, 7′) initial; (**a**), (1–6) distributed in the volume and (**b**), (1′–4′) in the surface of PE films. CI concentration in the films: (1, 3, 5, 1′–3′) 0.09 wt%; (2, 4, 6) 1.0 wt% and (4′) 1.4 wt%. Temperature: (1, 2, 1′) 25°C; (3, 4, 2′, 4′, 5′) 50°C and (5–7, 3′) 60°C; m and m_θ -amounts of G-2 desorbed from the films in time τt and $\tau t = \infty$, correspondingly

polymer matrix, VCI evaporation is retarded perceptibly [3,5,6,102,108,109]. The evaporation rate of VCI from polymer films is correlated with the elasticity of the inhibitor vapors [24,102] and their concentration in the material (Figs. 2.31a and 2.32) [5,6,24,102,110]. It is also a function of temperature (Fig. 2.33) and production process of the inhibited film (Figs. 2.31a and b; 2.33 and 2.34) [3,5,6,24,109,111].

It is seen from Fig. 2.31a and b [109] that the sublimation rate of VCI (G-2) diminishes both in the bulk and in the surface layer of PE films. The kinetic curve of the initial VCI sublimation has a linear character. At the same time, all desorption curves of G-2 from the extruded PE films (Fig. 2.31a) display parabolic dependencies of the $m/m_0 = a\tau t^{0.5}$ kind and obey Boltzmann's solution of the diffusion equation in a semi-infinite medium [110]. Therefore, it is possible to anticipate that the VCI desorption rate from the film carrier is limited by its diffusion. At the initial moment of diffusion, the surface concentration of the diffusant in the film is equal to that in the volume, although a concentration gradient is formed with time. So, diffusion of VCI in the films within a wide time and temperature range is described by the relation

$$m/m_0 = (4/h)(Dt\tau/\pi)^{0.5} , \qquad (2.2)$$

where D is diffusion factor of the CI and h is the film thickness.

It has been found [109] that the dependence ln $(D) = f(C)$, where C is the G-2 content in PE, has a linear behavior. With increasing G-2 content from 0.09 to 1.0 wt%, the diffusion factor increases 5 -10 times. It has been

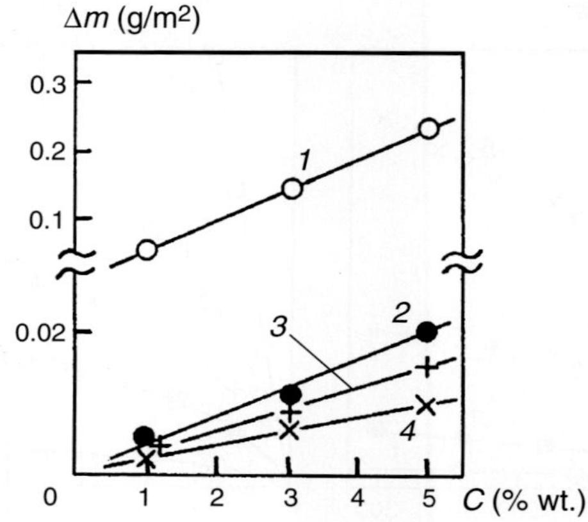

Fig. 2.32. The amount of PHC (Δm) isolated from PE films in 200 hours at 25°C depending on the initial concentration (C) of these VCI in the films: (1) Tr; (2) Tet; (3) KPhTet; (4) ATr

also established that the CI sublimation rate depends linearly on the PHC concentration in PE films (Fig. 2.32). Concentration dependencies of the diffusion rate are commonly related to the effect of VCI on the polymer structure. The diffusants may alleviate interactions between macromolecules and activate mobility of macrosegments via the polymer plastification effect and breakage of intermolecular hydrogen and other chemical bonds [102, 103, 109].

Evaporation of VCI speeds up with increased temperature. According to previous work [109], diffusion factors of G-2 in PE films obey the Arrhenius equation (Fig. 2.33) and the activation energy of diffusion decreases from 135.5 to 104 kJ/mole as the inhibitor concentration in the film bulk increases from 0.09 to 1.0 wt%.

The CI desorption intensity from the films can be regulated by combining CI with the polymer base. G-2 introduced into the porous surface layer of PE films [30] evaporates much faster than when distributed across their volume. The kinetic constants that characterize the CI diffusion process in the polymer matrix are higher by 1–2 orders in the former case and the activation energy is lower by roughly 30-35 kJ/mole [109] (Figs. 2.31, 2.33).

Introduction of Pl noticeably intensifies isolation of inhibitors (Figs. 2.34–2.36). This is probably connected with changes in the polymer structure at plastification that promote the formation of a branched system of pores and channels over which the Pl, together with CI, diffuses to the film surface via syneresis and volatility mechanisms, and is then transported to the surface of the protected metal [3, 5, 6, 104, 111].

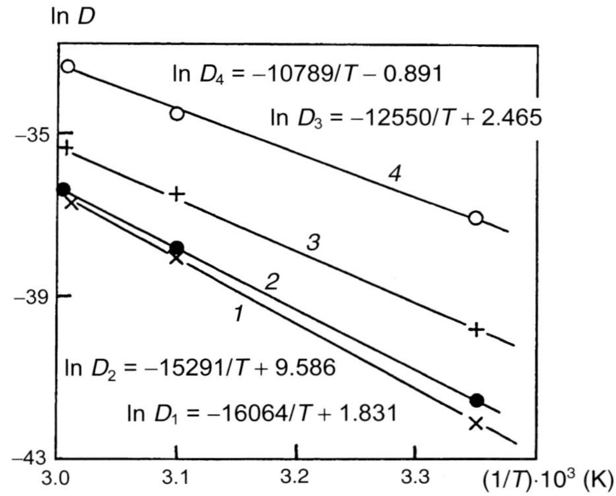

Fig. 2.33. Temperature dependence of diffusion factor of G-2 impregnated in PE films: (1–3) in the volume and (4) in the surface layer. CI concentration: (1, 4) 0.09 wt%; (2) 0.16 and (3) 1.0 wt%

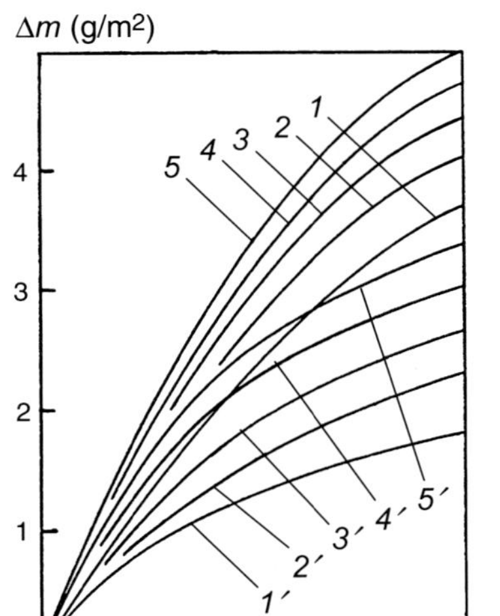

Fig. 2.34. Liberation kinetics of bis-(2-ethylhexyl)-sebacate (1–5) and its blends with MTtet (1′–5′) from PE films at 20°C. Initial concentrations of components in the film: MTtet – 5 wt%, Pl: 1, 1′ – 10; 2, 2′- 15; 3, 3′ – 20; 4, 4′ – 25; 5, 5′ – 30 wt%

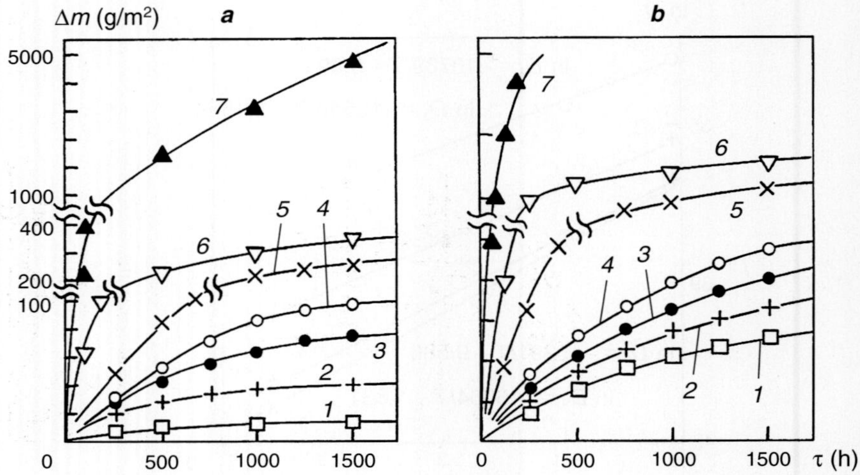

Fig. 2.35. Evaporation of CI at 20°C: (a) in the initial state; (b) from blends with mineral oil in 1:3 ratio. Russian CI grades: (1) NDA; (2) VNH-5; (3) M-1; (4) MSDA; (5) VNHL-20; (6) IFHANGAZ; (7) IFHAN-1

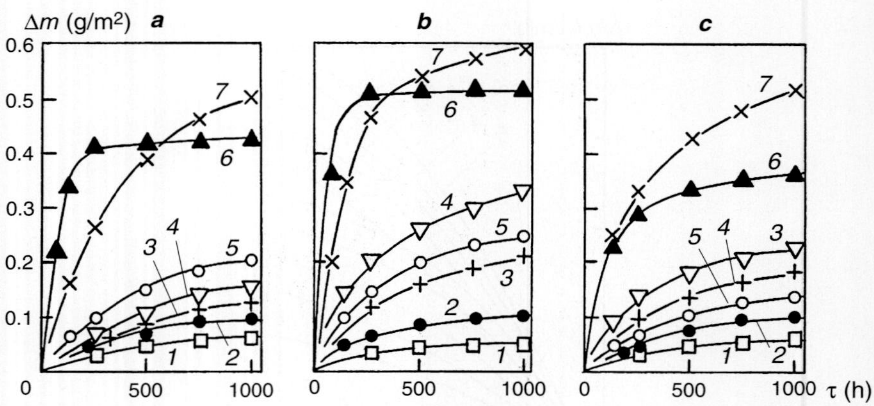

Fig. 2.36. Liberation of CI at 20°C from polymer films of the following composition (in wt %): (a) 98 PE + 2 CI; (b) 88 PE + 2 CI + 10 oil; (c) 83 PE + 2 CI + 15 oil. CI: (1) VNH-5; (2) M-1; (3) NDA; (4) IFHABGAZ; (5) MSDA; (6) IFHAN-1; (7) VNHL-20

An efficient accelerator of PHC diffusion in PE films is believed to be bis-(2-ethylhexyl)-sebacate, which perfectly plasticizes PE and dissolves this CI.

The kinetic curves of solid and liquid CI evaporation in a free state from solutions in the mineral oil (M), and from films of extruded double PE-CI and triple PE-CI-Oil MO blends are illustrated in Figs. 2.35 and 2.36 [3]. The comparison of kinetic dependencies shows that the immobilization of CI in the polymer matrix retards their evaporation rate by 2 to 3 orders due to, most probably, the limited diffusion velocity of the inhibitor in the polymer (Figs. 2.35 and 2.36a). Introduction of mineral oil into the extruded blend (Fig. 2.36b) speeds up CI evaporation from the polymer carrier by several times. This is evidently connected with formation of polymer gels where CI diffuses faster than in the initial polymer and localizes mainly in the liquid phase. Besides, the relaxation mechanism of CI desorption takes place in the gels via evaporation of CI from the liquid isolated on the surface as a result of syneresis. VCI are liberated even more quickly when impregnated in the surface layers of plasticized films (from either the gaseous or liquid phase) [3].

The data above proved that the structure of the film carrier and the CI liberation rate could be regulated by varying the composition and process regimes. It is critical for the CI to be liberated at a velocity able to create a required concentration on the metal surface for its transfer into a stable state prior to the onset of the intensive corrosion process.

This requirement has been taken into account in an algorithm [5, 6] aimed at optimizing formulations of wrapping films based on plasticized PE contained in PHC in the form of VCI. The algorithm is based on the information on isolation kinetics of CI from films of different compositions and includes the design of CI flow balance. Being dissolved in a condensed electrolyte layer on the metal surface or found as a vapor in the air gap between the film and metal, it is spent on the formation of an adsorption protecting layer on the metal surface. A part of the CI is kept by the exudate and evaporates in the environment.

Permeability to liquids and gases is one of the significant properties of inhibited polymer films intended for packaging and preservation of hardware. In particular, *vapor-penetrability* of the films may perceptibly effect the formation of condensed water layers on the protected metal surface. Of no less importance for film materials is their *oxygen-penetrability* since the limiting stage of atmospheric corrosion of metals under adsorption layers of moisture is a mated reduction reaction of oxygen diluted in water [112]. The *permeability of the films to electrolyte solutions* may affect corrosion kinetics, particularly via removal of the products of electrochemical reactions.

Vapor- and oxygen-penetrability values of PE modified by both solid (NDA, G-2) and liquid (IFHANGAZ) CI have been studied [113, 114]. The extruded polymer composition was either impregnated by 0.5–3.5 wt% of CI or brought into contact at the outlet with a mixture of CI and Pl. It was found that small additions of CI (0.5–2.5 wt %), independent of their method of introduction, decreased the vapor- and oxygen-penetrability of PE films (Fig. 2.37, Table 2.7). This is most probably because CI particles function as active structure-forming agents in respect to the polymer [115]. Serving as crystallization centers, they promote PE crystallization leading to lower gas

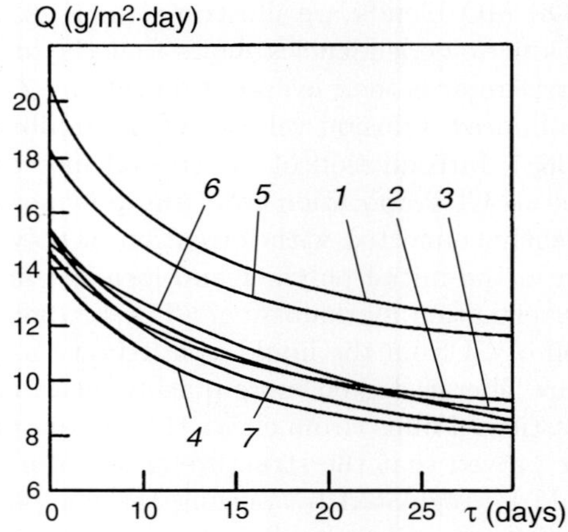

Fig. 2.37. Kinetic curves of vapor-penetrability of LDPE-based films: (1) without CI; (2–4) with volume distribution and (5–7) with surface distribution of inhibitors; (2) G-2; (3, 4, 6) and (7) NDA; (5) IFHANGAZ. CI concentrations: (2) 0.5; 3 and (6) 1; (5) 3; (7) 2.5 wt%

Table 2.7. Vapor- and oxygen-penetrability of inhibited LDPE films 100 ± 5 mm thick

Content of Modifying Agents in the Film, wt%	Vapor-Penetrability, g/(m^2·days)	Oxygen-Penetrability, g/(m^2·days)	Distribution Pattern of Modifiers
Without a modifier	13	0.45	Volume
G-2 (0.5)	10	0.44	– "–
NDA (1)	9	0.20	– "–
G-2 (1.5) + DOP (2.0)	24	0.63	– "–
G-2 (0.5) + DOP (0.5) + Al (0.25)	8	0.32	Surface
G-2 (1.0) + DOP (0.5) + Al (0.25)	6	0.33	– "–
AFHANGAZ (3.0)	12	0.28	– "–
IFHANGAZ (3.0) + MO (3.0)	10	0.45	– "–
IFHANGAZ (3.0) + MO (7.0)	8	0.25	– "–

penetrability. Slight elevation of the degree of PE crystallinity in the presence of 0.5–1 wt% of NDA has been supported by the X-ray diffraction structural analysis [115]. This phenomenon is also in agreement with model representations on the permeability of filled polymers put forward by Nilsen [116]. Since the diffusant molecules have to bend around the filler particles evenly spread over the polymer matrix, their path through the film extends and results in decreased penetrability, whose relative value can be described by the relation

$$P_{\mathrm{f}}/P_{\mathrm{n}} = V_{\mathrm{p}}/[1 + (l/2d)V_{\mathrm{f}}]\,, \tag{2.3}$$

where P_f and P_n are the penetrability of filled and non-filled polymers, respectively, V_p is the volume share of the polymer, l and d are the mean length and thickness of filler particles and V_f is the volume share of the filler.

Increase of the concentration of solid powdery CI impregnated into the extruded composition above 2.5–3 wt% results in elevated gas-penetrability of PE films, evidently due to the accumulation of micro- and macrodefects [113].

In contrast, the presence of CI based on low-molecular amines of NDA or G-2 type hampers gas-penetrability of the polymer films as a result of atmospheric aging [114]. Vapor-penetrability of PE films increases by 1.5–2 times after exposure to UV radiation for 200 hours or to climatic factors for two months. Under the same conditions, the vapor-penetrability of inhibited films grows, but by only 10–20%. Apparently, CI of the named type function in this case like a stabilizer that hampers photo and thermal-oxidative aging of PE.

The penetrability of film materials depends on the degree of plastification of the polymer base. The character of the observed changes is a function of the nature of the Pl and the laws of its physical-chemical interactions with the polymer. Vapor and oxygen-penetrability of inhibited PE films increase at impregnation levels of up to 10 wt% of polar Pl (DOP [113] or bis-(2-ethylhexyl)-sebacate [5, 6]) and decrease with similar amounts of non-polar Pl (mineral oil) [113]. It is apparent that, in the first two cases, structural plastification hinders molecular interactions and elevates the flexibility of macromolecules. Both factors should evidently contribute to penetrability growth [115]. On the contrary, introduction of a non-polar Pl (mineral oil) into the non-polar polymer (PE) brings about antiplastification [107, 117]. This phenomenon is accompanied by raised ordering of structural elements and, as a consequence, lowered penetrability of the polymer. Nevertheless, at high concentrations (about 30–40 wt%) of mineral, oil vapor-penetrability of inhibited films increases 2–4 times as compared to the initial PE [104]. This elevated vapor-penetrability is attributed [104] to the high solubility of water vapor in the inhibiting liquid.

In any case, single-layer inhibited polymer films cannot substantially restrict the rate of corrosion processes by limiting their vapor or oxygen-penetrability. This becomes especially clear in comparison with the mean rate of steel corrosion in atmosphere ($1.9\,\mathrm{g}/(\mathrm{m}^2 \cdot \mathrm{days})$ [112]). The amounts of water and oxygen needed for the corrosion process to run at the indicated rate calculated by the equation

$$4Fe + 3O_2 + 2H_2O = 2Fe_2O_3 \cdot H_2O\,, \qquad (2.4)$$

are 0.3 and $0.8\,\mathrm{g}/(\mathrm{m}^2 \cdot \mathrm{days})$, correspondingly.

Water and oxygen penetrate through the majority of single-layer inhibited polymer films in amounts sufficient for corrosion generation at a rate almost similar to unprotected metal surfaces.

An efficient means of limiting the penetrability of polymer films to corrosion-active substances is impregnation of fillers able to absorb these substances in the polymer matrix. Data on hampered vapor- and oxygen-penetrability of inhibited PE films containing aluminum powder are presented in Table 2.7 [113]. The powder particles can interact both with atmospheric oxygen and water vapor forming stable compounds that localize in the film. As a result, the penetrability of the films for these reagents is lowered. If compounds able to bind aggressive components of the air were used as fillers, it is possible to protect metals from atmospheric corrosion. Engineered Materials Inc. produces PE films filled with copper having covalent bonds with the polymer. Copper molecules interact actively with hostile atmospheric gases passing through the film and hamper their permeation into the packing space. These films render protection against atmospheric corrosion for non-ferrous metals for 15 years.

Multilayered films [29,34,36,40] no doubt present a more serious obstacle for corrosion-active matter. The resistance of a multilayered film to penetration ($1/P$) of some diffusant equals the sum of penetration resistances of the separate layers ($1/P_{\mathrm{i}}$) [115], leaving aside interfacial barriers between the layers

$$1/P = \sum 1/P_{\mathrm{i}} \tag{2.5}$$

Inhibited films containing layers of high-barrier and metallized polymers display elevated protective properties. This is connected with their parallel capabilities of simultaneous kinetic and diffusive regulation of corrosion processes.

One of the reasons for corrosion hampering in this case is the low ionic conductivity of the polymer films. Some of the reagents used, as well as the products of electrochemical reactions, represent ions in aqua solutions of electrolytes. Therefore, diffusive retardation of their supply and removal from the reaction zone may hamper both anodic and cathodic processes [112].

For instance, in previous work [104] the penetrability of PE films with contact CI towards hydrogen and chlorine ions has been estimated. Film samples were separated in the cells with distilled water and 0.1 N solution of HCl (a model of a hostile medium). Neither H^+ nor Cl^- ions were detected on the other side of membranes after seven months of endurance.

The penetrability of plasticized PE films containing MTet VCI towards iron ions was estimated in [5,6] visually judging by the color of 0.01 N aqua solution of Na_2SO_4 with immersed steel electrodes sealed with the inhibited film. The samples were kept in the electrolyte at room temperature for three months. The Na_2SO_4 solution did not change its color from the yellow-green hues characteristic of bivalent and trivalent iron solutions during the specified time. The film material averted the removal of iron cations from the zone of the electrochemical reaction and small amounts of corrosion products in the form of brown iron oxides remained beneath the film.

The efficiency of inhibited polymer films depends on the tightness of their packages (protecting shell). Reliability of sealing depends, in turn, on a combination of the barrier and strength characteristics of the film material, as well as ability to fuse.

Irregularities of the deformation and strength properties of the polymer films upon the introduction of contact or volatile CI are conditioned by the aggregate state and physical-chemical characteristics of the inhibitors, their concentration distribution in the polymer matrix and the presence of plasticizers.

With increasing content of solid dispersed VCI in extruded or hot-pressed inhibited PE films their breaking stress σ_b under tension drops monotonously together with the relative elongation at rupture ε [5, 6, 24, 114]. According to [114], introduction of 0.5–2.5 wt% of NDA or G-2 decreases σ_b by 5–18%. This is evidently connected with the growing defectiveness of the polymer crystalline lattice upon filling with CI particles. Note that, after two months of endurance, the rupture strength of the indicated inhibited films decreases by only 4–5% under the effect of atmospheric factors [114]. This is proof of the ample resistance of inhibited materials to climatic aging.

Impregnation of CI into PE films at co-extrusion with a polymer affects the strength of the films but differently. The dominating factors here are the chemical origin, melting point T_m and dispersion degree of the solid CI. Films containing 2 wt% of highly dispersed NDA are characterized by higher σ_b values compared to the initial PE, whereas those with a similar amount of PhTet, which has a lower dispersion and higher T_m, display a lower σ_b value (Table 2.8).

Table 2.8. Strength parameters of hose inhibited LDPE-based films

Concentration of Modifying Agents in the Film (wt %)	Breaking Stress Attension, MPa	Relative Elongation Rupture, %	Modifier Distribution Pattern in the Film
Without a modifier	20.5 ± 2.9	170	Volume
PhTet (2)	16.0 ± 1.1	125	– " –
NDA (2)	25.8 ± 0.9	–	– " –
PhTet (0.5) + DOP (3.7)	19.0 ± 1.1	–	Surface
NDA (0.6) + DOP (5.7)	20.6 ± 0.7	–	– " –
M-1 (0.6) + MO (1.7)	20.2 ± 1.9	82	– " –

A decrease in $\Delta\sigma_b$ was observed [5,6] upon volume filling of PE films with PHC CI (Fig. 2.38). The T_m of the modifiers varied from 120°C (Tr) to 352°C (KPphTet) within the studied series as follows Tr < MTet < Tet < ATet < ETet < KPhTet. It can be concluded that fusing of PHC particles during heat pressing of their powder compositions with PE ($T_{press} = 150 - 160$°C) promotes a more even distribution of the inhibitors within the polymer and lowers the defectiveness of the film material.

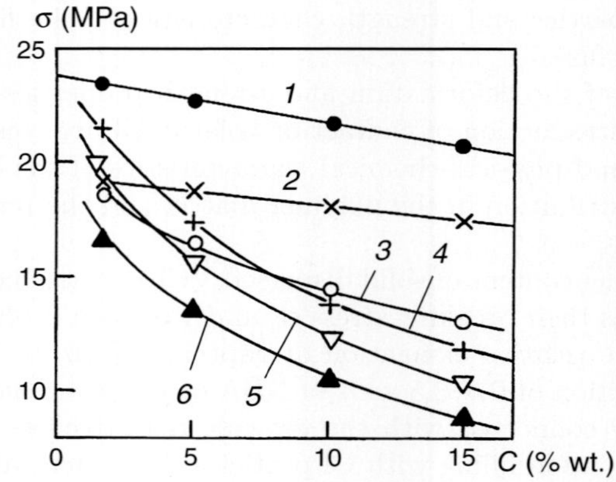

Fig. 2.38. Breaking stress at tension of HDPE-based films versus their concentration of solid CI: (1) Tr, (2) MTtet, (3) Tet, (4) ATet, (5) ETet, (6) KPphTet

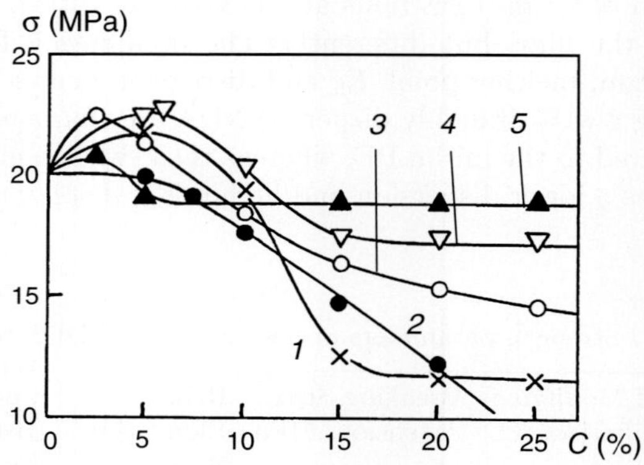

Fig. 2.39. Breaking stress at tension of HDPE-based films versus liquid CI content: (1) IFHANGAZ, (2) GRM, (3) MSDA, (4) VNH-5, (5) IFHAN-1

The dependence of the breaking strength under tension of PE films on evenly distributed liquid CI content has a more complex character (Fig. 2.39) [3, 24].

The introduction of liquid CI in amounts less than 5 wt% affects σ_b negligibly and may even slightly elevate it. According to X-ray diffraction structural analysis, this also leads to the growth of crystallinity of the film materials and stresses in the crystalline lattice of PE at 5–10 wt% [24]. CI particles evidently behave like additional crystallization centers in PE. Increasing the concentration of CI above 7–10% noticeably impairs the strength of inhibited films.

Liquid oil-soluble CI, e.g. sulfated and nitrated mineral oils, and salts of synthetic fatty acids, may render a plasticizing effect on PE thus lowering the degree of crystallinity and the strength of the modified films [106]. Plastification is accompanied by declining intermolecular interactions, which facilitates conformal transformations of macromolecules.

Along with this, plastification induces growth of highly elastic deformations on the background of reduced elasticity modulus and other strength characteristics of the polymer materials [103]. Impairment of breaking strength under tension of LDPE-based films takes place as the concentration of the non-polar plasticizer bis-(2-ethylhexyl)sebacate increases. The decrease in σ_b reaches 25% at Pl concentration of about 20 wt% [5,6].

Variations in the physical-mechanical properties of polymers are very often non-monotonous because of ambiguous changes in their structure on plastification [3]. More fundamental changes in the concentration dependencies of some physical-mechanical and structural parameters of PE plasticized by mineral oil take place at 10–30 wt% content (to a less degree at 50–60 wt%) (Fig. 2.40) [3,107]. Maximum values of deformation and strength characteristics of PE-based films plasticized by mineral oil and incorporating contact CI (Vital and GRM) are within similar concentration limits of Pl: 60 wt% of PE, 20–30 wt% of oil and 10–20 wt% of the inhibitor (Fig. 2.41) [117].

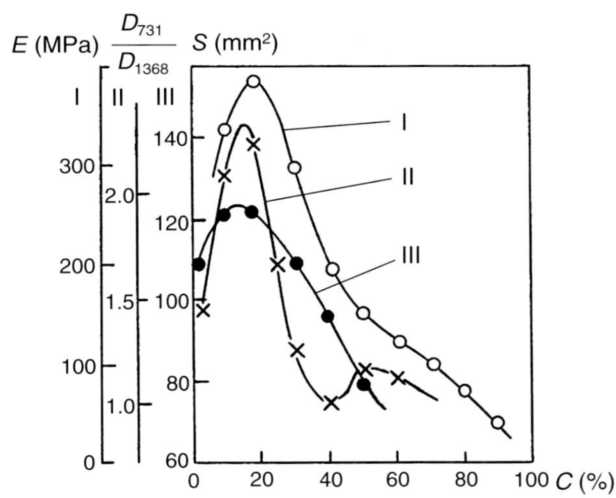

Fig. 2.40. Variations of physical-chemical and structural characteristics of film materials based on PE as a function of mineral oil content: (I) elasticity modulus (E), (II and III) crystallinity degree (D_{731} and D_{1368} – optical density of absorption bands at 731 and 1368 cm^{-1} frequencies in IR spectra of the materials; S – melting peak areas on thermograms)

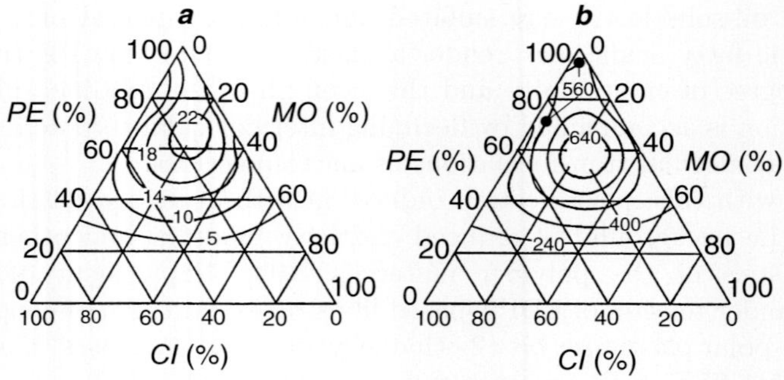

Fig. 2.41. Isolines of equal values: (a) breaking stress at tension, MPa and (b) relative elongation at rupture, %, for film materials based on compositions PE + mineral oil MO + CI (Vital)

The effect of extreme changes in properties of non-polar polymers as a result of accelerated crystallization in the presence of non-polar Pl is called antiplastification, which has been considered previously in this chapter.

Estimates of the effect on strength characteristics of inhibited films of combining CI with the polymer binder show that surface modification of film materials is preferable to bulk modification.

Thermodiffusive modification of the inner surface of the film hose during blowing does not result in any strength deterioration, whether the CI are used as a solution (M-1) or suspension (PhTet, NDA) in Pl (DOP) (Table 2.8).

The strength characteristics of inhibited PE films subjected to model atmospheric aging tests with exposure to UV radiation and temperature drops have been studied elsewhere (Table 2.9) [24]. A cycle of light-induced aging consisted in the exposure of film samples to a UV lamp with a power of 750 W for six hours at 50–60°C temperature and endurance without irradiation for 18 hours at 20°C. One cycle of temperature aging corresponded to endurance of the samples for six hours at 60°C followed by three-hour endurance at −40°C and for 15 hours at 20°C. It was found that impregnation of PE-based films with CI reduced their stability and strength parameters under conditions of temperature change and UV radiation.

Weldability of polymer films does not in fact worsen on volume filling with CI. It was proven in [24] that the strength of welds obtained on heat fusing of PE films modified by blends of mineral oil and either contact (Vital, MSDA) or volatile CI (IFHAN-1, VNHL-49) was lower by just 1–2% than the analogous parameter of usual PE films. The volume- and surface-modified films presented in Table 2.8 fuse satisfactorily in a packing machine with lower and upper fusing board temperatures of 120–125°C and 200–210°C, respectively. The strength of the weld varied, depending on the composition and method, from 7.0 to 13.0 N/cm. On thermodiffusive filling of the hose films, a relief structure is formed in their inner (modified) surface with a

Table 2.9. The effect of climatic factors on breaking stress (MPa) of inhibited film samples

Production Method	Composition	Number of Cycles of Photo-Aging				Number of Cycles of Temperature Aging	
		0	45	90	180	90	180
Extrusion	LDPE	14.2	13.0	11.1	–	12.3	7.3
	LDPE + MO + MSDA	16.5	5.9	–	–	11.0	13.3
	LDPE + MO	19.8	20.0	4.6	–	15.5	14.4
Hot-pressing	LDPE + MO + IFHANGAZ	21.0	13.6	–	–	14.9	14.1
	LDPE + MO + IFHAN –1	21.5	19.7	–	–	15.2	15.0
	LDPE + MO + VNH –5	21.1	11.7	4.1	–	13.4	13.4
	LDPE + MO + MSDA	22.3	20.2	–	–	16.3	14.2
	LDPE + MO + GRM	20.3	6.25	–	–	13.1	13.0

characteristic size of projections and depressions about 10 – 100 μm. The exudate concentrates largely in the relief depressions, while projections are devoid of the liquid phase, which guarantees strength of the welds during preservation and packing operations.

The compatibility of the barrier material and CI carrier in the polymer films promotes their high *Anticorrosive on characteristics* of polymer inhibited films are promoted my synergetic actions of two factors: barrier film property and IC specific action.

Test results carried out using an accelerated technique are presented in Table 2.10 [3, 24]. Resistance to corrosion of samples of different metals has been tested when tightly sealed in single-layer hose inhibited films. The films were produced by the "liquid in the hose" method (see Sect. 2.2.4, Fig. 2.5e. The most efficient at protecting iron-based alloys against corrosion turned out to be films containing VCI of the VNHL-49, IFHAN-1, MSDA types; in protecting aluminum – IFHAN-1, VNHL-20; copper – IFHANGAZ; silver – IFHANGAZ and VNHL-20. When extrapolated to field conditions, the named results certainly prove the capability of these films to protect the indicated metals safely for five to seven years.

Single-layer PE films containing solid VCI (G-2 and NDA) additions outdo conventional PE films by a factor of 10–15 in terms of their protective properties [72]. The corrosion rate of the steel samples insulated by PE films modified by Pl and CI and placed in a 1 N solution of Na_2SO_4 (simulating a harsh condensed medium of atmospheric corrosion) is lowered by three to five times [104, 118].

An original approach to the optimization of the composition of polymer films modified by CI using three-component composition-property charts is described in [24, 104]. It presumes calculations of coordinates and mapping

Table 2.10. Anticorrosion characteristics of PE-based films depending on CI nature and metals protected*

Oil Solution of CI (1:1)	Metal	Corrosion of Samples (in Severity Numbers) After Tests (in number of cycles)							
		10	15	20	25	30	40	50	70
IFHANGAZ	St 3	0	0	1	1	2	4	–	–
	Al	0	0	0	0	1	2	3	5
	Cu	0	0	0	0	0	1	2	8
	Ag	0	0	0	0	0	0	0	0
MSDA	St 3	0	0	0	0	0	2	3	–
	Al	0	0	0	1	1	3	–	–
	Cu	0	0	1	2	2	2	3	–
VNHL-20	St 3	0	0	0	0	0	0	1	–
	Al	0	0	0	0	0	0	–	–
	Cu	0	1	1	3	–	–	–	–
	Ag	0	0	0	0	0	0	1	–
VNHL-49	St 3	0	0	0	0	0	0	0	2
	Al	0	0	0	0	0	1	1	–
	Cu	0	0	2	3	–	–	–	–
IFHAN-1	St 3	0	0	0	0	1	3	–	–
	Al	0	0	0	0	0	2	–	–
	Cu	0	0	0	2	2	3	–	–

*One test cycle corresponds to endurance of samples at 55°C and 95% relative humidity for 8 hours, and at room temperature with the same humidity for 16 hours. A severity number of (1) corresponds to corrosive damage less then 0.2% of the whole metal surface, (2) 0.2–0.5%, (3) 0.5–1%, (4) 1–2.5%, (5) 2.5–5%, (8) 25–50% of the surface.

of isolines of the limiting properties of the films that seem essential in given operation conditions (Fig. 2.42).

The isolines restrict the regions on the chart of compositions that guarantee the needed service characteristics of the material.

The inhibited polymer films may simultaneously present a barrier for electrochemical and microbiological corrosion. Table 2.11 and Fig. 2.43 illustrate data from [5, 6, 54] characterizing the capability of PE films containing CI of the PHC series to inhibit vitality of microscopic fungi and bacteria whose metabolic products evoke corrosion damage of metals.

As can be seen, the inhibited film materials are able to display a prolonged electrochemical and microbiological activity thanks to the gradual liberation of CI. In addition, they can represent a diffusive and mechanical barrier for harsh media. The corrosion rate of hardware packed in these films is lowered by tens to hundreds of times. Their fusing capability and resistance to atmospheric aging make the film packages tight and reliable. A

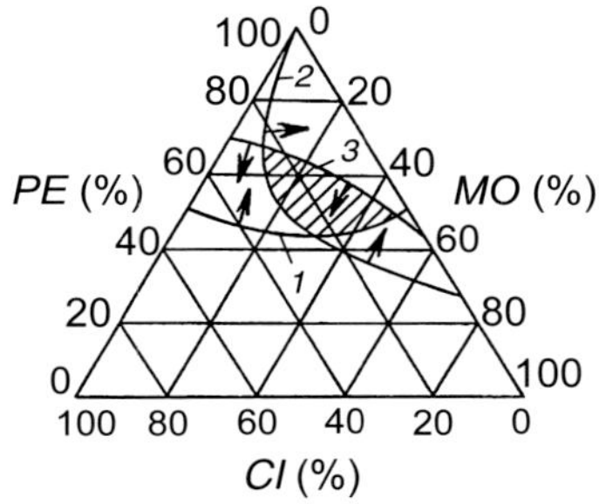

Fig. 2.42. The range of compositions of PE + mineral oilMO + oil-soluble CI film materials promoting the required deformation, strength and anticorrosion characteristics. Isolines: (1) breaking stress at tension $\sigma_b = 10\,\mathrm{MPa}$; (2) relative elongation at rupture $\varepsilon = 500\%$; (3) corrosion rate of steel beneath the film in 1 N aqua solution of Na_2SO_4, $i = 0.1\,\mathrm{mg/(m^2{\cdot}h)}$. Arrows indicate desired alterations of film properties

Table 2.11. Fungicidal effect of PHC-containing HDPE-based films

CI	Width of Zones Inhibiting Fungi Growth Around Film Samples, mm*		
	Aspergillus	*Penicillium*	*Trichoderma*
Without a CI	0	0	0
Tr	20–25	20–24	23–26
ATr	6–8	5–9	5–10
NTr	3–5	4–6	3–7
Tet	14–16	15–16	17–19
ATet	4–7	6–8	6–9
NaPhTet	0	3–4	6–8

*The fungi have been cultivated for 48 hours at 29°C on Chapeck-Dox's agar medium

combination of properties intrinsic to inhibited polymer films gives grounds for their application in the preservation of hardware for prolonged periods without additional reconservation.

2.3.3 Anticorrosion Mechanism

Inhibited polymer films are distinguished from barrier-type films by their electrochemical activity. This property arises from the ability of these films to liberate CI from the polymer matrix and transfer it to the mated metal surface. The CI strongly affects the kinetics of electrochemical reactions leading to corrosion damage of metals. The inhibiting mechanism of corrosion can be treated in this case as follows.

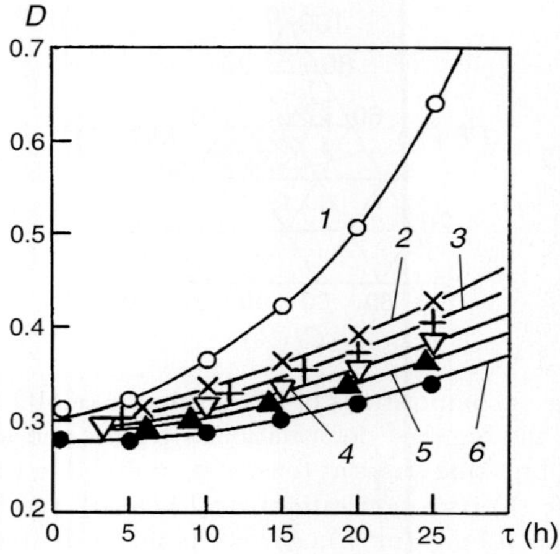

Fig. 2.43. Variation kinetics of the optical density of suspensions of Thiobacillus ferrooxidants bacteria cells in a liquid nutrient medium upon placing PE films into the media: (1) the film without CI, (2–6) films inhibited with PHC, (2) Tr, (3) NaPhTet, (4) ATr, (5) NTr, (6) ATet

With this aim, the kinetics of the variation of the polarizing resistance (R_p) of a steel electrode has been studied [118] in the vicinity of its stationary potential when the electrode is in contact with an inhibited polymer film in aqua solutions of electrolytes (HCl and Na_2SO_4). The films were produced by hose extrusion of PE-based compositions with Pl (mineral oil) and oil-soluble CI.

It is common knowledge that there exists an inversely proportional relation between the polarizing resistance of the electrode system and corrosion rate, which significantly increases R_p upon contact with the electrolyte [119].

Comparing curves of $R_p - t\tau$ (Fig. 2.44) [118], we conclude that, when there are no adhesive contacts of the inhibited film material and the electrode is absent, R_p and the kinetics of its variations will be defined by the penetrability and deformation of the film material along with the adsorption and hydrophobic properties of the CI.

The initial values of the polarizing resistance of the electrode in contact with modified PE films are 1.5–2 times higher than those of unmodified ones and are in agreement with the hydrophobic properties of the CI. Hydrophoby rises through the series MO, SIM, Ukramin, AKOR. The kinetic curves of $R_p - t\tau$ show two characteristic sections. The slope in the first second (initial) is dependent upon the Pl and CI content in the film material. The inclination of the second section remains constant for all samples. Consequently [118], R_p attenuation on the first section is the result of the contact area of the electrolyte and metal surface enlarging with time as the adsorbed CI molecules are removed from it. The variation velocity of R_p in the second section is

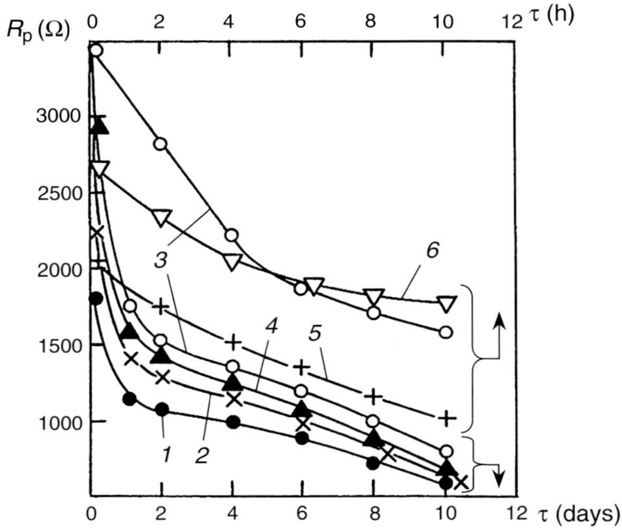

Fig. 2.44. Variation of polarizing resistance (R_p) of steel electrode in 0.1 N solution of HCl under film materials: (1) PE film; (2–6) PE films modified by mineral oil (MO) and contact CI (wt%); (2) 50 MO; (3) 45 MO + 10 AKOR; (4) 50 AKOR; (5) 45 MO + 10 SIM; (6) 45 MO + 10 Ukramin

connected with the diffusive penetrability of the films to corrosion-activating H^+ and Cl^- ions, which differs substantially for various PE-based materials.

Against the background of mechanical screening of the electrode that was roughly similar for all studied samples we can observe quite evident electrochemical activity of inhibited film materials related to the absorption and hydrophobic abilities of isolated CI.

Using a polarization transducer [120], averaged values of the polarizing resistance and corrosion rate of the steel electrode in the electrolyte in contact with inhibited PE films have been estimated. In Fig. 2.45 isolines of constant corrosion rate values of the electrode (isocorrs) are presented as a function of film composition. The inhibited films are arranged in order of growing inhibiting efficiency in acidic and neutral media depending on the hydrophobic properties of the contained CI: SIM, AKKOR, Ukramin. Impregnation of the mineral oil retards the corrosion rate of steel in neutral electrolyte solutions but augments corrosion in acidic media. This is probably due to chemical resistance and raised permeability of plasticized films, which is exhibited to the greatest degree in acidic media. The corrosion rate of steel samples under the films did not exceed $0.5\,\mathrm{mg/(m^2 \cdot h)}$, inflicting no damage to the metal.

The structure of adsorptive VCI-Tet coats formed on steel samples upon packing into a tetrazole-containing PE film has been studied [5] and [6]. The samples were exposed to periodical temperature differences from 55 to

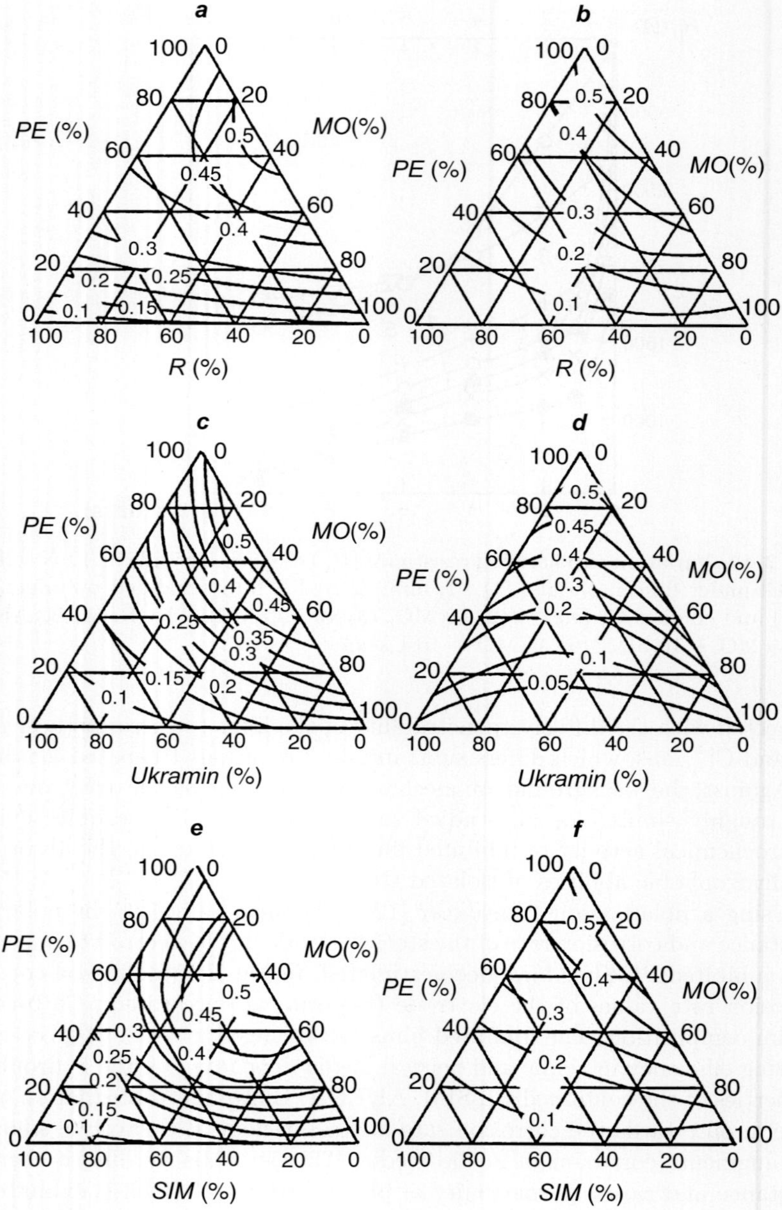

Fig. 2.45. Steel isocorrs under PE films modified by mineral oil (MO) and CI: (**a**) and (**b**) AKOR; (**c**) and (**d**) Ukramin; (**e**) and (**f**) SIM; (**a**), (**c**) and (**e**) in 0.1 N solution of HCl; (**b, d**) and (**f**) in 1 N solution of Na_2SO_4. Test duration – 14 days

20°C at 95% relative humidity, favoring moisture condensation on the metal surface. Upon removal from the packages the surface of the steel samples has been studied by IR spectroscopy (IRS), using the method of attenuated total reflectance (ATRMVTIR), X-ray photoelectron spectroscopy (XPES), X-ray structural analysis (XSRDA), atomic-force microscopy (AFM) and scanning electron microscopy (SEM).

It has been detected that a phase chemisorbed protective layer is formed on the steel sample surfaces. A protective layer similar in chemical composition is formed on samples exposed to a tetrazole-containing electrolyte (0.05 M aqua solution of Na_2SO_4), simulating the condensed medium during atmospheric corrosion.

IR- and REXPE-spectra of the protective layer differ significantly from the spectra of a free Tet and the steel substrate (Figs. 2.46 and 2.47, Tables 2.12 and 2.13).

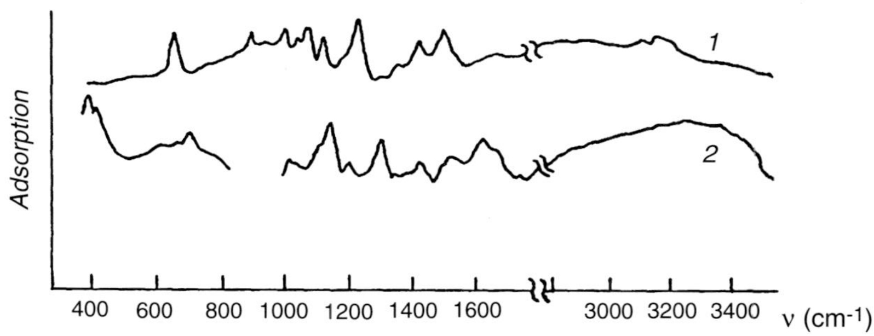

Fig. 2.46. IR spectra (MVTIRATR): (1) free Tet, (2) adsorptive protective layer of Tet on steel surface

Tet adsorbed on the steel surface does not show any characteristic of the absorption bands in the initial spectra, others shift to the region of high and low frequencies while some new bands appear. These changes are not typical of physical absorption processes [121, 122]. Therefore, a supposition has been put forward that a chemisorbed coordination compound (CC) Tet-Fe is formed on the steel surface. IR spectra of the protective Tet layer display a very strong absorption band at 390 cm^{-1} related to valent oscillations of the Fe-N link [121]. The shift to the high-frequency region and cleavage of the band in IRS spectra at 660 cm^{-1} corresponding to torsion oscillations of the heterocycle suggest that the heterocycle participates in the coordination [121, 123]. The presence of a clear-cut band at 1320 cm^{-1} of the protective layer spectrum indicates the formation of the Tet-Fe complex along the nitrogen atoms of the ring [124].

RE XPE spectra of protective coats differ essentially from those of the free Tet and steel substrate (Fig. 2.46, Table 2.13). They possess signals of Nl*s* and Cl*s* electrons in the link-energy regions (E_l) typical of Tet. The Cl*s* band

Table 2.12. The analysis of IR spectra of Tet, its sodium salt monohydrate (NaTet) and CC formed on steel surface at Tet adsorption

Observed Frequencies ν, cm^{-1}			
Tet 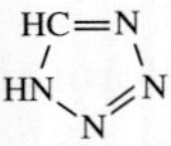	NaTet 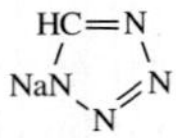	CC "Tet-Fe"	Bands Attributive to [124–128]:
–	3300	3300	ν (O-H associated)
	v.str.br.	str.brs.	ν (O-H)
3150 str.	3110 str.	–	ν (C-H)
3050 avm.	–	–	ν (N-H)
2300–2900 avm.	–	–	ν (N-H associated)
–	1690 av.m	1680 } str.	δ (HOH)
–	–	1640 } br.	
1520 str.	–	1550 av.m	ν, δ (rings)
1450 av.m	1435 av.m	1460 av.m	ν (rings), δ (CH)
–	–	1350 weak. }	ν, δ (corrdinated rings)
–	–	1320 str. }	
1255 v.str.	1290 avm.	–	ν, δ (rings); δ (CH)*1, δ (NH)*
–	1215 strms.	1215 av.m	–
–	1160 str.s	1155 str.s	ν, δ (rings)
1140 str.	–	–	ρ (CH)
–	1135 av.m	–	–
1080 str. }	–	–	ν, δ (rings), δ (NH)
1050 av.m }	–	–	
1015 strs. }	–	–	
–	1070 av.m }	1110 armsh }	ν, δ (rings), δ (CH)
–	1025 str.s }	1065 w }	
–	1015 str.s }	1030 av.m }	
935 av.m	–	–	ρ (NH)
–	910 str.s	–	ρ (CH), κ (rings)
905 av.m	–	–	δ (rings)
–	702 str.s	–	δ (rings)
660 str.s	–	700 } str.	κ, δ (rings)
–	–	680 }	
–	665	635 m	ρ (HOH)
–	v.str.br.	av.br.	
–	–	420 s	ν (Fe-O)
–	–	395 vs	ν (Fe-N)
	–	3434 tr.	
		395 v.str.	ν (Fe-N)

*In IR spectra of Tet
vs = very strong; s = strong; m = medium; w = weak, sh = shoulder

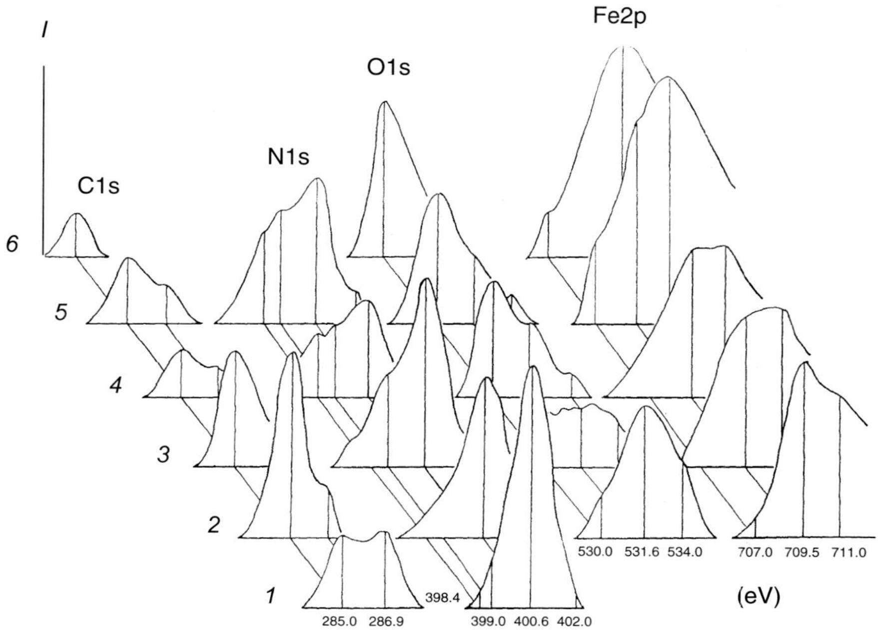

Fig. 2.47. RE XPE spectra: (1) free Tet; (2–5) adsorptive protective layer of Tet on steel surface: (2) after ion cleaning of the surface; (3) at 1.5 nm depth; (4) at 3 nm depth; (5) at 6 nm depth; (6) steel substrate

Table 2.13. Analysis of XPE spectra of Tet, steel substrate and adsorptive Tet-Fe protective layer

Energy Level	E, eV			Attribution of Bands
	Tet	Steel Substrate	Adsorptive Film	
C1s	285.0	285.0	285.0	Hydrocarbon impurities
	286.9	–	–	in C_5 cycle
N1s	–	–	398.4	Tet – physically sorbed
	–	–	399.0	N_1 coordinated
	400.2	–	–	N_1, N_4 in Tet
	400.8	–	400.6	N_2, N_3 in Tet and CC
	–	–	402.0	N_4 coordinated.
O1s	–	530.0	530.0	Fe_2O_3
	–	–	531.6	H_2O coordinated
	–	–	534.0	H_2O coordinated
$Fe2p_{3/2}$	–	707.0	–	Fe
	–	–	709.5	Fe (II) coordinated
	–	711.2	711.0	Fe_2O_3

is not changed, while N1s is broadened and cleaved into a few peaks. It can be concluded that coordination links of Tet with the metal are formed along the cycle nitrogen atoms. In addition, a high-energy component (709.5 eV) appears in the E_l region of Fe2$p_{3/2}$ electrons near the line (711.0 eV) corresponding to oxidized ferrum (Fe_2O_3) [129]. This component evidently belongs to the Fe(II) ion brought into the coordination link with the electron donor ligand (there are 710.3 eV Fe2$p_{3/2}$- electrons in FeO [129]). In addition, we can see new energy bands associated with the 1s level of oxygen in the protective layer spectra at 531.6 and 534.0 eV that are most likely indicative of the participation of water molecules in the coordination with the metal. Thus, the actually observed XPE spectra lines of Fe2$p_{3/2}$, O1s, N1s and C1s, with maxima at 709.5; 531.6 and 534.0; 402.0; 400.6 and 399.0; 286.9 correspondingly, represent the spectral characteristic of Tet-Fe CC.

The comparison of IR spectra of the initial and adsorbed Tet as well as that of its monohydrate of soda salt (Table 2.12) indicates that the ligand in Tet-Fe CC has a tetrazolate form. Particularly, the adsorption bands v (NH) at 3050 cm^{-1} and δ (NH) at 935 cm^{-1} are absent in the spectra of NaTet and CC compounds incorporate the tetrazolate-anion as a structural fragment. Along with this, the intensity of a number of bands related to mixed valent-deformation oscillations ν, δ (of the ring) and δ(NH) (1255, 1080, 1050, 1015 cm^{-1} in Tet and 1215, 1110, 1065, 1030 cm^{-1} in CC) attenuate.

Based on the quantum-chemical calculations of the electronic structure of a deprotonated Tet molecule [130], whose electronic density is concentrated in the nitrogen atoms of the N_1 and N_4 cycle, their participation in coordination seems most probable. This conclusion is supported by the detailed IRS and XPES analyses [131–133]. No anions of the outer sphere were found in the CC by XPES, which presupposes the anion form of the ligand.

It has been also established that Tet's anion is a bidentate ligand. Each ion of Tet links simultaneously with two neighboring Fe(II) ions and forms a single ion (basic valence) and single donator-acceptor (side valence) link, as a result of which polymer-type bridge structures are formed in the CC. Thanks to these structures in particular, the adsorptive Tet-Fe films are resistant to organic solvents and chemical agents.

In addition to Tet, CC contains water molecules in the inner coordination sphere. Their presence is clearly exhibited in the IR spectra (Fig. 2.45, Table 2.11) – ν (OH) 2700–3600 cm^{-1}, δ (OHO) 1620, 1640 cm^{-1}, ρ (OHO) 635, 680 cm^{-1}, ν (Fe-O) 440 cm^{-1} [121, 133] and XPE spectra (Fig. 2.46, Table 2.12) in O1s lines at 531.6 and 534.0 eV. Broadening and a shift into the low-frequency region of the ν (OH) 2700–3600 cm^{-1} band relative to the ν (OH) 3150-3640 cm^{-1} band of normal crystallized water proves that the aqua ligand participates in the hydrogen bond [121, 131]. Apparently, molecules of H_2O promote cross-linking of the neighboring chains in Tet-Fe via hydrogen bonds between the atoms of water hydrogen and those of cyclic nitrogen not involved in the coordination.

Based on the above-cited investigation results [5,6,131–133], the following conclusions on the *molecular structure of the surface Tet-Fe CC* have been derived:

(1) oxidation degree of the central ion of plus two (E_l of Fe2$p_{3/2}$ in 709.5 eV XPES [129]),
(2) atomic relationships of the elements as $C_N/C_{Fe} \sim 8.6$; $C_C/C_{Fe} \sim 2.5$; $C_O/C_{Fe} \sim 2.4$ (calculations are based on CC line intensity in XPES [129]),
(3) bidentate-bridge function of the tatrazolate ligand with N_1 and N_4 coordination centers;
(4) participation of aqua molecules in coordination,
(5) coordination number equal to 6 (acc. to the characteristic v value (Fe-N) equal to 395 cm^{-1} in IR spectra [135]) and, consequently, octahedral structure of the coordination node,
(6) presence of hydrogen bonds between hydrogen molecules of water and non-coordinated atoms of cyclic nitrogen (N_2 and N_3).
The complete structural formula of CC $[Fe(CN_4H)_2(H_2O)_2]_m$ and its structural model illustrated in Fig. 2.48 comply fully with the conclusions

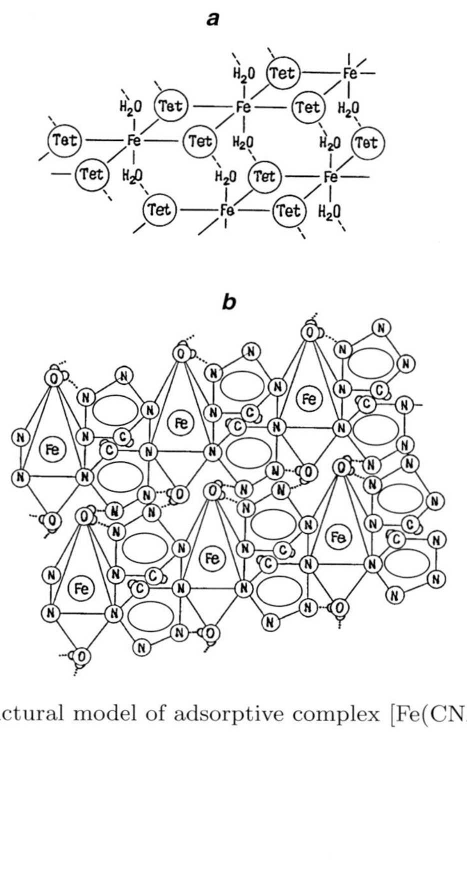

Fig. 2.48. Structural model of adsorptive complex $[Fe(CN_4\ H)_2(H_2O)_2]_m$

detailed above when the CC is electron neutral and sterically unrestricted.

Thickness of the adsorptive protecting layer under study estimated versus time (~25 min) of etching with argon beam ions (at 0.3 nm/min velocity) until appearance of distinct Fe^0 reflexes in XPE spectra ($Fe2p_{3/2}$ 707.0 eV) and disappearance of Nl*s* reflexes made up 7–8 nm. It has been experimentally verified that the method of XSA method is insensitive to the layers of this thickness. The sought thickness has been estimated by AFM from the crack thickness recorded in a profile section of the protective layer (Fig. 2.49a, which constituted 8–10 nm and matched up well with XPES data.

Proceeding from the analysis of AFM images of the surface and profile sections of the adsorptive protecting layer (Figs. 2.49a and b, a conclusion has been made that Tet-Fe molecules form clusters of with a characteristic size of 250 × 300 nm in the 3–5 nm plane across the layer thickness on the metal surface [136].

The clusters presented in Fig. 2.49b acquire the direction of the main orientation.

The microstructure and surface microtopography of adsorptive protecting layers Tet-Fe have been studied by SEM [136]. Micrographs of steel plates presented in Fig. 2.50 show the samples before and after their endurance in humid atmosphere under a PE film with and without a CI.

The steel samples c and f do not exhibit any signs of corrosive damage after contact with the inhibited film material. The samples b and e, on the other hand, have a porous surface layer structure resulted from accumulated corrosion products. The microstructure of sample c is in many respects similar to that of the initial substrate a and displays a characteristic relief left upon grinding with an abrasive. However, the ridges and grooves of the abrasive scratches on sample c surface are less pronounced and the acute angles of the ridges are smoothened. The differences become still more vivid on detailed examination of Figs. d and f. Abrasive scratches on surface d look like parallel alternating ridges, whereas on surface f they acquire the character of small-scale waves without clear anisotropy. The observed microstructure peculiarities in c and f can evidently be related to the formation of adsorptive Tet-Fe CC on the steel surface sharing the polymer pattern. Irregularities of the elemental and phase composition of the protective Tet layer across its thickness (Figs. 2.51, 2.52), established by XPES [132,137], have clarified the mechanism of structure formation and structural anisotropy.

The steel surface is initially covered by an oxide film of Fe_2O_3 about 1.5 nm thick. The thickness has been estimated by XPES using the intensive Fe2*p* lines of the substrate at 711.2 and 707.0 eV (Fig. 2.47) and confirmed by the absence of Fe_2O_3 signals in radiographic analysis of the substrate.

The adsorptive protective layer consists predominantly of Tet-Fe CC with inclusions of Fe_2O_3. The concentration of the inclusions gradually reduces

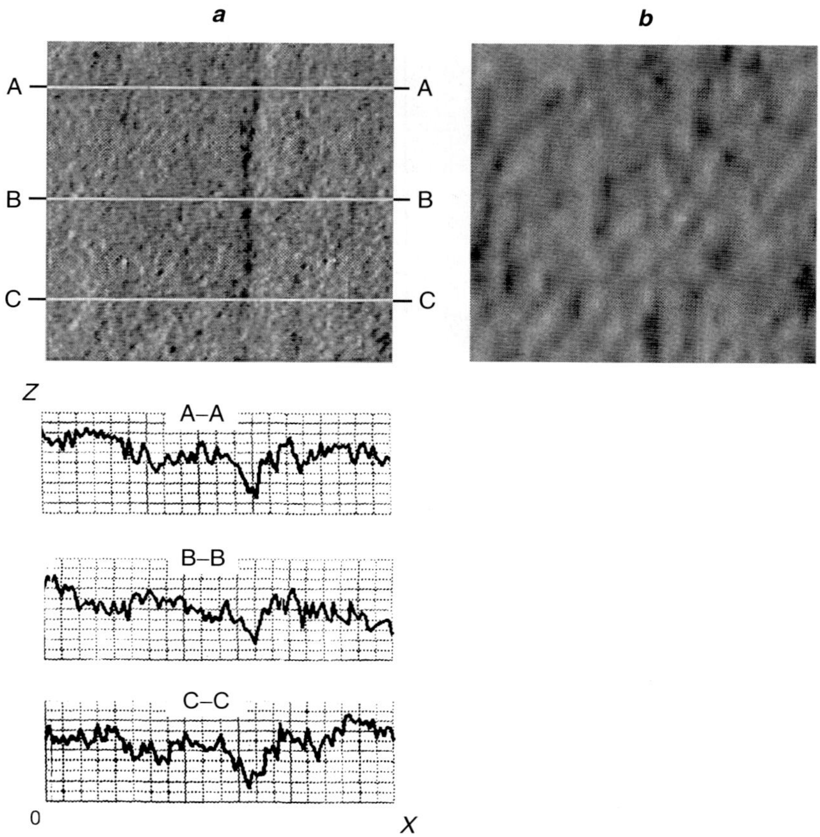

Fig. 2.49. AFM images of adsorptive protecting layer of Tet on steel and its profile sections. Scaling factor along the vertical line – 3.5 nm, scanning field size – $7.2 \times 6.4\,\mu m^2$ (**a**) and $1.4 \times 1.6\ \mu m^2$ (**b**)

with distance from the substrate and is insignificant compared to the adsorptive layer surface (Fig. 2.51).

The metal surface probably acquires an electrochemical inhomogeneity, i.e. energy discontinuity [38], on contact with condensed moisture inhibited by Tet. The initial acts of Tet adsorption take place on active centers with maximum positive charge density. These centers are concentrated on anodic areas of the Fe surface, where Fe(II) ions are generated. Subsequently, two processes take place simultaneously on the sample surface:

(i) Fe_2O_3 structures grow and become ordered on areas free from CC,
(ii) Multi-layered cross-linked polymer CC chains are forming and start to occupy still greater portion of the steel surface as long as the complexing reaction is thermodynamically preferable to oxide-formation.

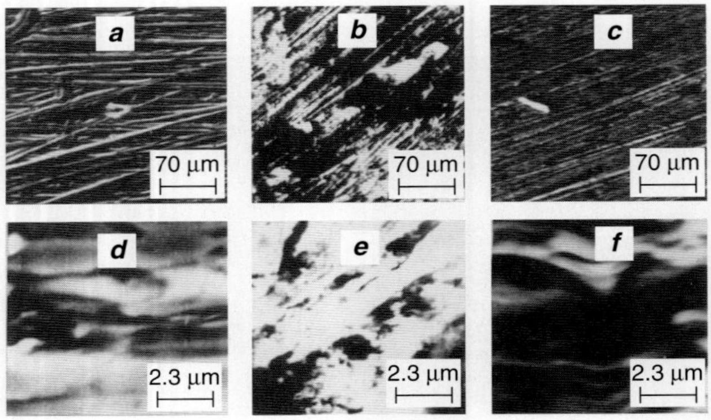

Fig. 2.50. Surface microstructure of steel 10 kp before (**a, d**) and after endurance in humid atmosphere at temperature gradient under a PE film without (**b, e**) and with Tet (**c, f**): (**a–c**) 70 μm marker; (**d–f**) 2.3 μm marker

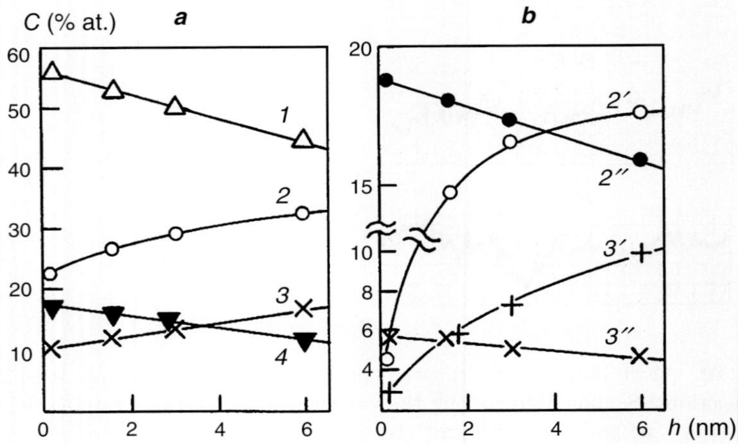

Fig. 2.51. Variations of elemental composition (C, % at) across depth (h, nm) of the adsorptive protecting Tet layer on steel. (**a**) gross composition: (1) N, (2) O, (3) Fe, (4) C; (**b**) O and Fe elements bound into iron oxide (2′, 3′) and CC (2″, 3″)

The corrosion process as a whole and the anodic one in particular are retarded still more perceptibly as the adsorptive layer grows. The concentration of free Fe(II) ions, representing centers of Tet coordination, diminishes. As a result, the adsorptive layer growth ceases upon reaching a certain thickness.

The considered experimental data on molecular constitution and morphology of chemisorbed protective Tet layers on a steel surface suggest that the mechanism of this CI is of the activation-suppressing type (with a prevailing screening effect) in the conditions under study.

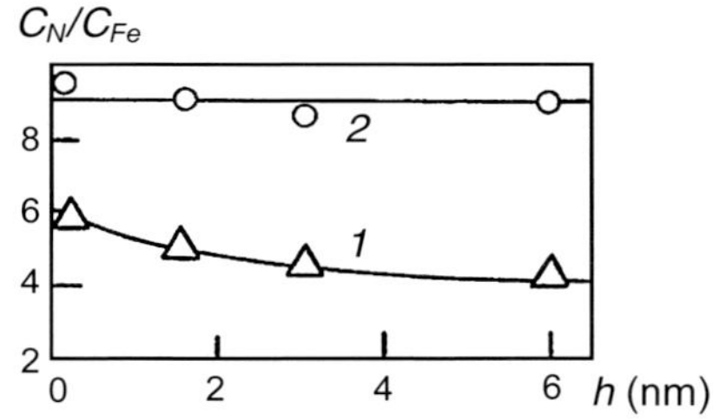

Fig. 2.52. Change of atomic ratio C_N/C_{Fe} through depth of protective layer Tet on steel versus relative total amount of iron (1) and combined in the complex iron (2)

The protective mechanism of PE films modified by Tet as a whole consists in the following. The CI (Tet) is liberated from the polymer matrix, is then transferred to the metal surface, dissolves in the moisture-condensed layer and is chemisorbed onto the metal surface, forming a protective layer that suppresses corrosion centers and hampers (together with the film barrier) the feeding of reagents and removal of products from the zone of electrochemical reactions. Formed on the metal surface phase chemisorbed protective layer is 8 nm thick and consists of CC clusters of the $[Fe(CN_4H)_2(H_2O)_2]_m$ composition. The tetrazolate anion behaves as a bidentate-bridge ligand having coordination centers in nitrogen atoms N_1 and N_4 of the cycle and represents, together with aqua molecules, a coordination node FeN_4O_2 with a distorted octahedral configuration. The polymer chains in CC are cross-linked with each other by hydrogen bonds. The structural anisotropy of the protective layer is attributed to Fe_2O_3 inclusions.

The films described should liberate amounts of CI sufficient to form protective adsorption layers on the metal surface before the initiation of intensive corrosion processes.

Single-layer inhibited films can neither present a insurmountable barrier to diffusive permeation of corrosion-active agents from the outside, nor they can avert infiltration and condensation of water vapor on the metal surface, but are able to hamper the removal of ion products from the zone of electrochemical reactions. The corrosion protection barrier factor is drastically improved in multilayered film materials.

2.4 Quality Control

Like any other kind of commercial goods, inhibited polymer films should comply with certain quality parameters and requirements indicated in standards and technical specifications.

Such characteristics as the external view of the films, their linear dimensions, deformation and strength parameters, and CI concentration are subjected to obligatory checking. The protective properties of the films are evaluated as necessary.

The *appearance* of the films is examined in order to exclude various defects in the form of ruptures, punctures, bends, fracture, foreign inclusions, surface soiling, etc. A defect peculiar to inhibited films produced by joint processing of VCI and polymer melts is gas inclusion arising from foaming of the polymer binder by CI vapors. This may occur when temperature regimes for molding are not strictly observed and may lead to violation of film continuity. Films containing powder CI sometimes acquire defects as a result of agglomeration of inhibitor particles. Uneven distribution of liquid-phase CI in the polymer binder results in areas of excess exudation of inhibiting liquid. Multilayered and combined film materials should be devoid of any delamination resulting from disruption of the adhesive bonding of the constituent layers.

The linear dimensions of the films refers to their width and thickness. The thickness of the films is an important parameter as it reflects stability of molding regimes and defines their strength. The measurement procedure for linear dimensions of films is specified in both national and international standards [139–142]. Relative changes in linear dimensions of heat-shrink films under elevated temperatures are characterized by a shrinkage parameter [143].

The most indicative *deformation and strength characteristics* of inhibiting films are the breaking tensile strength and relative elongation at rupture [1]. The methods for their determination also have respective standards [144–146]. Films intended for wrapping bulky or sharp-edged hardware have to meet the requirements on resistance to rupture [1, 147–151] and puncture [1, 152].

Control of CI concentration and its agreement with specifications guarantees the required protection properties of the films. The content of CI in the films is estimated on the basis of specific procedures, which consist of isolation of CI from the polymer matrix (via evaporation, extraction or chemical interaction with a technological medium), and quantitative determination by chemical or physical-chemical methods. The estimation procedure is chosen individually for each CI kind or a group of inhibitors, depending on the film structure and production method.

In particular, the content of NDA in PE films produced by either joint extrusion of VCI and the polymer or hot-rolling of the inhibitor powder on the film base is estimated by weighing. The CI is removed from the polymer matrix by sublimation upon when kept in a thermostat at 100°C until reaching a stable constant mass.

A method of determining G-2 inhibitor content in PE films is proposed in [109], during which this CI (hexamethylenimine and m-nitrobenzoate) is desorbed from the film samples by subjecting them to a flow of air heated to 60°C. The air is directed onto the G-2 absorber filled with distilled water. The amount of absorbed G-2 is estimated by analyzing the aqua solution for the content of acidic and amine constituents of the inhibitor. The acidic component (m-nitrobenzoic acid) is estimated for non-plasticized films by spectrophotometry. For this purpose, the optical densities of aqua solutions of G-2 are determined at 215 and 269 nm wavelengths. Since the presence of Pl impedes spectrophotometry, the absorbing solution for films plasticized by dibutyl or dioctyl phthalate is analyzed using the presence of the amine group (esters of the phthalic acid show absorption maxima at 225 and 275 nm wavelengths).

The method of defining IFHANGAZ content in a PE film consists in grinding a film sample, extraction of the VCI from the polymer using a solvent (chlorobenzene or methylisobutylketone) followed by potentiometric titration of CI components converted into an organic extract (secondary and tertiary amines) by solution of $HClO_4$ in a waterless acetic acid [24, 153].

The essence of a derivatographic method of analyzing CI content in polymer films [3, 24] involves measurement of the area (S) of endothermic peaks of CI evaporation recorded on DTA curves of the films (Fig. 2.53a). Each relative value of S corresponds to a certain CI concentration (C) in the film found by the calibration $C = f(S)$ (Fig. 2.53b). It is expedient to employ this method under constant derivatographic conditions for films whose components do not produce overlapping peaks on DTA curves.

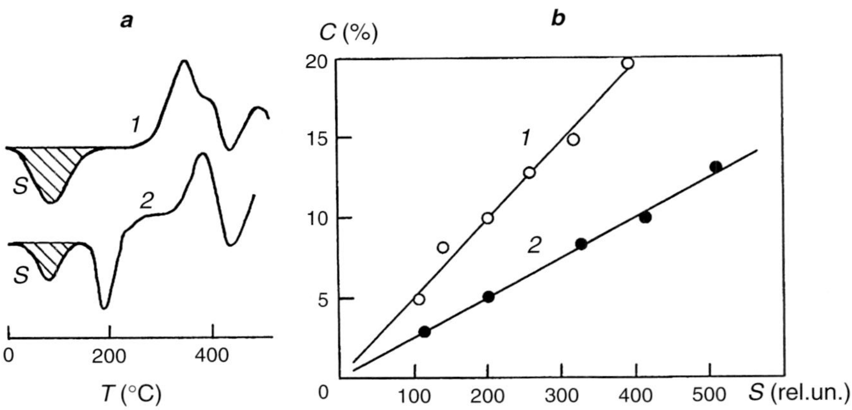

Fig. 2.53. Derivatographic analysis of CI content in polymer films. (**a**) DTA curves: (1) MSDA inhibitor; (2) PE film with 5 wt% of MSDA. (**b**) calibrating dependencies of CI concentration (C) in PE films upon area (S) of endothermic peaks of inhibitor evaporation on DTA curves of film samples: (1) MSDA; (2) IFHANGAZ

The protective properties of inhibiting films are estimated periodically. With this aim, standard as well as purpose-developed methods of accelerated test can be used (electrochemical, microbiological, model). These tests may be time consuming and are rather laborious.

A new method of express-estimate of the protective properties of inhibited films involves the application of chemical indicators forming colored products on interaction with the inhibitors. For instance, a number of VCI, representing compounds containing amine and azomethine groups, serve as modifiers of polymer films and represent the. For visual control of CI content in a film package modified by the mentioned inhibitors (e.g. dicyclohexylamine, VNHL-49, VNHL-20 or IFHANGAZ-1) it is possible to use bindon (anhydro-bis-indandione-1,3) [154,155]. The agent containing an indicator is applied as strips on the film hose surface during molding. The coloring intensity of the allochromatic strips depends, under all other conditions being equal, on the concentration of CI vapors in the packing space. It is possible to estimate the service characteristics of the films and time during which they render protective effect using the color indicators whose lower sensitivity threshold to CI vapor content in the air is $0.1\,mg/m^3$.

2.5 Fields of Application

Inhibited polymer films render a promising, efficient and supple means of protecting hardware against atmospheric corrosion. The operations of preservation and packaging of hardware can be joined into a single technological process that furnishes a highly efficient anticorrosion shield during transportation, in process and during warehouse storage. In addition, inhibited film materials can be used as independent means in sealing systems of metal structures, machines and mechanisms.

The world packaging market offers a wide range of inhibited films intended for the protection of various ferrous and nonferrous metal products from all types of corrosion as well as a multi-metal corrosion protection. Packing in inhibited films presents an alternative to anticorrosion agents that are costly and insufficiently adapted to production processes (rust preventives, desiccants, CI solutes in organic solvents and other) and eliminates the operations of applying and removal of lubricating, polymer or other rustproof coatings. The method concerned is highly productive, can be easily automated and is ecologically friendly and safe compared to traditional methods of fighting corrosion (inhibiting oils, air, primers, coats, paper etc.). The majority of inhibited films are transparent to allow the contained item to be seen. A high-quality film packing elevates the competitiveness of the goods for sale, makes them attractive and can also be a method of advertising.

Inhibited film materials have found application in anticorrosion protection of a broad spectrum of commercially produced commodities, including steel structures, rolling and wire mills, steel forged parts, tools, accessories, knife sets, kitchen and heating utensils, boilers, metal sheets, wire and springs, spares for cars and agricultural machinery, motors, diesels, electric motors, instruments, construction, mining and ship-building equipment, textile looms and metal-working machines, printing equipment, ball bearings, air-engines and their spares, measuring apparatuses, medical facilities and instruments, various kinds of weapons, household appliances and a multitude of other metal articles.

In confirmation of the above list, the Northern Technologies International Corporation, which is a large-scale manufacturer of inhibited film materials, informs with pride that among its customers are such world-known companies as Ford Motor Co., Ryerson Steel, General Motors, Buik, Delco Remy, Kone Oy, Valmet Group, Wartsila Diesel, Power Train Group, Cadillac Motor, John Deere, Caterpillar Tractor, General Electric, Chrysler Corporation, Detroit Diesel, Minnesota Mining & Manufacturing (3M), Overseas Packing, Texas Instruments, Universal Packaging, Volkswagen, BMW, Merñedes, Nissan, Toshiba, Toyota, Mitsubishi, MAN, Isuzu, Honda, Suzuki, Komatsu, Renault, Pont-a-Mouson, Fiat/Lancia and Alfa Romeo.

2.5.1 Methods of Packing into Inhibited Films

The simplest method of preserving hardware consists in their *wrapping and sleeving* (or *covering*) in inhibited films.

One, or a few articles, can be wrapped in the film using a wrapping machine. The machine functions [1] include feeding of the film wrap, forming and closure of the package. The wrapping material from the unwinding unit fitted with a strainer is transmitted to the cutter and bundling machine. The film sheeting placed on the packed article envelopes it and the film ends are folded up and tucked-in. The package is then heat sealed or spliced.

Sleeving or covering of the hardware presupposes the use of *packages or bags* [13, 18, 155, 156]. The simplest package and bag technology is their manufacture from a hose or tubular film during which plane-folded hose cut-offs are cross heat-sealed and cut into bags. Films extruded as a plane sheet are first folded and then heat-sealed. The sheets are folded in the longitudinal direction to form a hose, transversely as loops or envelopes or into W-shapes [1, 157, 158].

To make bags of the inhibited films tight, they are heat-sealed or glued together after packaging using contact, thermal-pulse bonding or HF current methods. The methods and equipment employed for sealing usual films can also be used for inhibited films [1, 11, 34]. The sealing temperature and seam strength of the inhibited films is dependent largely on the nature of the polymer base and only insignificantly on the CI type if it is present in small amounts [24]. For example, the sealing temperature for Zerust^{TM} LDPE-based inhibited films of different composition ranges within 170–180°C (depending on the equipment, time and pressure). Surface modification of film by CI commonly lowers the sealed package strength. Nevertheless, it is possible to ensure the required strength of the seam by an appropriate choice of the heat-sealing regime, usually by elevating the sealing temperature (see Sect. 2.3.2). In particular, the Russian State Standard 12303-83 [158] indicates that the seam strength of PE and PVC-based bags should surpass their

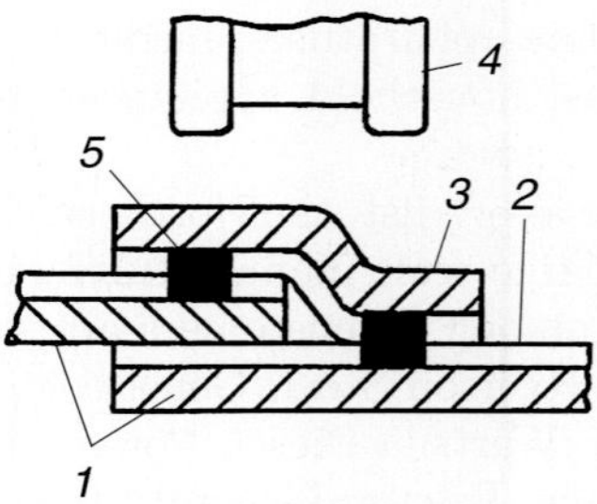

Fig. 2.54. Heat-sealing scheme of a rustproof paper laminated with PE: (1) inhibited paper, (2) PE coat, (3) strap, (4) sealing roller, (5) seam

rupture strength value by a factor of 0.7. The seam strength of the bags from combined films depends on the mass of the hardware packed; so, at $m < 3.5\,\text{kg}$, it is 2–6 N/cm; and with $3.5\,\text{kg} < m < 7.5\,\text{kg}$, it is 7–10 N/cm.

Polymer films are heat-sealed to form either lap or fin seams resisting shear and delamination loads, respectively. The fin seal is less strong than the lap seal [34]. It is recommended to seal PE-coated inhibited paper bags with a lap seam and strap the butt with the same material [3] (Fig. 2.54).

For packing small-piece and piece goods it is convenient to use flat bags of inhibited films fit with a tang-slot closure. A tang of the required profile is made on the inner side of the mated wall and can be tightly inserted between a pair of slots on the other wall of the bag. The bags can be uncapped many times and a short-term uncapping does not impair the anticorrosion effect of the package since the protective adsorption layer of CI on the packed article surface is able to self-restore by replenishing the inhibitor from the film.

Inhibiting covers should be kept out of direct sunlight, airflows, excess heat and mechanical loads to preserve their protective properties. This is why it is better to store the covered hardware in boxes, cases or other types of containers.

Wrapping in thermoshrinking inhibited films seems to be a more efficient means of covering hardware as compared to the previous type. The shrink-wrapping presupposes shrinkage of the film at heating to skin tight around the package charge. As a result, CI isolated from the film easily reaches the protected metal surface, improves reliability and appearance of the package. If the polymer film contains a VCI, its evaporation from the polymer matrix presents an additional film shrinkage factor. Films containing a plasticizer as a CI carrier shrink depending on the plasticizer content. In particularly, PE-based film containing above 40 wt% of mineral oil shrinks intensely (see Sect. 1.4) and accelerates the liberation of CI from the film.

Fast assimilation of inhibited films in engineering has lead to the application of *combined packages*, some types of which are illustrated schematically in Fig. 2.55 [3].

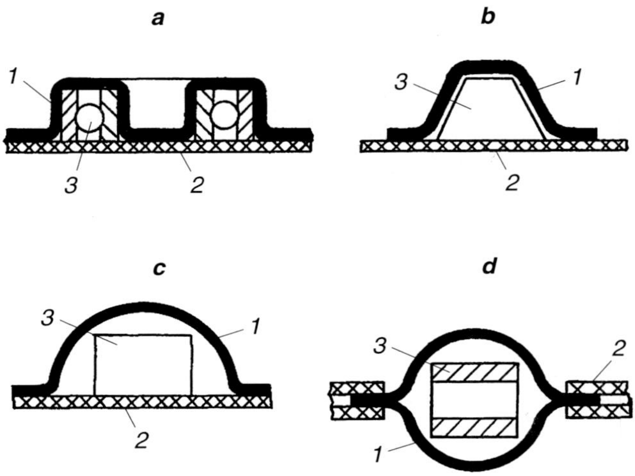

Fig. 2.55. Combined packages of inhibited polymer films: (**a**) skin pack; (**b**) blister strip pack; (**c**) blister bubble pack; (**d**) stretch package. (1) polymer film, (2) bottom card, (3) protected article

Skin packaging is a combination of a polymer film sleeve with a bottom liner (cardboard or sheet polymer). The article is placed on the card liner and is wrapped into the inhibited shrinkage sleeve (Fig. 2.55a). The film ends are fixed on the card, which can also be made of an inhibited material. During heating the film shrinks skin tight around the article. Skin packaging is used for small-size machine parts, spares, tools etc.

A variant of skin packaging employs either single film or bag-in-bag packaging in a thermal vacuum former. Various types of hardware, including air filters for internal combustion engines, wrenches, high-carbon steel drills etc. can be packed by this method. During packaging an article is placed on a bottom liner, the covering sleeve is modified by CI in places of contact, the space between the bottom and the sleeve is evacuated and the film is sealed using e.g. the thermal-pulse bonding.

An automatic machine combining film modification by CI with packing is shown in Fig. 2.56 [160].

The machine operates as follows. The substrate film 5 is fastened between frames 7 and 8 and is heated by heater 4 to a viscous-flow state. Chuck 2 travels upwards and air is exhausted through its breathable surface. The substrate film imitates at curing the profile of the chuck. Film 10, fed by unit 9, is subjected to embossing by needle roller 11 and treatment by CI. The inhibiting liquid is applied onto the film from vessel 12 using castor 13 fit with a porous pad on its outer surface. The CI is applied onto the film areas over the packed article contours. The inhibited film covers articles 6 on substrate 5, the air is evacuated and the pack is sealed. The machine economizes

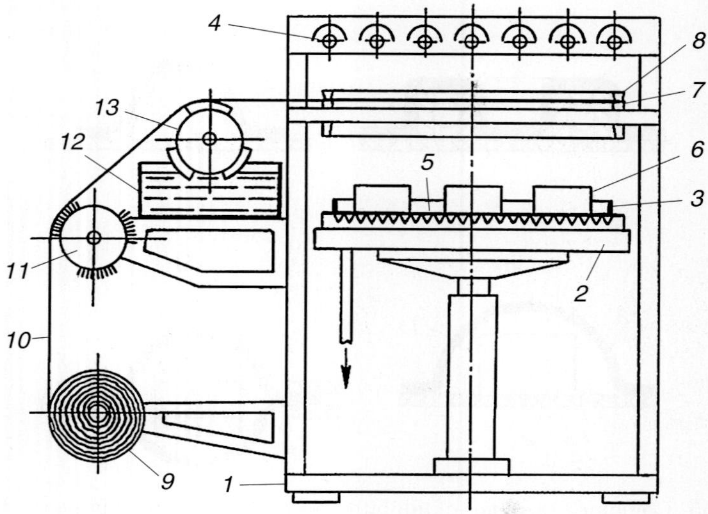

Fig. 2.56. A skin packaging machine using inhibited polymer films: (1) base, (2) vacuum chuck, (3) pallet, (4) heater, (5) substrate film, (6) packed articles, (7) and (8) frames for film fastening, (9) film feeding unit, (10) film, (11) needle roller, (12) CI vessel, (13) CI applicator castor

CI consumption and produces high-quality seams between articles. Fig. 2.57 shows metal articles packed by the method described above.

The skin packaging method is advantageous for simultaneous packing of products of different shape and size, it is distinguished by high production standards, rather high efficiency and package quality.

A blister pack consists of a rigid (cardboard or other) bottom liner and a sleeve of a transparent polymer material. The sleeve can either reproduce the packed article contours (blister strip pack, Fig. 2.55b) or be of regular spherical shape (blister bubble pack, Fig. 2.55c). The sleeve is produced by thermoforming and is bonded to the bottom liner by welding. For the blister pack, commonly inhibited film materials and sheets up to 0.15–1.0 mm thick are used. Mainly small-size hardware and tools are packed by this method. The blister pack adds an attractive appearance to the goods, protects reliably against various effects and enables direct observation of the contents by customers.

Flow packing is achieved by the application of inhibited polymer melt onto the article being packed [3, 24]. When cured the polymer layer forms a thin film on the article. For tight covering the article is placed on a porous substrate, which is evacuated on melt spraying. Another variant of the flow pack consists in the insertion of the packed part into a protective cap and placing into a press mold. Pressurized polymer is poured over the part to form a highly strong and tight pack. The flow method is extremely economical as

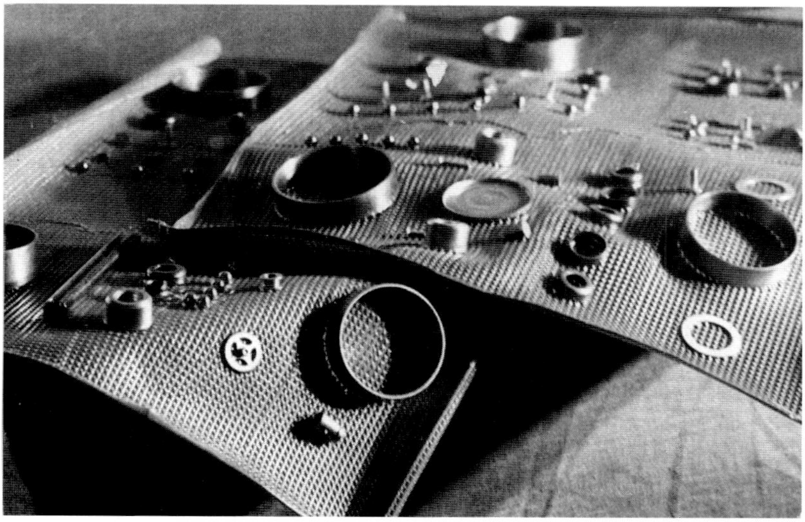

Fig. 2.57. Hardware packed into inhibited PE film by a skin packaging method

polymer excess can be reused and it does not necessitate additional sealing and bonding of the sleeve to the substrate; its application is restricted by the size of the goods and a short-term temperature effect on the packed article by the polymer melt. A modification of the flow method is a cocoon pack, in which the film shell is formed by polymer spraying on a metal or fabric framework with a charge inside [157]. The cocoon pack is advantageous for protection against atmospheric, crevice and bacterial corrosion of large hardware and structures, especially during carriage by sea and in tropical climate conditions [24].

Inhibited stretching film is used in stretch packages, a variant of which looks like (Fig. 2.55d) a blank of two cards (or sheet material) having a window reproducing the article shape. The article is covered from two sides by the stretching film whose ends are heat-sealed or bonded by some other method between two cards.

Various types of packing, wrapping, bundling and stacking machines are used for packaging hardware in inhibited films, along with numerous devices for pasting and heat-sealing, extrusion, molding and thermal vacuum aggregates that are described at length in literature on the subject [1,157,161,162].

As has already been shown, engineering goods are preserved in films containing contact CI by sleeving in shrink and stretch films, skin packs, etc. Single-ply films of this type, in which CI are liberated from both sides, have found application in anticorrosion gasket materials, during stacking or winding of metal sheets and plates or winding of foil in rolls (films with VCI are equally applicable with this aim [155, 163, 164]).

Film materials with VCI furnish an unlimited choice of packaging methods and hardware kinds. The protective effect is exerted by CI vapors without the obligatory film contact with the product, although the closer the film to the part surface the more efficient the rendered protection. Complete tightness of the package is compulsory only for prolonged storage for a few years. VCI-modified films are especially efficient for rustproof protection of complex-profile articles as CI vapors can penetrate in any hard-to-access places. Articles packed in such films are fit for operation immediately after unpacking. The presence of an invisible monomolecular or nanosize phase layer of CI on the article surface does not affect the properties of the metal in any way. This is highly important for preserving electronic instruments where slightest changes in conductivity, permittivity or dimensional stability may lead to malfunction of the equipment. Under certain conditions traces of VCI disappear two to three hours after unpacking the hardware from the inhibited film.

Perfection of inhibited packing films goes hand in hand with modern developments in the packaging industry, maintaining high standards of efficiency, multifunctionality and ecological safety [165].

2.5.2 Application of Inhibited Films in Sealing Systems

Inhibited polymer and combined film materials display corrosion-inhibiting properties together with good barrier and strength characteristics that can be efficiently employed in sealing systems as well.

An important field of application of inhibited polymer films is *insulation of anticorrosion coatings* applied on metal structures by either hot or cold methods. In the latter case inhibited film materials with an adhesive layer are used, into which the contact CI is commonly impregnated [58, 166].

A modification of the laminated anticorrosion film coat used on trunk Pipelines is shown in Fig. 2.58 [4, 165]. The film consists of two barrier PE-based layers, 1 and 2. The first of these has strip substrates 1 bonded by glue 4 to a cellular layer 3 containing a mixture of heat-resistant mineral oil and CI (sodium benzoate) 5. The second layer 2 is a separating layer and is weakly adhered to layer 3. Prior to application of the coating, layer 2 is removed and the film material is pasted over the metal pipeline, with inhibited layer 3 facing the surface being protected. Laminated materials of this type are usually based on very strong cross-linked or oriented films to make the insulation more reliable and reduce its penetrability to ambient matter [4].

An adhesive tape [58] intended for protecting metal pipelines against mechanical damage along with electrochemical and microbiological corrosion has been patented. The tape consists of a PE substrate, intermediate CEVA-based adhesive layer and a sticky bituminous-rosin layer containing volatile water-soluble CI of the 5-R-tetrazole and 3-R-5R′-triazole series. When evaporating from the sticky layer or dissolving in the water solute (after ingress

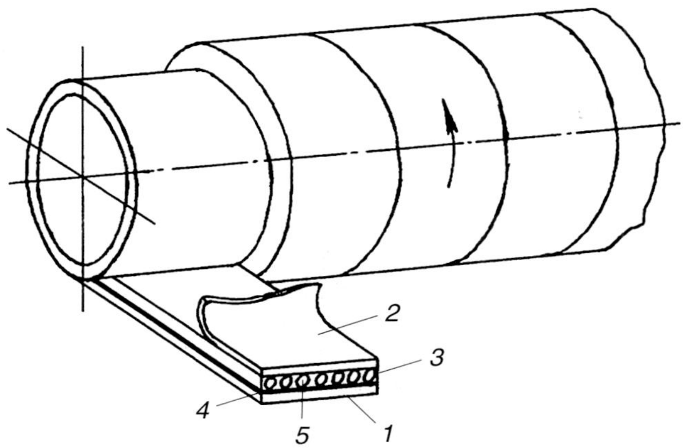

Fig. 2.58. Rust-inhibiting of pipelines using a laminated film incorporating CI in its adhesive layer: (1) barrier PE layer; (2) separating PE layer; (3) inhibited cellular layer; (4) adhesive layer; (5) cells containing CI

of moisture on depressurization of the adhesive layer) the CI form an adsorptive protecting phase film on the metal surface and thus avert local corrosion.These CI are able to suppress vitality of mould fungi and bacteria, including hydrosulfuric ones and thus protect metal structures against biological corrosion.

Insulation coats based on inhibited films with an adhesive layer will be discussed in more detail further in Sect. 3.5.3.

Inhibited films are also employed in *shell-type seals.* A shell seal means a thin-walled lining that separates the sealed medium from the ambient space; hollow seals of contact gaskets also belong to this category [167].

Use of polymer films in shell-type sealing systems has a number of advantages. The polymer shell walls can be made thin enough to regulate stiffness and the deformation properties of the seals' volume by creating excess pressure inside them. Moreover, it is possible to regulate heat regimes of the seals by injecting auxiliary media into the shell hollows, to keep lubrication conditions in movable joints at the required level, transport different matter that hamper crevice corrosion and neutralizing harmful effect of the sealing media on operation of tight joints and so on. Already developed methods of polymer film orientation reinforcement may impart sufficiently high strength to sealing shells [3].

Plasticized polymer films are known to possess an unusual property of self-healing. Their sealing capacity is not impaired noticeably even by 100 μm diameter punctures. The data presented in Fig. 2.59a prove that the electrolyte (0.1 N solution of HCl) cannot penetrate through the PE film plasticized by the mineral oil even in the presence of puncture defects and the effect of mechanical vibration. Testing of a plasticized PE film has shown

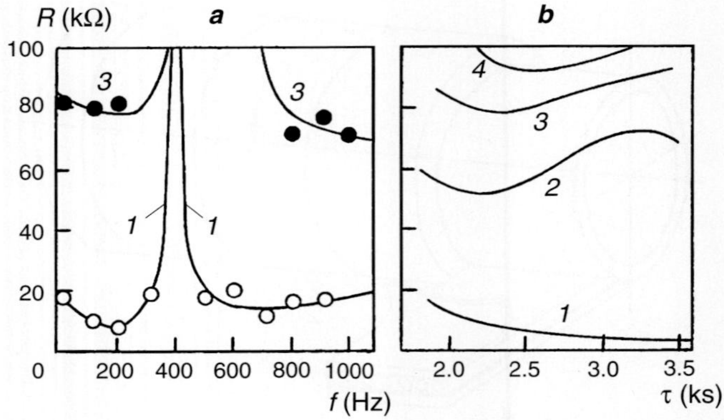

Fig. 2.59. Changes in electrical resistance R of the electrolyte PE film with a puncture (100 μm diameter) – substrate system depending on (**a**) frequency of mechanical oscillations of the sample, f, and (**b**) time at $f = 0$. Plasticizer content in the film, wt%: (1) 0; (2) 30; (3) 40; (4) 50

a considerable improvement of electrical resistance (R) of the electrolyte–film sample–substrate system (curve 3) in contrast to a non-plasticized film (curve 1).

This phenomenon can be explained in terms of the variation kinetics of the electrical resistance of systems with punctured films (Fig. 2.59b). In contrast to non-plasticized films, for which R diminishes with time, the resistance of the plasticized samples increases as a result of, firstly, incrementing oil content in the film, and secondly, filling of the puncture hole in the film with a hydrophobic mineral oil through syneresis [72].

Sealing shells are mostly applied in insufficiently stiff and weakly loaded *immovable joints.* They are usually made as flat or hollow tubular seals fastened along the packed hole outline (lids, hatches, car and instrument bodies, doorway and window openings, etc). It should be emphasized that the introduction of CI into sealing compositions is strongly advisable since this is found in the narrow gaps and clearances where the corrosion process intensity is the highest.

Polymer films and shells are widely applicable in *separating baffles of sealing devices.* Polymers are also preferable as materials for membranes and flexible baffles separating unlike or differently loaded media intended for the conversion of pressure into motion or vice versa. The membranes should meet the requirements of high deformability, chemical resistance and perfect sealing properties.

There are also cases when the medium being sealed is found between the contacting inhibiting film material of the membrane and metal surface. If the medium is gaseous, VCI can be used to protect the metal. When the sealed medium is a liquid, then it is expedient to impregnate a CI into the membrane, which is either soluble in this medium or is able to form a dispersion or emulsion with it. In particular, a composition for sealing materials based on PE and containing a CI (a mixture of potassium and triethanolamine salts of unsaturated fatty acids), Pl and emulsifier forms an inhibitor emulsion in contact with the water solution of the sealed medium. Inhibitor liquid incorporating a mixture of CI and Pl isolates on the seal surface via syneresis. CI transported by the sealing medium protects against corrosion of the seal parts not in immediate contact with the inhibited material [168].

Seals of movable joints should first of all make the sealant fill microasperities of the conjugated part to avert transfer of the sealing media on motion. This can be achieved in two ways using sealing shells: i) by augmenting the contact pressure in the sealed joint when forcing media into the shell hollow, or ii) by raising the deformability and, consequently, sensitivity of the sealant to tracing of the counterface via thinning the shell walls. The shell walls can be imparted with the properties of semipermeable membranes with this aim by making use of an original technical decision described in [72]. Media containing CI, lubricants and other matter raising the service characteristics of the seal are forced into the shell under low pressures. As a result, it is possible to achieve the metered liberation of CI into the contact zone of movable seal parts. In addition, sufficiently thick shell walls can be made of an inhibited material containing lubricants and other necessary antifrictional components [167].

The complexity of calculations and designing as well as labor intensity of some shell-seal designs restricts their fields of application. However, their versatility makes them unique and applicable in almost all types of contact seals.

It should be underlined in conclusion that inhibited polymer films are an efficient rust-inhibiting means of protecting hardware and metal structures against atmospheric, hydrosulfuric, acidic, microbiological and other types of corrosion. The numerous film types and their production processes fit any customer's preference in line with the chosen preservation method, transportation or storage conditions and anticorrosion terms.

References

1. W.A. Jenkins and K.R. Osborn. *Plastics Films: Technology and Packaging Applications.* Lancaster, PA, Technomic., 1992.
2. V.E. Gul (ed.) *Polymer Film Materials*, Moscow, Khimiya, 1976.
3. L.S. Pinshuk and A.S. Neverov. *Polymer films containing corrosion inhibitors*, Moscow, Khimiya, 1993.
4. A.P. Kondratov, A.P. Gromov and V.N. Minin. *Encapsulation in Polymer Films*, Moscow, Khimiya, 1990.
5. A.V. Makarevich. *Physico-Chemical and Technological Principles of Developing Active Film and Fibrous Materials Based on Thermoplastics.* Dr. Sci. thesis, Minsk, 2000.

6. I.V. Tsarenko. *Modified by Polynitrous Heterocycles Polymer Materials and Their Protective Mechanism.* Ph.D. thesis, 2000.
7. U.S. Patent 5855975, B 32 B 5/16. *Anti-corrosion plastic film containing recycled resin.* B.A. Miksic and J.M. Foley, 1999.
8. EP Patent 0990676, C 08 L 23/06, 67/00 3/00, B 65 D 65/46, C 08 K 3/00, 5/00. *Biodegradable vapor corrosion inhibitor products.* C. Chandler, S. Bradley and B.A. Miksic, 2000.
9. G.W. Ehrenstein. *Polymer Materials: Structure, Properties, Applications.* Munchen, Carl Hanser Verlag, 2001.
10. R. Vincenti. *Elsevier's Dictionary of Plastics and Polymers.* Amsterdam, Elsevier, 1998.
11. V.A. Kabanov, M.S. Akutin, N.F. Bakeev et al. (Eds.), *Encyclopedia of Polymers.* In 3 Vols., Moscow, Soviet Encyclopedia, 1972–1977.
12. Verpackung mit eingebautem inhibitor. *Werkst. und Korros.*, 1995, Vol. 46, No. 10, s. 603.
13. Nove antikorozivni folie Zerust. *Plasty a kauc.*, 1997, Vol. 34, No. 11, p. 347.
14. EP Patent 0566789, C 23 F 11/02, B 65 D 81/26/ Vapor phase corrosion inhibitor material. B.A. Miksic and M. Tarvin, 1993.
15. Zerust bags. *British Corros. J.*, 1997, Vol. 32, No. 3, p. 158.
16. Folie antykorozyjne. *Ochr. Koroz.*, 2000, Vol. 43, No. 4, p. 101.
17. Transparenz schutz. Korrosionschutz fur transport und lagerung. *Produktion.*, 1997, Vol. 36, No. 40, p. 15.
18. DE Patent 3938125, B 29 D 9/00, B 32 B 5/18, 27/08, 27/32. Two- or three-layer air cushion film – with integrated corrosion protection based on polyethylene or related materials. A. Trinkl and H. Oldenburg, 1991.
19. Japan Patent Application 5-45420, B 32 B 27/32, 27/18. Anticorrosion film for metal and metal ware packaging. Mitsui Du Pont porichemikaru K.K., 1993.
20. Japan Patent 4147852, B 32 B 27/28, B 32 B 25/08, C 08 L 25/04. *Wrapping film.* Watanabe Takeshi, et al., 1992.
21. Japan Patent Application 62-6507, B 32 B 27/18, 27/28. *Anticorrosion film.* Nitto Electric Industrial Co., 1987.
22. U.S. Patent 6054512, C 08 K 5/34. *Corrosion inhibiting thermoplastic alloys.* S. Bradley, T. Nelson, B. Miksic and M. Kharshan, 2000.
23. Ya.M. Zolotovitsky, L.S. Pinchuk and G.V. Rechits. Polymer anticorrosion film. *Plastic Masses*, 1986, No. 7, pp. 51–53.
24. G.V. Rechits. *Development and Investigation of Plastics Containing Corrosion Inhibitors for Packing Engineering Products.* Ph.D. thesis, Gomel, 1987.
25. Japan Patent Application 60-43306, B 32 B 27/30, H 01 B 7/28, 17/62. *Anticorrosion protective film.* Tatsuta densen K.K., 1985.
26. Japan Patent 11071472, C 08 J 5/18, B 32 B 27/18, B 32 B 27/20, B 65 D 65/40. *Film for metal wrapping.* Y. Otsuka, and T. Fukaya, 1999.
27. Japan Patent 10100300, B 32 B5/32, B 05 D 7/24. *Formation of corrosion-proof and waterproof film.* K. Ganai, 1998.
28. U.S. Patent 5863642, B 32 B 7/02, 5/08, 27/32. *Water resistant and vapor phase corrosion inhibitor composite material.* E. Henderson, L. J. Reed, K. and D. Jagielski, 1999.
29. JP Patent 11179838, B 32 B 15/08, B 32 B 27/38, B 65 D 65/40. *Corrosion-inhibitor laminated film for soft packaging and corrosion inhibitor soft package.* K. Ono, and K. Okita, 1999.

30. U.S.S.R. Patent 827496, N 08 J 5/18, B 32 B 27/32. *Method for production of inhibited polyethylene film.* O.R. Yurkevich, Ya.M. Zolotovitsky, L.V. Zaborskaya and L.S. Golubova, 1981.
31. Belarus Patent 1943, C 08 J 5/18. *Method of production of polymer film.* L.S. Pinchuk, A.V. Makarevich and I.V. Tsarenko, 1997.
32. L.I. Antropov, E.M. Makushin, and V.F. Panasenko. *Metal Corrosion Inhibitors*, L.I. Antropov, (ed.), Kiev, Tekhnika, 1981.
33. U.S. Patent 5715945, B 65 D 85/84. *Vapor phase corrosion inhibitor package utilizing plastic packaging envelopes.* C. Chandler, 1998.
34. D.F. Kagan, V.E. Gul and L.D. Samarina. *Laminated and Combined Film Materials*, Moscow, Khimiya, 1989.
35. U.S. Patent 5422187, B 32 B 9/00. *Vapor corrosion inhibitor-desiccant material.* B.A. Miksic, J.M. Foley and T-Z. Tzou, 1995.
36. DE Patent 3735811, B 32 B 15/08, 27/06, B 65 D 65/40, D 21 H 5/22. *Packaging material, in particular for products sensitive to corrosion.* Bosch GMBH & CO KG W, 1989.
37. Japan Patent 1199927, B 29 C 63/02, B 32 B 7/10, 15/14, E 01D 19/02, E 02 D 31/06. *Corrosion protective processing.* Sh. Katayama, 1986.
38. DE Patent 39334273, B 65 D 65/38, C 08 K 5/17. *Anti-corrosion packaging material for metals – comprising combination of thermoplastic polymer and active mixture of salicylic acid-octadecyl amine salt and benzoic acid-diethylamine salt.* J. Bouzek and P Bouzek, 1991.
39. Japan Patent 6009912, B 32 B 27/02. *Corrosion-resistant sheet material.* Kitamura Hirosi and Ivadzaki Hirobumi, 1994.
40. U.S. Patent 5840381, A 47 G 19/22. *Corrosion inhibiting laminate sheets and containers.* R. Ohtsuka, 1998.
41. Japan Patent 7316314, C 08 J 5/18, 7/04; B 32 B 27/18, 27/20; B 65 D 65/40; C 08 K 3/40, 5/17; C 09 D 5/08. *Film for packing metal.* Y. Otsuka, et al., 1995.
42. EP Patent 0639657, C 23 F 11/02, B 65 D 81/26. *Vapor phase corrosion inhibitor material.* B.A. Miksic and M Tarvin, 1995.
43. U.S. Patent 5958115, C09 K 3/00, C 23 F 11/00. *Corrosion-inhibiting composite material.* H. Boettcher, G. Hahn, G. Reinhard and K.-H. Kallies, 1999.
44. U.S. Patent 4124549, C 23F 11/14. *Corrosion inhibiting plastic film.* K. Hashiudo and K. Imanishi, 1978.
45. U.S. Patent 4973448, B 65 D 67/00, C 23 F 11/02. *Vapor phase corrosion inhibitor product and method containing desiccant.* R.R. Carlson, N.E. Malwitz, B.H. Hill and Ph.J. Martin, 1990.
46. U.K. Patent Application 1263553, B 65 B 55/00. *A method of producing inhibited film*, 1978.
47. U.S.S.R. Patent 859178, B 29 D 7/02, B 29 D 7/04, B 29 F 3/04. *A method of manufacturing hose polymer film and an extrusion head thereof.* O.N. Kopyl and B.G. Karabut, 1979.
48. WO Patent application 86107009, B 32 B 3/14, 5/16. *Anticorrosion film for metals containing bivalent iron.* D.M. Weil, 1986.
49. S.P. Papkov. *Jelly-like state of polymers.* Moscow, Khimiya, 1986.
50. U.S. Patent 4913874, B 05 D 7/26, B 29 C 55/28. *Process of providing tubular polymeric anticorrosion film.* L.S. Pinchuk, E. N Aleshkevich, V.A. Goldade, et al., 1990.

51. U.S.S.R. Patent 1669929, N 08 J 5/18. *Polymer antirust film and a method of its production.* L.S. Pinchuk, V.A. Goldade, I.M. Vertyachikh, et al., 1991.
52. EP Patent 0373255, B 05 D 7/02, B 29 C 47/00, B 65 D 81/26, C 23 F 11/02. *Corrosion-inhibiting polymer film, method of its production and apparatus for carrying out the method.* L.S. Pinchuk, V.A. Goldade, I.M. Vertyachikh, et al., 1990.
53. France Patent 2714386, C 08 J 5/18. *Elastomeric film and a method for its production and application.* G. Riess and A. Cheymol, 1994.
54. I.V. Tsarenko, A.V. Makarevich, D.A. Orekhov. *Microbicidal properties of polymer films modified by five-membered polynitrogen heterocycles. Bioprocess Engineering*, 1998, Vol. 19, No. 6, pp. 469–473.
55. V.E. Gul, and V.P. Diyakonova. *Physico-Chemical Foundations of Manufacturing Polymer Films*, Moscow, Vysshaya Shkola, 1978.
56. Czech Patent 188739, B 65 D 81/26. *Anticorrosion wrapping film.* D. Mityska, V. Novak and V. Volsik, 1981.
57. Japan Patent Application 55-29509, C 09 J 7/02, 3/14. *Production of antirust adhesive tapes. Nagau Tatsuhiko, Vatanabe Ioity*, et al., 1980.
58. Belarus Patent 2668, C 09 J 7/02. *Adhesive tape.* L.S. Pinchuk, A.V. Makarevich, I.V. Tsaremko and E.I. Parkalova, 1999.
59. L. S. Pinchuk,V. A. Goldade, A. V. Makarevich. Structure, properties and use of active polymer films. *Proc. 12-th World Conf. on Packaging*, Warsaw, Poland, 2001.
60. U.S.S.R. Patent 1407024, C08 L 23/06, C 08 K 5/34/ 5/35. *Polymer composition for manufacture of polymer wrapping film.* L.S. Pinchuk, G.V. Rechits, E.I. Parkalova et al., 1988.
61. U.S.S.R. Patent 1815970, C 08 L 23/06. *A composition for wrapping film.* L.S. Pinchuk, G.V. Rechits, S.V. Bezrukov, et al., 1992.
62. Belarus Patent 1813, C 23 F 11/00. *A method of protecting ferrous and non-ferrous metals against corrosion.* A.V. Makarevich, I.V. Tsarenko, L.S. Pinchuk, et. al., 1997.
63. Belarus Patent 394, C 08 J 3/18. *A method of producing composite articles on the base of thermoplastic polyamides or polyolefines.* V.A. Goldade, A.S. Neverov and L.S. Pinchuk, 1995.
64. D.D.R. Patent 259967, C 08 L 23/06, B 65 D 81/26. *Composition of an antirust element V.A. Goldade.* A.S. Neverov and L.S. Pinchuk, 1988.
65. D.D.R. Patent 263417, C 08 L 23/06, C 08 K 9/00, 5/16. *A polymer composition.* V.A. Goldade, Y.M. Zolotovitsky, A.S. Neverov, et al., 1989.
66. U.K. Patent 2187466, C 08 L 23/06, C 08 K 9/00. *Anticorrosive material.* V.A. Goldade, Y.M. Zolotovitsky and A.S. Neverov, 1986.
67. U.S.S.R. Patent 504663, B29D 7/00. *A method of producing films from plasticized thermoplastics.* V.A. Belyi, A.S. Mikhnevich, A.S. Neverov and L.S. Pinchuk, 1976.
68. Indian Patent 1207125, C 08 J 5/18, B 29 D 7/02. *Process for producing tubular inhibited film.* I.M. Vertyachikh, Yu.I. Voronezhtsev, V.A. Goldade, et al., 1989.
69. U.K. Patent 2188274, B 29 C 55/28, C 08 J 5/18, C 08 L 23/06. *Process for producing tubular inhibited polyethylene film.* I.M. Vertyachikh, Yu.I. Voronezhtsev, V.A. Goldade, et al., 1987.

70. D.D.R. Patent 265290, C 08 J 5/18, B 29 D 7/02. *A method of producing hose inhibited film.* I.M. Vertyachikh, Yu.I. Voronezhtsev, V.A. Goldade, et al., 1989.
71. Switzerland Patent 670216, B 29 D 23/22. *A method of manufacturing inhibited polyethylene film.* I.M. Vertyachikh, Yu. I,. Voronezhtsev, V.A. Goldade, et al., 1989.
72. V.A. Goldade, A.S. Neverov, and L.S. Pinchuk. *Low-Modular Composite Materials Based on Thermoplastics*, Minsk, Nauka i Tekhnika, 1984.
73. U.S.S.R. Patent 1669929, C 08 J 5/18. *Polymer antirust film and methods of production.* L.S. pinchuk. V.A. Goldade, I.M. Vertyachikh, et al., 1991.
74. U.S. Patent 5028479, B32B 5/14, 3/26. *Polymer anticorrosion film.* L.S. Pinchuk, V.A. Goldade, I.M. Vertyachikh, et al., 1991.
75. U.S.S.R. Patent 1721968, B 29 C 47/20, 47/06. *A device for production of hose polymer anticorrosion film.* S.Ya. Liberman, S.V. Bezrukov, G.A. Goncharov et al., 1991.
76. U.S.S.R. Certificate on Industrial Prototype 25535. *A device for inhibiting polymer films*, 1987.
77. U.S.S.R. Patent 1669745, B 29 C 47/20, B 29 L 23/00. *Extrusion head for production of hose polymer anticorrosive film.* S.Ya. Liberman, I.M. Vertyachikh, V.A. Goldade et al., 1991.
78. U.S.S.R. Patent 1723808, C 08 J 5/18. *A method of manufacturing polymer anticorrosion film.* L.S. Pinchuk, S.V. Bezrukov, I.M. Vertyachikh et al., 1991.
79. U.S.S.R. Patent 1510324, C 08 J 5/18, C 08 L 23/06, 1/00, C 08 K 5/09. *A method of producing wrapping polymer film.* L.S. Pinchuk, E.I. Parkalova, G.V. Rechits et al., 1989.
80. U.S.S.R. Patent 1762548, C 09 J 7/02. *Method of production of polymer anticorrosion film.* S.V. Bezrukov, Yu.I. Voronezhtsev, V.A. Goldade et al., 1992.
81. Belarus Patent 1617, B 29 C47/20, 47/26. *Extrusion head for production of hose polymer materials.* L.S. Pinchuk, A.V. Makarevich, V.A. Goldade and Yu.V. Gromyko, 1997.
82. U.S. Patent 5139700, C 23 F 11/13. *Vapor-phase corrosion inhibitor material.* B.A. Miksic, N. Oaks and M. Tarvin, 1992.
83. U.S.S.R. Patent 974027, F 16 L 59/10. *Laminated film coating.* Yu.I. Blinov, V.M. Shtanko and B.P. Pashenin, 1979.
84. U.S. Patent 4290912, C 23 F 11/02, 11/12, 11/18. *Volatile corrosion inhibiting article.* F.P. Boerwinkle and D.A. Kubik, 1981.
85. V.N. Kestelman. *Physical Methods of Modification Polymer Materials.* Moscow, Khimiya, 1980.
86. G.M. Sessler (ed.) *Electrets*, Berlin, Springer-Verlag, 1987.
87. V.A. Goldade, and L.S. Pinchuk *Electret Plastics: Physics and Materials Science.* Minsk, Nauka i Tekhnika, 1987.
88. V.N Kestelman, L.S. Pinchuk, and V.A. Goldade. *Electrets in Engineering: Fundamentals and Applications.* Boston-Dordrecht-London, Kluwer Academic, 2000.
89. U.S.S.R. Patent 870192, B 29 B. *A method of manufacturing films and sheets of polymer materials.* V.A. Goldade, A.S. Neverov and L.S. Pinchuk, 1982.
90. U.S.S.R. Patent 899357, B 29 D. *A method for production of laminated metal-polymer materials.* V.A. Belyi, V.A. Goldade, A.S. Neverov and L.S. Pinchuk, 1981.

91. I.M. Vertyachikh, V.A. Goldade, A.S. Neverov and L.S. Pinchuk. *Effect of electrical field of a polymer electret on vapor sorption of organic solvent, Vysokomolec. Soedinenia*, 1982, Vol. B24, No. 9, pp. 683–687.
92. U.S.S.R. Patent 1762538, C 08 J 3/28, 7/02. *A method of producing polymer anticorrosion film.* V.V. Snezhkov, S.V. Bezrukov, Yu.I. Voronezhtsev et al., 1992.
93. U.S.S.R. Patent 1826498, C 08 J 5/18, B 29 C 47/06. *A method of manufacturing polymer anticorrosion film and a device for implementation.* S.Ya. Liberman, G.A. Goncharov, G.V. Rechits et al., 1992.
94. U.S.S.R. Patent 1300773, B 29 C 47/20, 47/06, 47/78. *Extrusion head for production of a hose polymer film.* S.Ya. Liberman, V.A. Goldade, L.S. Pinchuk, et al., 1986.
95. U.S.S.R. Patent 1822068, B 29 C 47/20, 47/06, 47/78. *Extrusion head for production of hose polymer film.* S.Ya. Liberman, I.M. Vertyachikh, V.A. Goldade, et al., 1992.
96. A.P. Merkulov. *Vortical Effect and its Application in Engineering.* Moscow, Mashinostroenie, 1969.
97. U.S.S.R. Patent 1352903, C 08 J 5/18, N 08 L 23/06. *A method of producing polymer film.* A.S. Neverov, A.M. Prishchepov, V.P. Rusov and L.S. Pinchuk, 1987.
98. U.S.S.R. Patent 1669744, B 29 C 47/20, 47/06. *A production method for hose polymer anticorrosion film and an extrusion head for its implementation.* G.V. Rechits, V.A. Goldade, S.Ya. Liberman, et al., 1991.
99. U.S.S.R. Patent 1820592, B 29 C 47/20, 47/06. *Extrusion head for production of hose polymer anticorrosion film.* S.Ya. Liberman, S.V. Bezrukov, L.S. Pinchuk and G.V. Rechits, 1992.
100. U.S.S.R. Patent 1304295, B 29 C 47/20. *Extrusion head for production of hose polymer film.* S.Ya. Liberman, V.A. Goldade, L.S. Pinchuk and G.V. Rechits, 1992.
101. Yu.S. Lipatov. *Polymer Reinforcement.* Tironto-Scarborough, ChemTec, 1994.
102. L.V. Zaborskaya, L.S. Golubova, Ya.M. Zolotovitsk, and O.R. Yurkevich. The effect of polymer carrier on the efficiency of volatile corrosion inhibitors. *Protection of Metals*, 1985, Vol. 21, No. 2, pp. 301–305.
103. P.V. Kozlov, and S.P. Papkov. *Physico-Chemical Bases of Polymer Plastification.* Moscow, Khimiya, 1982.
104. Ya.M. Zolotovitsky, A.S. Neverov, L.S. Pinchuk and G.V. Rechits. Anticorrosion inhibited materials based on plasticized polymers. *Protection of Metals*, 1985, Vol. 21, No. 2, pp. 244–248.
105. Ya.M. Zolotovitsky, V.A. Goldade, A.S. Neverov and V.S. Uss. Preservative materials based on inhibited polyethylene. *Protection of Metals*, 1980, Vol.16, No. 5, p. 630-633.
106. V.A. Belyi, V.A. Goldade, Ya.M. Zolotovitsky, A.S. Neverov and L.S. Pinchuk. Inhibited polymer materials for protection of metals from corrosion. *Proc. USSR AS*, 1982, Vol. 267, No. 5, pp. 1163–1166.
107. A.S. Neverov. On polyethylene plastification by coolants. *Vysocomolecul. Soedinenia*, Vol. B31, No. 2, pp. 141–145.
108. Ya.M. Zolotovitsky, A.M. Dorfman and A.M. Golubova. Desorption of volatile corrosion inhibitors from the polymer matrix. *Protection of Metals*, 1992, Vol.28, No. 1, pp. 151–156.

109. A.M. Dorfman, Ya.M. Zolotovitsky, A.M. Lyakhovich, et al. Desorption of volatile corrosion inhibitor from polyethylene. *Protection of Metals*, 2001, Vol. 37, No.1, pp. 14–18.
110. A.E. Chalykh. *Diffusion in Polymer Systems.* Moscow, Khimiya, 1987.
111. Ya.M. Zolotovitsky, L.S. Pinchuk, G.V. Rechits, et al. Plasticized polymer materials containing a complex of corrosion inhibitors. *Protection of Metals*, 1987, Vol. 23, No.1, pp. 93–97
112. I.L. Rozenfeld and V.P. Persiantseva. *Inhibitors of Atmospheric Corrosion.* Moscow, Nauka, 1985.
113. N.I. Domantsevich, Ya.M. Zolotovitsky, O.N. Kopyl and G.V. Rechits. The effect of corrosion inhibitors on penetration of vapor and oxygen in LDPE films. *Plastic Masses*, 1987, No. 10, pp. 32–33.
114. Ya. Zolotovitsky, and N. Domantsevich. Anticorrosion polymer films. *Packages and Packaging*, 1992, No. 6, pp. 4–5.
115. S.A. Reitlinger. *Permeability of Polymer Materials.* Moscow, Khimiya, 1974.
116. L. Nielsen. The penetration of filled polymers. *J. Macromolec. Sci.*, 1967, Vol. 1A, pp. 929–942.
117. A.V. Neverov. Antiplastification in polyethylene-mineral oil systems. *Trans. Belarus AS, Ser. Phys.-Techn. Sci.*, 1978, No. 4, pp. 70–73.
118. V.A. Goldade, Ya.M. Zolotovitsky, A.S. Neverov, and L.S. Pinchuk. Estimation of protective efficiency of inhibited film materials by the polarization resistance method. *Protection of Metals*, 1982, Vol. 18, No.6, pp. 946–949.
119. I.I. Antropov, M.A. Gerasimenko, and Yu.S. Gerasimenko. Determination of corrosion rate and inhibitor efficiency by the polarization resistance method. *Protection of Metals*, 1966, Vol. 2, No. 2, pp. 115–121.
120. L.I. Antropov, V.M. Babenkov, E.A. Budnitskaya, et al. Corrosion rate meter P5035. *Protection of Metals*, 1976, Vol. 12, No. 2, pp. 234–238.
121. K. Nakamoto. *Infrared and Ramon Spectra of Inorganic and Coordination Compounds.* New York-Chichester-Toronto, John Wiley & Sons, 1986.
122. A.V. Kiselev and V.I. Lyguin. *IR Spectra of Surface Compounds and Adsorbed Matters.* Moscow, Nauka, 1972.
123. G. Vos, J.G. Haasnoot and W.C Groenereld. Complexes of 1,2,4-triazoles. Part XVI. *Z. Naturforsh*, 1981, Vol. 32 B, No. 7, pp. 802–808.
124. R.D. Holm and P.L. Donnelly. Spectral studies of metal ion interaction with tetrazole and tetrazolate anion. *J. Inorg. Nucl. Chem.* 1966, Vol. 28, pp. 1887–1894.
125. M.M. Sokolova, V.V. Melnikov, V.A. Ostrovsky, et al. Oscillation spectra and structure of tetrazole derivatives. *J. Gen. Chem.*, 1975, No. 8, pp. 1744–1749.
126. N.I. Sushko, N.A. Matveeva, N.I. Makarevich, et al. The analysis of oscillation spectrum of tetrazole. *J. Appl. Spectrosc.*, 1990, Vol. 53, No. 2, pp. 323–327.
127. O.A. Ivashkevich, P.N. Gaponik, N.I. Sushko and V.A. Krasitsky. The analysis of oscillation spectra of tetrazole salts. *J. Appl. Spectrosc.*, 1994, Vol. 61, No. 1–2, pp. 45–49.
128. L.L. Garber, L.L. Sims, and C.H. Brubaker A vibrational analysis of the tetrasolate ion and preparation of bis(tetrazolato)copper (II). *J. Am. Chem. Soc*, 1968, Vol. 90, pp. 2518–2523.
129. V.I. Nefedov. *X-ray Electronic Spectroscopy of Chemical Compounds.* Moscow,. Khimiya, 1985.

130. V.A. Ostrovsky, G.B. Yerusalimsky and G.B. Scherbinin. Investigation of five-membered nitrogen-containing henetrocycles by the methods of quantum chemistry. II. Structure and aromatic properties of azoles. *J. Inorg. Chem.*, 1995, Vol. 31, No. 9, pp. 1422–1431.
131. I.V. Tsarenko. Adsorptive tetrazole-containing films on steel surface. I. Molecular structure. *Surface, Physics, Chemistry, Mechanics*, 1995, No. 10, pp. 35–43.
132. I.V. Tsarenko and A.V. Makarevich. Anticorrosive metal protection by adsorption films containing tetrazole. *Application of Surface and Interface Analysis: Proc. of 6th European Conf. Montreux, Switzerland*, 1995, pp. 224–227.
133. I.V. Tsarenko and A.V. Makarevich. A structural model of polymer complex with protective property formed on corrosive liquid/steel interface. *Polymer-Solid Interfaces: From Model to Real Systems: Proc. of the 2-nd Int. Conf.*, Namur, Belgium, 1996, pp. 467–482.
134. Yu.N. Kukushkin. *Chemistry of Coordination Compounds.* Moscow, Vysshaya Shkola, 1985.
135. I.I. Plyusnina. *Infrared Spectra of Minerals.* Moscow, Publ. of Moscow Univ., 1977. 175 pp.
136. I.V. Tsarenko, S.A. Chizhik and A.V. Makarevich. Morphology of protective films formed on steel in inhibited by tetrazole aqua media. *Protection of Metals*, 1999, Vol. 35, No. 1, pp. 1–4.
137. I.V. Tsarenko. Adsorptive tetrazole-containing films on steel surface. II. Structural features. *Surface: Physics, Chemistry, Mechanics*, 1995, No. 10, pp. 44–50.
138. L.I. Antropov. The formal theory of the effect of organic corrosion inhibitors. *Protection of Metals*,1977, Vol. 13, No. 4, pp. 387–399.
139. ISO 4592:1992. Plastics – Films and Sheeting – Determination of length and width.
140. ISO 4593:1993 Plastics – Film and Sheeting – Determination of thickness by mechanical scanning.
141. ASTM F 2251-03. Standard Test Method for Thickness Measurement of Flexible Packaging Material.
142. Russian State Standard 17035-86. Plastics. Methods of Determining Thickness of Films and Sheets.
143. Russian State Standard 25951-83. Polyethylene Heat-Shrinkage Film. Technical Specifications.
144. ISO 517-3:1995. Plastics – Determination of Tensile Properties – Part 3: Test Conditions for Films and Sheets.
145. ASTMD 882. Standard Test Method for Tensile Properties of Thin Plastic Sheeting.
146. Russian State Standard 14236-81. Polymer Films. Tensile Testing Method.
147. ISO 6383-1:1983. Plastics – Film and Sheeting – Determination of Tear Resistance – Part 1: Trouser Tear Method.
148. ISO 6383-2:1983. Plastics – Film and Sheeting – Determination of Tear Resistance – Part 2: Elmendorf Method.
149. ASTMD 1922-03. Standard Test Method for Propagation Tear Resistance of Plastic Film and Thin Sheeting by Pendulum Method.
150. Russian State Standard 26128-84. Polymeric Films. Method of determining Resistance to Tearing.

151. ISO 7765-2:1994. Plastics Film and Sheeting – Determination of Impact Resistance by the Free-Falling Dart Method – Part 2: Instrumented Puncture Test.
152. I. Gyenes. *Titration in non-aqueous media.* Budapest, Akademiai Kiado, 1967.
153. S.V. Bezrukov. Polymer film indicators of amine and corrosion inhibitors content in air. *J. Analytical Chem.* 1991, Vol. 46, No. 7, pp. 1348–1351.
154. Russian Patent 2050028, G 21 F 9/16. Method of Determination of Amine-Based Corrosion Inhibitors in Air. S.V. Bezrukov, E.A. Myshakova and L.S. Pinchuk, 1995.
155. S. Stoff. FachPack'91 bietet rundes programm: Keine wunsche offen. *Industrial – Anzeiger*, 1991, Vol. 87/88, pp. 30–32.
156. Vatababe Kadzuo. Antirusting of metals by tight packing. *Metals and Technology*, 1996, Vol. 66, No. 6, pp. 81–86.
157. A.A. Portyanko. *Preservation and Packaging of Engineering Goods.* Moscow, Mashinostroenie, 1972.
158. Russian State Standard 12302-83. Bags made of polymeric materials. General Specifications.
159. U.S.S.R. Patent 1202205, B 65 B 9/02. A method of packing articles in hot-tack film. V.A. Goldade, Ya.M. Zolotovitsky, L.S. Pinchuk and G.V. Rechits, 1985.
160. U.S.S.R. Patent 1253880, B 65 B 9/02. Packaging machine for articles packing in a polymer film. S.Ya. Liberman, V.A. Goldade, G.A. Goncharov, et al. 1986.
161. D.C. Moles, J.H. Briston. *Polymer Technology*, 2nd edn. New York, Chemical Publ. , 1979.
162. M.G. Solomenko, V.L. Shreder and V.N. Krivoshei. *Containers of Polymer Materials.* Ref. Book, Moscow, Khimiya, 1990.
163. Ship products has coil packing all wrapped up. *Anticorrosion Methods and Materials*, 1996, Vol. 43, No. 5, p. 36.
164. Japan Patent 62262309, H 01 B 7/28. Separating tapes for cables insulated with crosslinked polyethylene. Y. Ijiri and K. Mio, 1987.
165. U.S.S.R. Patent 947027, F 16 L 59/10. Laminated film coating. Yu.I. Blinov, V.M. Shtanko and B.P. Pashenin, 1982.
166. A.M. Dodonov and Ya.G. Muravin. Modern trends in consumer packaging. *Modern Packing*, 1992, No. 6, pp. 34–36.
167. L.S. Pinchuk. *Investigation and development of new engineering materials for sealing systems.* Dr. Sci. thesis, Minsk, 1981.
168. U.S.S.R. Patent 761 537. *A composition for sealing materials.* L.S. Pinchuk, K.M. Gilman, V.A. Goldade, et al., 1980.

3 Inhibited Coatings

Although polymer coatings have for a long time been used in anticorrosion techniques, the mechanisms of their protective effect have not been understood in sufficient depth so far. This lack of knowledge is explained by the high level of intricacy of the metal substrate-coating-medium junction as an electrochemical system and by the abundance of the phenomena that affect processes running in this system.

For a long time there existed a primitive representation of the corrosion inhibition mechanism by coatings as a physical barrier isolating a metal substrate from the hostile environment. It was believed that neither metal origin nor its chemical properties affected the corrosion kinetics of the substrate under a coating. Later on it has been found out that corrosion of metal substrates starts long before damage and peeling of the substrate surface. Besides, it has been established that the same coating protected alloyed steels (Cu, Cr, Ni) for longer periods than usual carbon steels. The period of protection turned out to be dependent on the finishing quality of the substrate surface. Passivation of the latter under the coating makes the protection last still longer. It was acknowledged finally that the polymer coating should be considered as a multifunctional element of a specific electrochemical system and that corrosion grows under the polymer film adhering to the substrate electrode.

The transfer processes of electrolyte ions in this system are not considered as macro-kinetic but at the level of molecular diffusion. The corrosion rate of the substrate depends largely on permeation of oxygen and aqua electrolytes through the coating, i.e. on its structure, charge state, swelling in electrolytes and so on. Adhesion of the coating to the substrate exerts an effect on the kinetics of electrode reactions thus bounding the access of reacting particles to active centers of the substrate and their removal from the reactive space.

Electrochemical processes of corrosion under the polymer coating are influenced by passivation and polarizing control of the metal substrate by active components of the coating. These components are passivating primers, metal fillers and conducting polymer binders (organic metals) of the coating. Chemical interactions between the polymer functional groups and active centers on the substrate surface lessen the intensity of the electrode processes. CI as coating components impose a specific action on the protective action too. These coatings alone are the subject of examination of the current chapter.

A huge body of theoretical and experimental data awaiting generalization has been accumulated in the recent two decades in materials science of polymer inhibited coatings. In this chapter we will present the investigation results of the mechanisms of structural formation and protective effect of polymer coatings containing CI. Apart from classifications of inhibited polymer coatings, impregnation technologies of CI have been described and their effect on adhesion and performances of the coatings has been explained. Technological and structural features of inhibited paint and varnish coatings, powder coverings, adhesive tapes, antifouling, antimicrobial polymeric coatings, etc. have been analyzed where the priority is given to modern trends in the materials science of polymer means of protecting against corrosion.

3.1 Polymer Coatings as Elements of Anticorrosion Systems

The viewpoint that a metal substrate can be fully insulated from the ambient corrosion medium using a polymer coating is erroneous. Physical-chemical properties of polymer materials available today as well as peculiarities of contemporary technologies used to form their coatings are incapable of averting totally the interaction between the metal, substrate and the corrosive medium. A continuous exchange of active particles and charge carriers is observed within the metal substrate-polymer coating-electrolyte system. A persuasive confirmation of this is that a standard electrode potential is unavoidably set up on a metal substrate protected by a coating shortly after the sample has been immersed in the electrolyte. Consequently, the polymer coating is not an obstacle for diffusion of water, oxygen and ions participating in electrochemical reactions. Therefore, the metal substrate-coating-electrolyte system can be characterized by the same parameters as a common electrochemical system with allowance for specific limitations imposed by the barrier parameters of the coatings. These parameters are stationary potentials, polarization characteristics, ohmic resistance, capacity, diffusion parameters and the ability for self-passivation.

3.1.1 Electrode Potentials

Electrode potentials of metallic substrates are more positive than the potentials of analogous electrodes without the polymer coatings. This is because water and oxygen inducing the cathodic process in neutral electrolytes penetrate through the coating much more easily than the metal substrate ions are removed. Since the permeability of the coatings for ions is very low, they hamper the anodic reaction of metal ionization.

In the case of the slightest defects in the coating corrosion elements appear on the metal substrate. As a result, areas of the substrate insulated by the coating from the corrosive medium function as the cathode and the naked areas of the substrate near the coating defects serve as the anode and acquire a more negative potential than the cathodic areas. The electrode potential difference between the anodic and cathodic areas of the substrate is a function of the ohmic potential drop on the cathode and can reach 200–300 mV. This is commensurate with the contact potential difference assumed dangerous for metal structures (0.25 V) [1].

Let us consider the mechanisms of electrode potentials settling on metallic substrates, illustrated in Fig. 3.1.

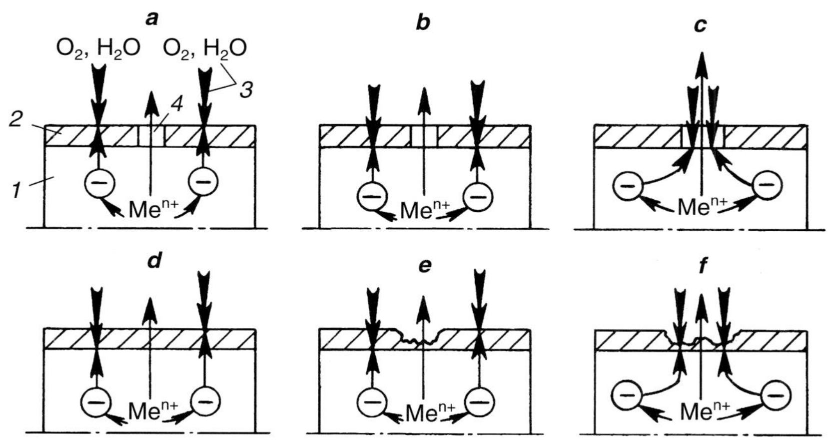

Fig. 3.1. Schemes of electrode potential settling mechanisms on metal substrates under polymer coatings [1]. (1) substrate, (2) coating, (3) corrosive medium, (4) and (5) coating defects (a pore or local thinning). See explanations in the text

(a) The polymer coating displays electron conductivity (organic metals), hence the cathodic process localizes at the electrolyte-coating interface. The anodic process concentrates in the coating defects, e.g. pores. Thanks to the large surface of the cathode, the current density on the anode (in the pore) is large enough. Therefore, the substrate polarizes and its stationary potential shifts to the side of positive values.

(b) The coating has no electron conductivity but allows through both oxygen and water easily. The cathodic process runs under the coating at its interface with the substrate while the anodic process is observed in the pores. The substrate stationary potential shifts to the positive side for similar reasons to the previous case.

(c) A larger portion of the substrate surface remains passive and is not involved in electrode processes. Both anodic and cathodic processes concentrate in the pores. Metal ions transfer into the electrolyte at a similar velocity to that with which oxygen and water approach the substrate. The difference in velocity of the two processes depends on the shift of the substrate potential.

(d) The coating has not got any pores. The reagents are transferred via diffusion in the polymer material. The cathodic process may run just as on the coating surface under it. The stationary potential shift value of the substrate metal is conditioned by the difference in diffusion velocities of the reagents. Penetrability of polymer coatings for metal ions is, as a rule, much lower than for water and oxygen. This is why the substrate potential shifts to the positive side.

(e) The coating is uneven in thickness. The substrate is more positive under the coating areas of standard thickness in contrast to the areas of deficient thickness. Under the former areas the cathodic process starts, under the latter – the anodic one. As a result of anodic polarization the stationary potential of the substrate becomes more positive.

(f) The coating is uneven in thickness. On the areas of standard thickness the coating totally insulates the substrate from the corrosive medium. Both the cathodic and anodic processes localize in the areas of the deficient coating thickness. Anodic polarization of the substrate promotes the positive shift of its stationary potential.

Filling of the polymer binder augments electrochemical inhomogeneity of the system. Uneven distribution of filler particles within the binder and their removal from the coating surface layer under service factors stipulates inhomogeneous distribution of the electrode potential over the substrate. Note that defects in filled coatings are more dangerous with respect to corrosion than those in non-filled ones.

3.1.2 Polarization Characteristics

Polarization characteristics of polymer coatings, i.e. polarizability of electrodes of the bearing coatings, are critical for protection of the coatings. The higher its polarizability, the better the coating preserves the electrode against corrosion. The most informative polarizing pattern of coatings is their polarization resistance. The classical method of recording polarization curves makes it possible to evaluate the protective ability of the coating. Named curves carry certain information on barrier properties and the passivating capability of coatings.

In Fig. 3.2, polarization curves are presented for HDPE coatings on carbon steel St.3 substrates recorded at 20°C. The coatings were formed by a vibrovortex method (see 3.3.1) under three temperature regimes: 1) $T = 170°C$ (preheating time of the substrate $\tau_1 = 30\,\text{min}$, time of coating aftermelting $\tau_2 = 5\,\text{min}$); 2) $T = 210°C$ ($\tau_1 = 20\,\text{min}$, $\tau_2 = 5\,\text{min}$); 3) $T = 260°C$ ($\tau_1 = 15\,\text{min}$, $\tau_2 = 5\,\text{min}$). Coating thickness was $150 \pm 10\,\mu\text{m}$. The polarization curves were recorded making use of a potentiostat and a silver chloride reference electrode in a 0.5 N solution of H_2SO_4.

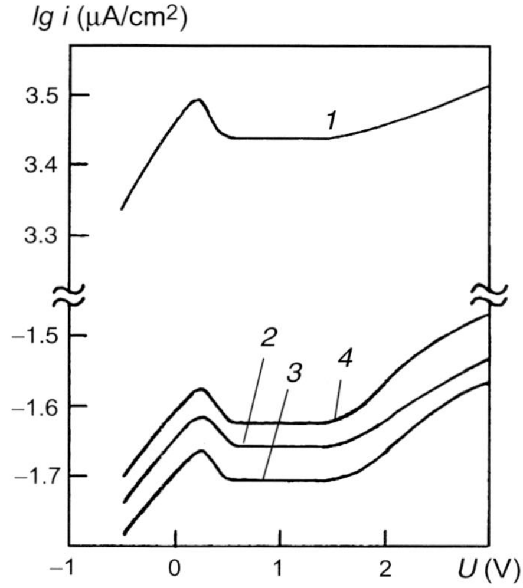

Fig. 3.2. Anodic polarization curves of steel. (1) unprotected electrode; (2), (3) and (4) electrode with HDPE coat formed at 170°C, 210°C and 260°C, respectively

Judging from the location of the polarization curves in Fig. 3.2, it is obvious that the coating significantly impedes electrochemical processes on the substrate electrodes. The similarity between the curves for the unprotected electrode and those with the coatings proves the hypothesis that the coating is incapable of ceasing metal solution totally. Apparently, the protecting mechanism of the coatings under study corresponds to scheme *f* in Fig. 3.1.

The formation temperatures of the coatings were chosen so as to correlate their adhesive and protecting action. A temperature dependence of adhesive strength of LDPE coatings is illustrated in Fig. 3.3 [2]. It is evident that adhesion is maximal for the coating formed at 210°C.

Adhesive strength of the coatings applied at 170°C and 260°C turned out to be identical. From the analysis of Fig. 3.2, it follows that the coatings with maximum adhesion display maximum protective ability too. Nevertheless, the protective properties of these coatings differ. The reasons for this difference may be [3]:

- the presence of passivating oxide layers on the steel substrate surface resulted from coating formation;
- the different permeability values of the coatings formed under different time and temperature regimes.

The potentiostatic method has been used to estimate the role of oxide layers for steel electrodes oxidized preliminary at time-temperature regimes similar to these of the coating formation. The following current density values in the passive region have been recorded: original electrode – 9.35–9.50 mA/cm^2;

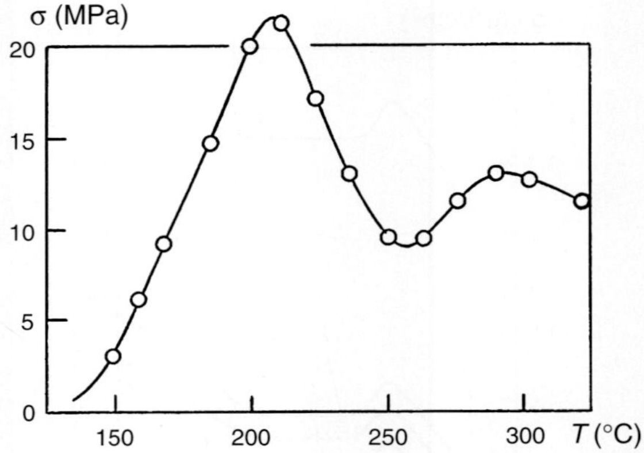

Fig. 3.3. Strength of HDPE – steel adhesive joints (normal detachment) versus their formation temperature

electrode at $T = 170°$ C – 7.75–8.20; at $T = 210°$C – 7.55–8.15; at $T = 260°$C – 8.40–9.00 mA/cm^2. The indicated differences are attributed to the structure of oxide layers forming on the substrate. Phase $\alpha - Fe_2O_3$ appears at $T > 130°$ C on the oxide film surface, being previously $\gamma - Fe_2O_3$. At $T > 190°$C, $\alpha - Fe_2O_3$ crystals start to grow and phase $\gamma - Fe_2O_3$ transfers starting from the substrate into Fe_3O_4. This process intensifies noticeably at $T > 210°$C [4].

The effect of the HDPE crystalline structure on the permeability of the coatings has been studied by X-ray structural analysis. The coatings formed at 170° C, 210°C and 260°C are characterized by a roentgen crystallinity degree 67%, 75% and 65%, correspondingly. With increasing crystallinity degree, the permeability of the coatings diminishes.

Similar results have been obtained for pentaplast coatings [5]. It is seen in Fig. 3.4, that minimum values of polarizing currents of the coated electrodes correspond to the coating formation regime of 240–255°C. Within the same temperature interval the coatings show maximum adhesive strength and the coating peels off around a hand-made puncture in the coating at the lowest rate.

The polarization current values in Figs. 3.2 and 3.4 match up with the electric resistance of the coatings.

3.1.3 Electrical Resistance

The electrical resistance of coatings is composed of the active (ohmic) and reactive (polarization) resistances. The latter is a resistance to alternating current through the capacity and inductance of the coating-substrate system. Ohmic resistance makes up an insignificant portion of the total electrical resistance of polymer coatings and characterizes electrolyte diffusion in the

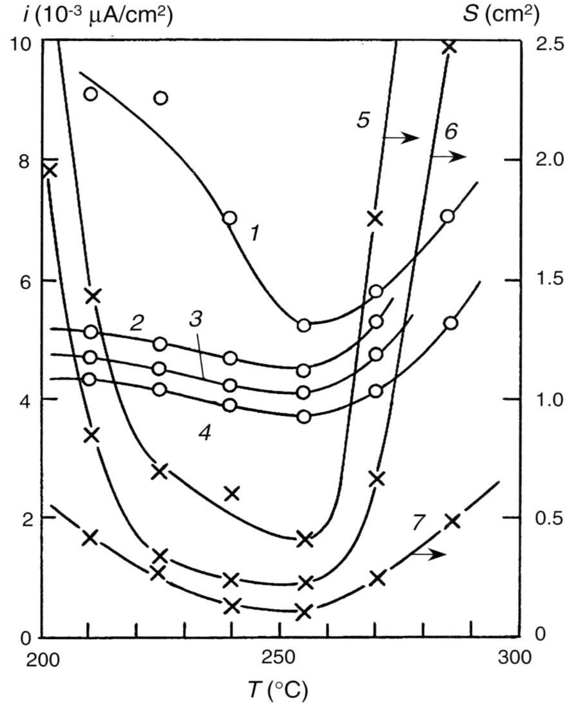

Fig. 3.4. Anodic current density (i, 1–4) in electrodes with pentaplast coatings (0.5 N solution of H_2SO_4, 20°C) and delamination area (S, 5 – 7) of the coatings round a puncture (pregnant solution of NaCl, 20°C) versus coating formation temperature (T). Curves (1) 4 correspond to polarization potentials 4.0; 3.0; 2.0 and 0 V. Curves 7, 6 and (5) after 75, 125 and 150 hours of exposition

coating. It is the polarizing resistance that largely conditions the protective properties of the coatings.

Alternating currents of variable frequency are commonly used to measure the total electrical resistance R_t of polymer coatings. On increasing the frequency to 2–25 kHz the ohmic constituent R_Ω of the total resistance can be estimated accurately enough. At low frequencies (400–1000 Hz) the recorded resistance value characterizes the polarizing resistance R_p of the coating.

Frequency dependencies of the electrical resistance of pentaplast coatings (deposition temperature of 250°C, coating thickness of $150 \pm 10\,\mu$m) recorded during exposure of the samples in 30% solution of H_2SO_4 at 20°C [5] are illustrated in Fig. 3.5. It is evident that the resistance of the system grew during the experiment. The resistance dependence on the electrical signal frequency intensified and then dropped as the frequency increased. The described resistance variations are rather unexpected since diffusion of electrolyte ions in the

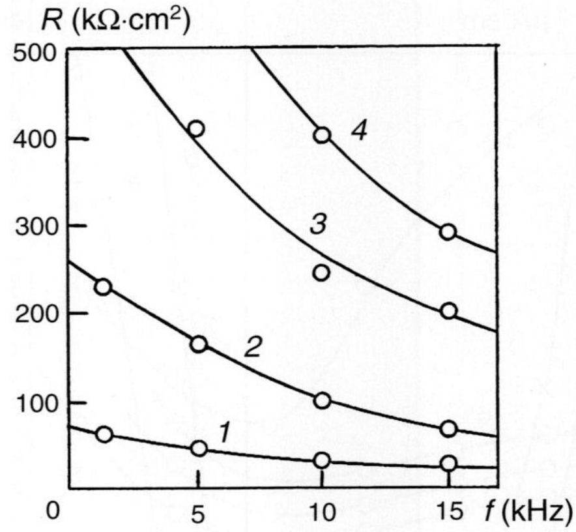

Fig. 3.5. Frequency dependencies of electrical resistance of the steel substrate – pentaplast coating – electrolyte system. (1) initial values; (2, 3 and 4) after 24, 48 and 72 hours of exposure in electrolyte

coating usually promotes a reduction of the ohmic resistance in the polymer film. The location of the curves in Fig. 3.5, proves that R_Ω (recorded mainly at $f < 1\,\text{kHz}$ frequencies) grows drastically during the experiment, much more than R_p (estimated reliably at $f \geq 15\,\text{kHz}$). The degree of crystallinity of pentaplast reduced after 75 hours of exposition in the acid from 30% to 20%. Consequently, the permeability of the coatings for the ions needed for the cathodic process increased. This fact gives grounds for the supposition that ions diffusing through the pentaplastic layer are reduced not only on the substrate surface but in the coating subsurface layer and at the interface with the substrate as well. Reduction of charged particles probably occurs with the participation of active centers (radicals, unsaturated links, polarization groups) of macromolecules bonded with the substrate via both adhesive and coordinative links. This is why the electrical resistance of the pentaplast surface layer increased at the interface with the substrate in the course of exposure in electrolyte.

Pentaplast is an atypical polymer, distinguished by its physical-chemical properties from the majority of thermoplastics used for anticorrosion coatings. This example is to show that the main function of polymer coatings in inhibiting electrochemical reactions on the substrate is only slightly governed by their initial ohmic resistance values. Factors that are far more significant are variations in the coating resistance during operation and the presence of active components actuating an additional resistance in the corrosive chain.

3.1.4 Capacity

The capacity of the coating-substrate system characterizes swelling (change in dielectric permittivity) of the coating in electrolytes and pore volume in the coating.

A polymer coating covering a metal substrate represents, in contact with an electrolyte, a capacitor in which the metal and electrolyte are the plates and the dielectric spacer is a polymer film adhesively bonded to the substrate. The capacitor value remains constant within sonic frequencies and is independent of the alternating electrical field frequency until the coating swells and its pores fill with electrolyte. The capacitor value diminishes as the coating swells and its pores are filled with electrolyte and become strongly dependent on the outer field frequency. By comparing the frequency dependencies of the coating-substrate system capacity in the original state to that during exposure in the electrolyte one can judge variations in the system as corrosion progresses on the substrate.

The frequency dependencies of a pentaplastic coating capacity (see electrical resistance values in Fig. 3.5) are presented in Fig. 3.6. After three days of exposure in the acid, the capacity of the system is seen to rise anomalously and then drop as the alternating field frequency grows [5]. Apparently, growth of the capacity can be attributed to the formation of a layer that we have already mentioned at the interface with the substrate whose electrical resistance is rather high. Such an improvement of the dielectric properties of the coating unavoidably leads to growth in the studied system capacity. Intensified frequency dependence of the capacity with extension of the exposure time in the acid is a result of liquid absorption by the coating.

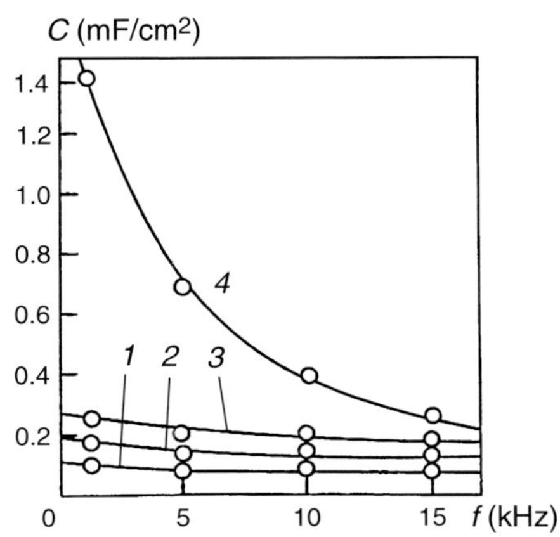

Fig. 3.6. Dependence of pentaplast coating – steel substrate system permittivity on alternating field frequency and exposure time in a 30% H_2SO_4 solution. (1) initial values; (2, 3 and 4) after 24, 48 and 72 hours of exposition

There is not any unambiguous dependence between the system capacity and corrosion rate of the substrate since generation of corrosion products can lead by itself to changes in the capacity [6]. The substrate-coating-electrolyte system capacity is considered an informative parameter for finding the volume of the electrolyte absorbed by the coating and determining its absorption mechanism.

R and C are varied by the capacitance-ohmic method. To estimate these parameters reliably it is necessary to choose the electrical circuit of the compensator unit correctly in the measuring instrument in the form of an alternating current bridge. This equivalent electrical circuit should adequately simulate the electrochemical scheme under study. The scheme is selected proceeding from the following considerations. The original samples represent an electrical capacitor with losses whose plates are the substrate and electrolyte, and a dielectric spacer that is the polymer coating. This capacitor can be represented by a parallel coupling of the electrical capacity with active resistance as the source of dielectric losses. In the case of a large amount of through pores in the coating and diffusive permeation of electrolyte to the substrate, the electrical circuit equivalent to the new state of the studied system will be a combination of capacitors and resistances coupled either in series or parallel.

The tangent of the angle of dielectric losses of the studied system is related to the capacity via the equation

$$\mathrm{tg}\delta = 1/2\pi fCR\ ,$$

where f is the alternating current frequency, and C and R are the capacitance and resistance of the system measured in a parallel circuit. The slight initial reduction of tgδ values with increasing f is a proof of growing dielectric losses due to a slowly settling polarization in the substrate. An abrupt drop of tg$\delta(f)$ dependence is evidence that the coating is highly permeable to the electrolyte.

3.1.5 Diffusion Characteristics

Diffusion characteristics of polymer coatings define to a large degree the metal substrate corrosion mechanism. The kinetics of the cathodic process on the substrate are dependent on the diffusion velocity of water, oxygen and electrolyte ions in the coating; that of the anodic process depends on the diffusion velocity of metal ions in the reverse direction. Sorption of vapors and moisture depends on the chemical nature and permolecular structure of the polymer. The main factors governing diffusion permeability of polymer films are the degree of macromolecular branching, their mobility and polarity of macromolecular fragments and diffusing molecules.

Vapor permeability parameters of films from linear polymers are given in Table 3.1

Table 3.1. Vapor permeability of polymer films

Polymer	Permeability for Vapors*, mg/(m^2·s)
Polyvinylidene chloride	27.7
Polyvinylidene chloride + Polyvinyl chloride (%):	36
92 + 8	33
80 + 20	83
60 + 40	300
40 + 60	392
10 + 90	
Polyvinylidene chloride + Polystyrene (%):	
60 + 40	395
50 + 50	530
Polyvinyl chloride	510
Polyvinyl chloride chlorinated	1020
Polyacrylonitrile	1060
Polyacrylonitrile + Polyvinyl acetate	3800
Polyacrylonitrile + Polystyrene (%):	
85 + 15	570
60 + 40	900
Polytetrafluoroethylene	27.7
Polychlorotrifluoroethylene	27.7
Polyethylene	100
Polyethylene + Polyvinyl acetate (94 + 6 %)	720
Polyethylene chlorinated (% Cl):	
33.3	725
46.6	350
64.0	463
Polystyrene	970
Polystyrene + Polyisobutylene (%):	
80 + 20	1390
50 + 50	1420
Polystyrene + Polybutadiene (90 + 10 %)	790
Natural rubber	1580
Synthetic isoprene	2770
Chlorinated natural rubber	265
Rubber hydrochloride	122
Gutta-percha	468
Gutta-percha hydrochloride	287

(continued)

Table 3.1. cont.

Polymer	Permeability for Vapors*, mg/(m^2·s)
Neoprene, grades:	
GH	1260
GG	1360
FR	2000
Neoprene chlorinated GN (61 % Cl)	300
Buna-rubber	1910
Thiokol (polysulfide) rubber	3850
Polyisobutylene	86
Polybutadiene	3930
Polyamide	1010
Polyester	2850
Polyetheramide	7600
Polymethyl methacrylate	3180
Polyethyl methacrylate	4380
Polybutyl methacrylate	4950
Polymethyl anilinoketone	6200

*Permeability is related to 50 μm thick films.

Model representations of the permeability of filled polymers have been put forward by L. Nielsen [7]. He suggests that the diffusing molecules in the polymer binder are compelled to bend around the filler particles, which lengthens their path. In addition, the presence of the filler diminishes the area occupied in a sample by the binder through which the diffusion stream passes. Furthermore, the models of diffusive permeation of media through filled polymer films have been elaborated to take account of the filling degree, shape, size and aggregation of particles, their orientation in the binder, properties of the surface layers at the binder-particle interface and other parameters [8].

Diffusive permeability of polymer coatings is affected considerably by electro-kinetic phenomena resulting from spontaneous electric polarization of the dielectric coating-conducting substance system. Electro-osmosis i.e. directed motion of ions and polar molecules along the coating micropores in response to the electromotive force leads to asymmetry of diffusive transfer of substances through the polymer coating. The reasons for spontaneous polarization of the coating-substrate system can be

- dissimilar potentials on the metal substrate areas under the coating, in a pore, or a region under the coating of different thickness;
- accumulation of ions of different polarity on the coating surface and at the interface with the substrate induced by diffusion transfer asymmetry;

- transformation of the coating structure during diffusion of media giving rise to layers that differ in their conductivity;
- heterogeneity of the polymer coating structure at the interface with the substrate and on the surface, which induces an inhomogeneous distribution of diffusing ions across the coating thickness;
- injection of charge carriers into the polymer coating from the metal substrate during thermal molding of the coatings from powder polymer materials.

There are examples supporting the effect of the named factors on the diffusion characteristics of the coatings.

Sorption of liquid media by polymer coatings is strongly influenced by the wetting and spreading processes of liquids over the coating surface. In this regard, capillary flow of water has been studied over the surface of polyethylene coatings on (low-carbon engineering) steel [9]. The coatings were formed from LDPE powder at $205 \pm 5°C$. Upon flashing, the surface was finished with a fluorolon matrix to reach an even surface roughness ($R_a = 0.5\,\mu m$).

The scheme of the experimental setup is presented in Fig. 3.7. The junction under study consists of discs 1 and the polymer coating 2 formed on the substrate 3. The substrate is made as a ring plate with a coating having a capillary puncture 4 in the center of its plane surface, which communicates with orifice 5 and clearance 6 of the junction. Orifice 5 is connected through tube 7 with the glass capillary 8 containing water (bidistillate).

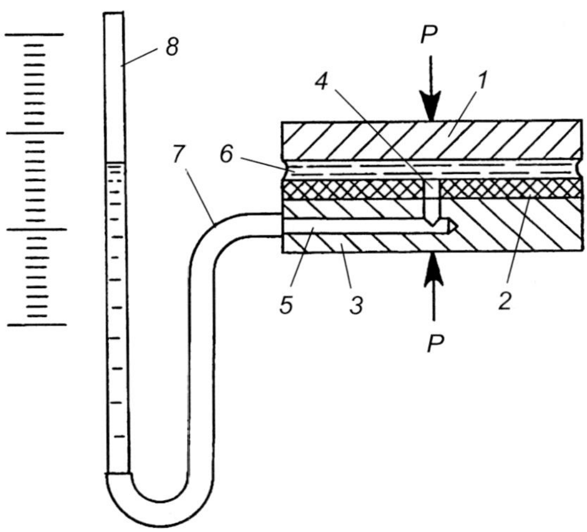

Fig. 3.7. A setup scheme for studying capillary flow of liquids over the polymer coatings surface

The initial water level h_0 in the capillary is set so as to form a convex meniscus in puncture 4 when disc 1 is offloaded. Once disc 1 is installed on the meniscus, water from capillary 8 flows to the junction under study and spreads under the capillary pressure effect over the surface under consideration. The new level h_t of water in capillary 8 and the height decrease $\Delta h = h_0 - h_t$ of the water column are recorded. The kinetic dependence Δh has the form of an exponent. The installation time of disc 1 before h_t measurement is taken is 5 min, corresponding to the exponent approach to a constant value.

The experimental results presented in Fig. 3.8 show that the substrate significantly affects the coating surface properties. The dependence $\Delta h(\delta)$ is distinguished by a high conjugation degree. Its correlation coefficient values are within 0.755 ± 0.102 at a significance level of 0.01.

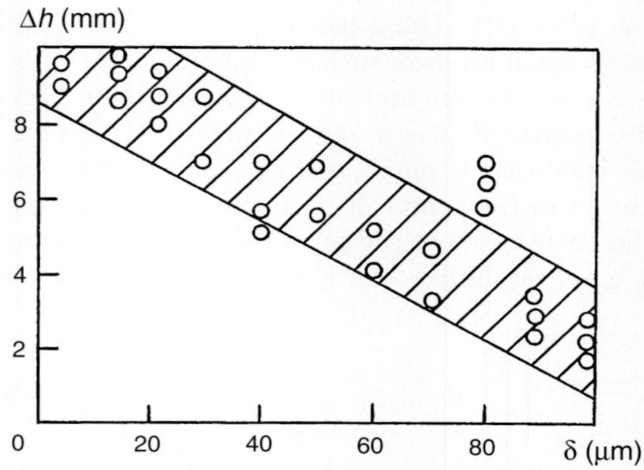

Fig. 3.8. Water level variation in a capillary as a function of polyethylene coating thickness over which the water spreads

The difference in coating properties can be attributed to the following reasons.

A. Dissimilar permolecular structure (consequently, different packing density of macromolecules and wetting parameters) of coating surface layers of different thicknesses. The surface energy of the substrate is known [10] to impose an influence on the formation of the permolecular structure of the coatings molded from thermoplastic melts. This influence is observed at a distance up to tens of μm from metal substrates. Beyond this distance, the influence of the substrate surface on the process of macromolecular ordering ceases. The thickness of the coatings studied (within 100 μm) complies with the range of the substrate action.

B. Electric polarization of the coatings during their finishing. The coating in a viscous-flow state is found in contact with the steel substrate and fluorolon matrix prior to setting, between which a potential difference exists $U = E_O - E_F$. Here E_O is the electrode potential of the steel in contact with the LDPE melt; E_F is the electrical potential caused by the charge of the matrix, which is close to zero. The thinner the coating, the stronger the polarizing field intensity and the higher the polarizing charge surface density in the set coating.

C. Uneven distribution of electrode potential over the substrate due to micropores and uneven coating thickness. This unevenness is more pronounced for thinner coatings. The following experiment has been carried out to prove the effect of electrical polarization of the coating on its diffusion characteristic [11]. 200 μm-thick polyvinyl butyral coatings sprayed by an electrical method onto an aluminum foil were exposed to a 5% solution of H_2SO_4. Adhesive strength was controlled by the peeling method, bearing in mind that adhesion in the experimental conditions was first of all conditioned by diffusive permeability of the coatings. Electrical polarization of the coatings was performed in the electric field of 20 kV/cm voltage at a temperature of 125°C.

The results of the experiments are shown in Fig. 3.9. The adhesive strength of the polarized coatings upon endurance in the electrolyte appears to be essentially higher in contrast to common ones. A short-term elevation of adhesion (for 24 hours) of the check coatings stored in air can be related to

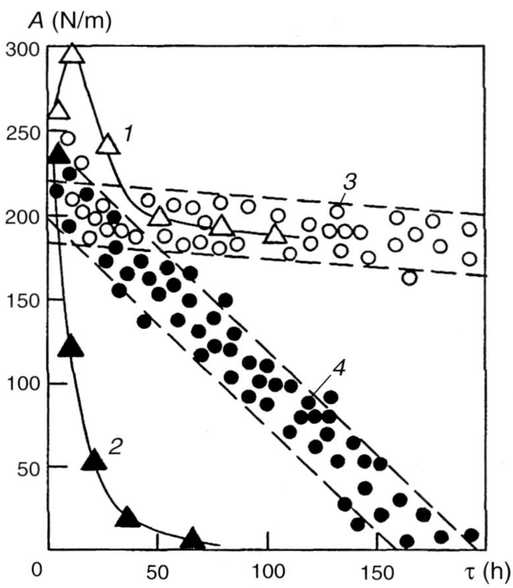

Fig. 3.9. Strength of adhesive PVB coating – Al substrate joints versus exposition time in air (1, 3) and in H_2SO_4 solution (2,4). 1 and (2) reference coatings, (3) and (4) polarized coatings

secondary crystallization of PVB. The exposure time of the coatings in a 5% solution of H_2SO_4 during which adhesive strength of the coatings fell twice as low makes up 8 hours for usual coatings and 75 hours for electret ones, in which the electric field is directed from the coating to the substrate. This is direct proof of the effect of electrical polarization of polymer coatings on their diffusion characteristics.

In conclusion it should be underlined that diffusion parameters of polymer films and coatings differ drastically. The diffusive permeability of polymer coatings is lower than the films acquire via adhesive interaction with the substrate. The diffusive parameters of the films and coatings equalize more or less only when adhesive bonds between the coating and substrate are broken. As has been mentioned in Sect. 1.2, rupture of adhesive bonds under the action of diffusion into the coating media depends on the degree of substrate polarization. The results cited in Fig. 1.8, prove unequivocally that peeling of the coatings can be reduced to a minimum by optimizing the electrical polarization regime of the substrates, so intrinsic barrier properties can be thus preserved.

3.1.6 Self-Passivation

The property of coatings to transfer metal substrates into a passive state (self-passivation) is realized during their formation or operation.

Substrate passivation during coating formation is achieved by treating the surface being covered with special technological fluids, i.e. passivators. The passivation operation precedes the application of the polymer layer to the substrate.

Passivators are inorganic substances possessing oxidative properties whose reaction products with metals form a passive film on the substrate surface, which shifts the corrosion potential of the substrate to the positive side by a few tens of volts. Like a depolarizer, the passivator generates a current on the anodic areas of the substrate of $i > i_{cr}$ density, where i_{cr} is the critical density of the passivation current. This means that the chemical composition of the passivating film on a metal substrate is the same whether the substrate is passivated by anodic polarization in an acid or is treated with solutions of chromates (CrO_4^{2-}), nitrates (NO_2^-), molybdates (MoO_4^{2-}), tungstates (WO_4^{2-}), ferrates (FeO_4^{2-}) or pertechnates (TcO_4^-).

Passivator ions acting as oxidizers are adsorbed on the substrate and reduced easily, thus enlarging the cathode surface area. An optimum passivator solution concentration should exceed some critical value and, the higher the passivator concentration, the easier it is adsorbed, and the smaller the anodic areas on the substrate will be. This promotes an increase in anodic polarization and total passivation of the substrate. If the passivator concentration is lower than some critical value, it initiates local corrosion of the substrate.

Self-passivation of metal substrates occurs in the course of application of a new generation of electronic polymer coatings called organic metals, described in Sect. 3.2.2.

Passivation of substrates during operation of coatings takes place when the passivators have been previously impregnated into the coatings as filler particles spread within a polymer binder or primer layer (contacting the substrate) applied from, e.g. an aqua solution of sodium silicate. Diffusion into the coating from the water environment dissolves the filler particles and carries the passivator ions to the substrate. As soon as the ions reach the substrate, they passivate the metal. The substrate transfers from the active to the passive state, even if the concentration of CrO_4^{2-} ions in the diffusive water is insignificant. In the case that the coating filling degree is consistent with the protective content of passivating ions in diffusive water, the electrode potential of the steel substrate shifts in aggressive electrolytes (0.1 N solution of Na_2SO_4) to the positive side (ennoblement) by 0.5–0.6 V [1].

It is expedient to fill the coating primer layer with passivator particles. The particles should be soluble enough in water to ensure the protective concentration of passive ions in the diffusion medium reaching the substrate. On the other hand, solution of filler particles impairs barrier properties as well as the adhesion of polymer coatings. At the same time, inadequate solution rate of the particles in the aggressive medium can lead to local corrosion of the substrate. Hence, the use of passivating primers requires decisions with respect to the passivator solubility and filling degree of the binder.

Only a few passivators are adequate for filling polymer primers [12]. For instance, red lead Pb_3O_4 with a structure of lead plumbate, which isolates PbO_4^{4-} ions, is considered to be the best filler for passivating primers. The solubility of zinc chromate in water is $2 \cdot 10^{-4}$ mole/l, which is sufficient for the creation of a protective concentration of CrO_4^{2-} ions ($>10^{-4}$ mole/l) on steel substrate surfaces under polymer coatings.

It should be noted in conclusion that there exist contradictory opinions about the chief factor that defines the protective properties of polymer coatings. A number of investigators consider adhesion to be the critical factor, others believe that the diffusion limitations that the coating brings into the corrosive system are critical, while others choose the electrical resistance of the coating, and so on. In our view, the best concept proceeds from the position that the property of polymer coatings for inhibition of corrosion of metal substrates is conditioned by the whole set of physical-chemical parameters of the coating-substrate system, which can be described by the following four characteristics [1]:

- Electrochemical and dielectric properties of the coatings
- Coating characteristics as a diffusion barrier for corrosion reagents and products

- The capability of coatings containing active fillers to passivate or protect the metal substrate via electrochemical mechanisms
- Mechanical and all above-cited adhesive properties of coatings.

These properties are interrelated and interdependent. For instance, raised permeability of the coatings inevitably results in lowered adhesion due to the propagation of corrosion processes. It is a matter of common experience that all service characteristics of polymer coatings unavoidably deteriorate in the course of exploitation. No perfect initial parameters can guarantee high protective ability of coatings. There is only one way out of this hopeless situation: aging of the coating should activate adaptation mechanisms to the working conditions. This means, first of all, that the coatings should contain active components that alter the coating-substrate system properties under the action of the corrosion system. On the agenda today is the development of smart coatings, which are able to compensate for the deterioration of the original parameters through respective changes in the structure and properties of the components of the electrochemical system.

3.2 Modern Trends in Coating Design

The purpose of polymer coatings as an element of the "coating – metal substrate – medium" electrochemical system is to avert or weaken the corrosive action of the ambient medium on the metal. This function is hampered by a number of factors, e.g. by impairing the initial physical-chemical properties of the coating during operation within the electrochemical system. Therefore, some compensation mechanism is needed to force the deteriorating properties to operate in coordination with aging and ensure anticorrosion protection of the substrate. This ideology is illustrated in Fig. 3.10.

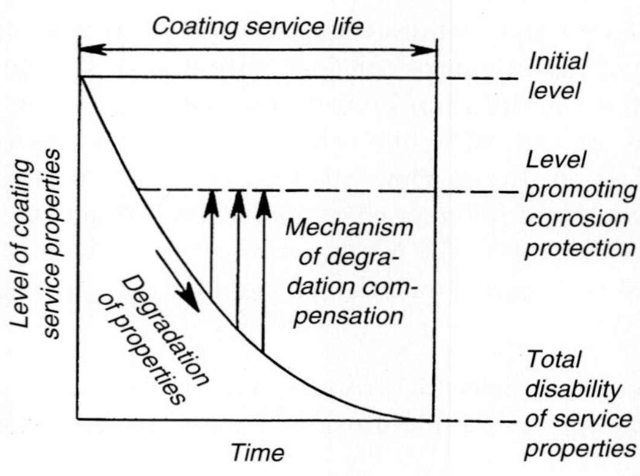

Fig. 3.10. Scheme of allowance for service characteristics of a polymer rustproof coating that are impaired during operation in the electrochemical system

As has been noted earlier, allowance for the properties lost by the coatings during operation can be realized by the introduction of active additives into the coating composition. These exert an effect of physical, chemical and biological origin on the electrochemical system that modifies the coating, medium or the metal substrate. Today's polymer anticorrosion coatings tend to acquire additional functional properties, such as electric field source, electrical charge carrier or regulator of electrode processes on the substrate. Three main types of polymer coatings, namely electret, conducting and inhibited, show this trend well., although they have different previous history, depth of scientific substantiation of their technical applications and usefulness in anticorrosive systems. In our view, the directions mentioned above match with the contemporary objectives of corrosion science and technology.

3.2.1 Electret Coatings

Electret polymer coatings are characterized by a spatial distribution of electric charge carriers in the polymer film and become sources of a permanent electric field. This field is, as a rule, directed normal to the substrate. A polymer coating covering a metal substrate is convenient for a polarizing charge to form. In such a system one of the polarizing electrodes is the substrate and the other is found in contact with the coating surface or above it (Fig. 3.11). In the first case, the polarization mechanism connected with dipolar orientation (a), ion and electron polarization of the macromolecules and the shift of spatial charges of the macromolecular segment dominate. Formation of a homocharge in the second case (b) is attributed to the injection of charges of the same sign as the electrode. The sign on the substrate surface is formed by the volume polarization mechanism and has a sign opposite to that of the substrate [13]. The case when the outside coating surface is charged less negatively than that adhesively bonded to the substrate is illustrated in Fig. 3.11b.

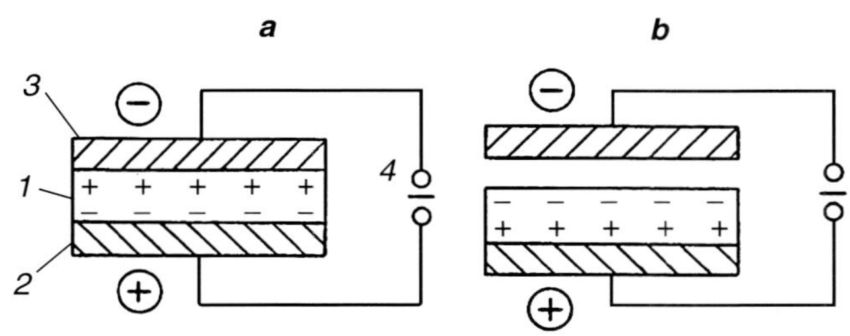

Fig. 3.11. Schemes of a heterocharge (**a**) and homocharge (**b**) formation in polymer coatings. (1) coating, (2) substrate, (3) electrode, (4) electric field source

The formation of electret coatings on flat surfaces in fact evokes no problems. They are formed on polarization of polymer coatings on curvilinear surfaces of complex shapes.

Liquid permeation through polymer coatings proceeds in the following stages: i) wetting and spreading over the coating surface, and ii) diffusion in the coating. These processes are verifiable by their own field in the electret coating.

Wetting dependence of pentaplast coatings by glycerin is presented in Fig. 3.12. The polarizing charge evidently generates an electric barrier in the coatings, overcoming which exhausts the motive force of the spreading liquid [13]. The velocity of liquid spreading over the electret coating surface is somewhat lower than over other non-electret ones (reference) and grows with increasing charge density σ_e on the electret (Table 3.2, Fig. 3.13).

The spreading motive force corresponds to the sorption capacity of the coating. The sorption intensity of electret coatings is less than that of the reference ones [15], therefore spreading over them is slower.

The sorption characteristics of electret coatings are described in Fig. 3.14 [15]. Sorption of liquids by electret coatings is seen to be noticeably lower than by reference ones (charge density 10^{-5}–10^{-6} C/cm^2). This difference increases

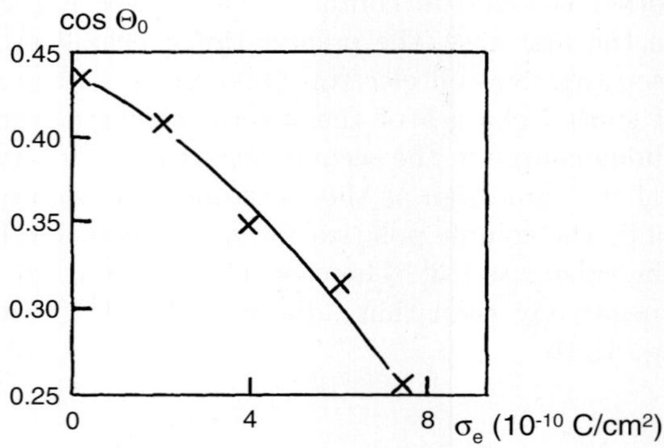

Fig. 3.12. Edge wetting angle of pentaplast coatings by glycerin as a function of the coating's polarization surface charge

Table 3.2. The initial spreading velocity (ν) of diethylene glycol over the surface of pentaplast coatings depending on their surface charge density (σ_e) [14]

Sample	σ_e, C/cm^2	ν, mm/s
Reference	0	0.28
Thermoelectret	$4.96 \cdot 10^{-10}$	0.14
	$2.78 \cdot 10^{-7}$	0.06
Corona electret	$3.60 \cdot 10^{-5}$	0.04

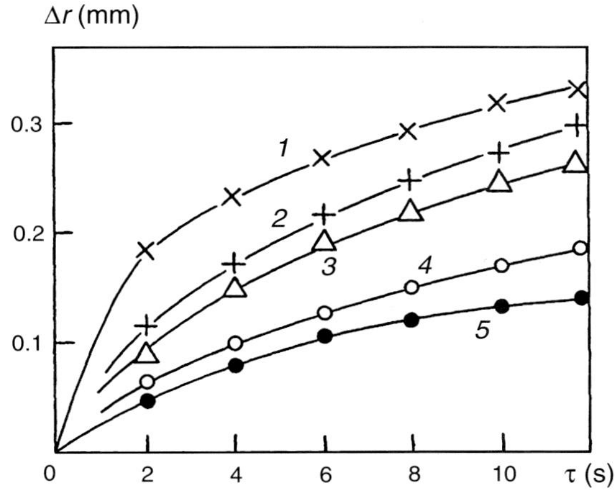

Fig. 3.13. Kinetic dependencies of a diethylene glycol drop radius (Δr) spreading over a pentaplast coating surface carrying electret charge (C/cm^2): (1) 0; (2) $1.16 \cdot 10^{-9}$; (3) $2.78 \cdot 10^{-7}$; (4) $3.6 \cdot 10^{-5}$; (5) $5.8 \cdot 10^{-5}$ [14]

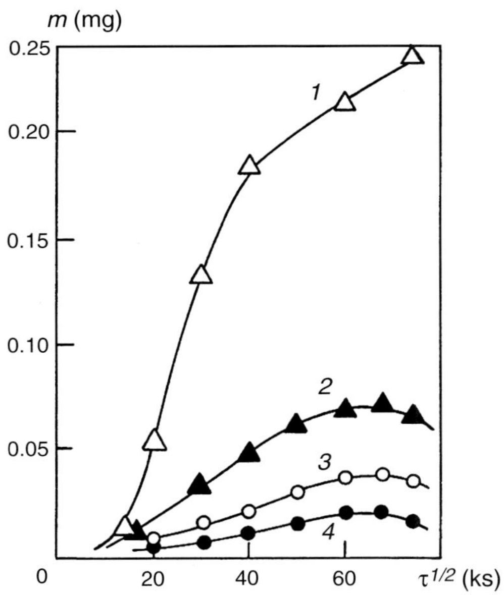

Fig. 3.14. Sorption of diethylene glycol vapors by non-polarized (1, 3) and electret (2, 4) coatings of polyvinyl butyral. Temperature: (1, 2) 50; (3, 4) 30°C

with growing temperature as a result of the elevating rate of conformal transformations in macromolecules ($T = 50°C$ corresponds to the glass transition temperature in polyvinyl butyral). Presumably, polar molecules of diethyelene glycol are attracted to the coating surface in its electrostatic field and form a layer of oriented dipoles that hamper further sorption. As soon as the layer is formed, the equilibrium concentration of the liquid establishes in the coating. Molecules of nonpolar liquids are evidently acquiring an induced dipolar moment in the electret field and interact with the coating following a similar mechanism.

Information on diffusive permeability of electret coatings is presented in Table 3.3. Based on these results, an analytical dependence between the liquid mass absorbed by the coating and the polarizing charge value has been derived [16].

Table 3.3. Quantity of the liquid *(Q)* diffused into a polyvinyl butyral coating and the diffusion constant *(D)* at different polarizing charge densities (σ_e) in the coating [16]

σ_e, 10^{-5} C/cm^2	0	0.05	0.18	0.98	2.00	2.94	3.51	4.80
Q, 10^{-7} kg/cm^2	1.51	1.48	1.38	1.02	0.76	0.57	0.44	0.29
D, 10^{-11} m^2/s	1.20	1.15	1.05	0.62	0.31	0.16	0.11	0.05

$$Q = \frac{2C_0}{\pi}\sqrt{D\tau} \cdot \exp(-\alpha\sigma_e) \ ,$$

where $C_0 = const$ is the liquid concentration on the coating surface; σ_e is surface density of the polarizing charge estimated by TSD; D is the liquid diffusion constant in the coating; and the factor α expresses the dependence of D on σ_e. For the data presented in Table 3.3, $\alpha = 0.34 \cdot 10^5$.

Electret polymer coatings peel more slowly from the substrates in electrolyte solutions around through pores in the coating and downgrading of the substrates accompanying exfoliation is less than in reference coatings. As is shown in Fig. 3.9, the adhesive strength of electret coatings with the substrates is higher upon exposition in various media as compared to non-electret coatings. Filling of the polymer binder with powder electret particles effects protective properties of the coatings similarly to electrical polarization [13]. A polarization model has been elaborated that visualizes the formation of a thermoplastic coating structure filled with electrets. The model is based on the electric charge relaxation of filler particles during coating formation and the generation of polarized surface layers at the binder-filler interface [17].

The described advantages of electret polymer coatings have not been called for until recently chiefly due to the lack of information and inaccessibility of electret techniques. It is taken for granted today that the electrically polarized state is a natural physical state of polymer dielectrics. The realization of merits that are intrinsic to polymer coatings may open new vistas in anticorrosion techniques. At present, a number of simple and low power-consumption polarization methods have been developed that combine molding and electrical treatment of polymer coatings in a single technological cycle [13].

3.2.2 Electroconductive Coatings

Polymers have been traditionally employed in anticorrosion techniques as dielectric materials. Investigations in protective ability and the applications of conducting polymers in anticorrosive systems such as, firstly polyaniline, and later polypyrrole, displaying intrinsic conductivity started in the early 1980s. The conductivity is induced by a specific molecular structure formed during synthesis, with a distinguishing feature of developed double links. An essential feature of each polymer is its highly conjugated π-system. When a (+) or (−) charge is placed in the π-system of electrically conductive polymers, it is this conjugation that enables charge movement (in the presence of an electric field), and the concomitant appearance of metallic properties. It can be seen from Fig. 3.15 that heteroatoms may be present in the backbone of the polymer, as long as conjugation is not interrupted [19].

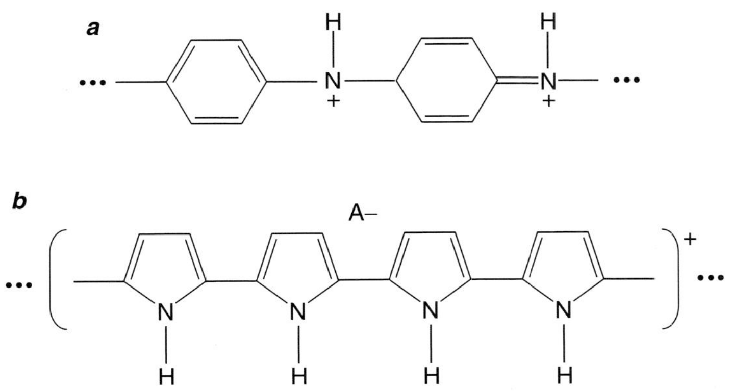

Fig. 3.15. Polyacetylene (**a**) and polypyrrole (**b**) structures. (A −) – additive

Polyaniline (PAni), which has been known for more than a hundred years, was modified in the 1980s. The conductivity of polyaniline was raised from 10^3 to 10^5 $Ohm^{-1} \cdot cm^{-1}$ at BASF Company (Nuarman H.) in conditions of high experimental reproducibility and strict control over polymerization. The indicated values surpass the conductivity of metals such as Fe and Pt, i.e. the dream of an organic metal has become a reality. Polypyrrole (PPy) in a doped state shows significant stability, while its conductivity is above 10^2 $Ohm^{-1} \cdot cm^{-1}$ [20]. In Fig. 3.16, the range of electrical conductivities possible with electrically conductive polymers is presented [19]. Note that

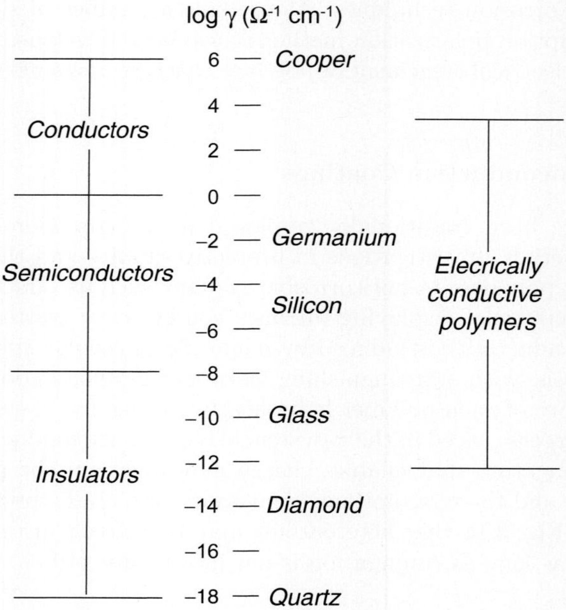

Fig. 3.16. The range of electrical conductivity (γ) for electroconductive polymers

(on a logarithmic scale) two thirds of the range available in nature can be attained.

In 1985, D. DeBerry found that PAni electrodeposited on passivated steel in a strong acid environment enhanced corrosion protection of the metal. ORMECON Chemie GmbH & Co (Germany) had used pure PAni coatings from organic dispersion, which were deposited on untreated metals (normal steel, stainless steel and copper). The corrosion potentials were shifted to the nobler region. The corrosion current was significantly reduced or even completely eliminated at comparable corrosion potentials [18]. A joint group from Los Alamos National Laboratory (LANL) and the National Aeronautics and Space Administration (NASA) stated in 1991 that the doped form of PAni showed excellent performance as a coating for preventing corrosion of carbon steel. When covered with an epoxy topcoat, the coatings have displayed excellent performance on immersion testing (3.5 wt % NaCl and 0.1 M HCl) [19]. The LANL–NASA groups were interested in developing a high-performance coating that could protect launch towers from acid vapor generated during space shuttle launches. They believed that interfacial contacts between metals and doped PAni would generate an electric field that would restrict the flow of electrons from the metal to the external oxidizing species, thus preventing, or at least reducing, corrosion. Protection of Al alloys was achieved using a double-strand PAni as a surface conversion coating. The effectiveness of the coating was tested by salt spray and immersion in salt and acidic salt solution [21].

PPy coatings were polymerized electrochemically from p-toluene lulphonate water–ethanol solution [22]. A very good corrosion protective effect of those coatings on mild steel has been observed previously. After drying in air, thin PPy coatings (~200 nm) were highly adhesive, whereas thicker films (~7.5 μm) could easily be peeled off the metal. AFM and XPES study of the interface between the metal and the polymer showed that a sulfur–rich layer was formed at the interface. This indicated that the counter-anions (p-toluene sulfonate) had accumulated on the surface of the metal. High-binding-energy iron and imine–type nitrogen were detected in a thin PPy coating. Moreover, thin films were substantially overoxidized. The number of anions incorporated into the PPy matrix varied as a function of film thickness [22].

Thin films were created after immersion of Mg alloy samples in a specially formulated emulsion of polyacrilic and PPy polymers. This resulted in better corrosion resistance of coated Mg foils in neutral aqueous NaCl solutions [23].

Air Product and Chemicals, Inc. (USA) have found out that (in the absence of any topcoat) while the doped electrically conductive form of PAni performed well as a corrosion-resistant coating for carbon steel, the undoped, nonconductive form of PAni performed better. This was determined by electrochemical impedance spectroscopy (EIS), immersion testing (3 wt % NaCl) and salt fog testing. Perfect performance of undoped PAni could be understood as follows. From the high pore-resistance value observed ($2{\cdot}10^{18}\,\Omega$) it can be concluded that PAni forms a dense, adherent, low porosity film on carbon steel. Such a film would restrict the access of components required for corrosion to occur:

$$2\mathrm{Fe} + \mathrm{O_2} + 2\mathrm{H_2O} \leftrightarrow 2\mathrm{Fe}^{2+} + 4\mathrm{OH}^- \,, \quad E^0 = 1.26\,\mathrm{V} \tag{3.1}$$

It has also been indicated that aniline itself is a corrosion-inhibiting agent. Thus, a PAni film coating is essentially a very dence collection of aniline molecules, held mechanically against a carbon steel surface. These molecules would effectively scavenge protons and create a basic environment at the surface (from reaction with water present). The sum of these effects is that the corrosion reaction (3.1) would be moved to the left, according to Le Chatelier's principle, thus fostering the maintenance of an unoxidized carbon steel surface. An undoped, nonconductive PAni coating would not be expected to alter the electronic state of the underlying metal or cause, as compared with doped PAni, the oxidation of the metal to form protective oxides [19].

EIS is commonly employed for the evaluation of coatings. Poor resistance values of electrodeposited PPy coatings have been demonstrated to be good predictors of long-term coating performance in corrosion prevention [24].

It was found (with the EIS technique) that PAni blending with a polyimide (prepared by solution blending) improved the corrosion resistance performance of polyimide. PAni can react with polyimide to form chemical bonds between these two polymers and these bonds keep these two polymers as a miscible system. The corrosion protection property of these coating increases on increase of the PAni component and the excellent anticorrosion effect emerges when the PAni content reaches 10–15% [25].

The poor solubility of intractable electrically conductive polymers has been a serious problem. The authors of [25] have explored a PAni-4- dodecylhpenol complex to improve the dissolution, and it has been shown to be an effective organic corrosion inhibitor. With the surfactant (dodecylphenol), PAni could be diluted into the coatings. The surfactant improves the wet adhesion property between the coating and the metal.

Clearly, electrically conductive polymers show good promise as an additive to improve the corrosion resistance performance of existing coating systems. The improved performance that can be achieved by blending PAni with an epoxide has also been observed by a group at Drexler University (USA) [19].

Multilayered coatings, consisting of combinations of PAni and PPy have been galvanostatically deposited onto both carbon steel and stainless steel [27]. Potentiodynamic polarization was used to access the ability of these copolymers to provide an effective barrier to corrosion in chloride environments. The performance of these multilayered coatings on carbon steel was not sufficiently better than single PAni coatings to justify their more complicated deposition procedures. However, in the case of stainless steels the new multilayered coatings proved to be significantly better than single PAni coatings, especially at protecting against pitting corrosion. SEM observations and adhesion measurements, along with the electrochemical data suggested that the ability of a conducting polymer film to act as an electronic and chemical barrier were more important in providing corrosion protection than its ability to act as a physical barrier [27].

A group at Ohio State University (USA) and the aforementioned group at Drexel University have reported that undoped PAni shows performance superior to doped PAni. The good performance of undoped PAni was attributed to the effective formation of a thin, protective layer of Fe_2O_3 on the surface of Fe. Several groups have reported, in agreement with the LANL–NASA group, that doped PAni performs very well and at least as well as undoped PAni. At the University of Texas (USA) it has been found that, when overcoated with an epoxy, both undoped PAni and doped PAni show very good performance in protecting carbon steel from corrosion. The PAni employed was the commercially available Versicon® product (doped form), now available from Monsanto Co., Inc. (USA). The explanation offered for the observed good performance was the presence of PAni, which caused the formation of a protective layer of iron oxides (Fe_2O_3 and Fe_3O_4) on the surface of the carbon steel. A group from Zipperling Kessler & Co (Germany) has found that coating carbon steel with a Versicon®, dispersion resulted in good corrosion protection [19].

Ormecon Chemie GmbH & Co. KG (Germany) now offers two commercial products comprising doped PAni: CORRPASSIV® and CORREPAIR® [28, 29] for resistance to mechanical stresses [18]. The potential distribution between two coating defects on a steel sample was measured (Fig. 3.17). On the left-hand side, the sample was coated with a conventional primer and on the right-hand side with the CORRPASSIV primer, both top-coated with a two-component epoxy lacquer (100 μm). The transition between negative and positive potentials marks the delamination front. The delamination zone is not an equipotential surface. More positive potentials are registered towards the delamination front. This is due to the changing reaction conditions within the delamination area, which depend on the distance to the defect. In Fig. 3.17, delaminating zones are clearly seen. The delamination area DL1 with a conventional primer is larger than the DL2 zone by a factor of two. Due to the PAni-containing coating the measured potential on the right is shifted to more positive potentials.

The three principal hypotheses so far advanced to explain the performance of electrically conductive polymers are:

- polymers contribute to the formation of an electric field at a metal surface, thus restricting electronically the flow of electrons from the metal to an oxidant [30];

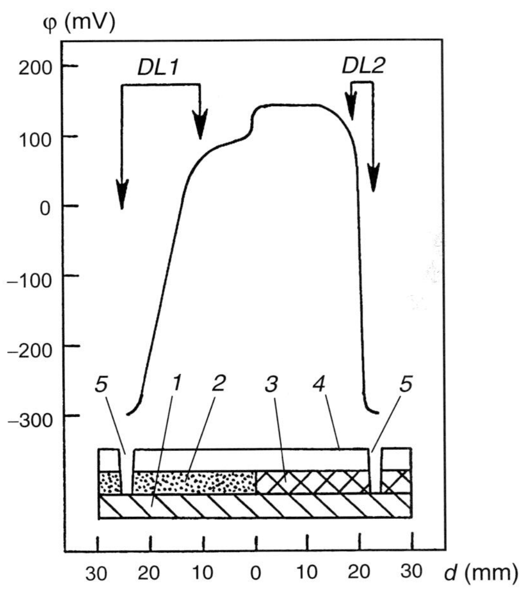

Fig. 3.17. Potential profile as measured with the Kelvin probe after an immersion time of 100 hours in 5% NaCl solution. (1) steel, (2) EP primer, (3) CORRPASSIV, (4) top coat, (5) defect, DL1 and DL2 – delaminating zones

- they form a dense, adherent, low-porosity film, and maintain a basic environment on the metal surface, thus restricting access of oxidants and forcing the corrosion reaction (3.1) in the direction of the unoxidized metal [30,31];
- conducting polymers cause the formation of protective layers of metal oxides on a metal surface [32, 33].

The actual mechanism may be a combination of two, or all three, hypotheses. Since most commercial coatings need to possess a number of properties in addition to corrosion resistance (e.g. abrasion resistance and good adhesion to the substrate), it is thought that the ultimate commercial use of conducting polymers will be as additives for existing coating systems [19]. This would help to minimize its probable relatively high cost. Recent research has demonstrated that PAni can be prepared in emulsion form [23, 26]. Such emulsions would seem to be ideally suited to blending with a water–based coating system. The use of conducting polymers as additives would also be fostered by their solubility in easily handled solvents. Recent research has shown the availability of such polymers [22, 25, 33, 34]. Thus, given the good performance already demonstrated for electrically conductive polymers and the potential for the availability of materials that are water processable and more easily handled, their future in corrosion-resistant coatings looks very promising.

3.2.3 Inhibited Coatings

Inhibited polymer coatings, i.e. coatings whose components are metal corrosion inhibitors, have a long previous history. Apparent successes in the application of inhibited coatings have intermingled with conspicuous failures. It is believed that the main reason for this was the contradictory effect of corrosion inhibitors on a combination of coating performance factors.

In the beginning, corrosion inhibitors were simply introduced into the polymer binder composition taking no notice of possible interactions between the components and the necessity for local concentration of the inhibitors on definite coating areas. Nevertheless, this approach was in many cases a success. For instance, they are said to improve substantially the protective ability of alkyd coatings by filling with acid phosphate coupled with chalk and zinc oxide [35]. The idea of adjusted consumption of corrosion inhibitors through their localization in porous particles of inorganic oxides was realized in [36]. Thus filled lacquer coatings show a high protective level at a lower inhibitor content. Coatings based on a liquid polyester gel, granted with a patent [37], contain an antioxidant evenly distributed within the binder, corrosion inhibitor and a rust converter. An analogous solution is employed for coatings based on high-temperature refinery waste – bitumen – with the use of an alkyl succinic acid mixture [38]. These components fulfill two functions: that of corrosion inhibitor and bitumen hardener.

Uniform distribution of inhibitors within the binder often resulted in an unpredictable outcome. Being most often a surface-active substance, corrosion inhibitors first affect the adhesion of the coating. Weakening of adhesion due to isolation of the inhibitor at the coating-substrate interface is used in easily peeling coatings intended for temporary protection of metal ware against corrosion. Thus, Alocit Group (GB) manufactures a thermoplastic material, Enviropeel, intended for such coatings. It protects steel products from atmospheric corrosion [39]. A 2–3-mm-thick coating containing a corrosion inhibitor is applied by spraying onto an article heated up to 160°C. The coatings are easily peeled off and can be reprocessed. After peeling, a thin layer of corrosion inhibitor is left on the article. Flange joints covered by the Enviropeel coating did not show any corrosion signs upon hot salt fog testing equivalent to 30-years service in conditions of the North Sea [40]. The coating composition for temporary anticorrosion protection [41] based on a perchlorovinyl enamel contains large amount (6–8%) of a corrosion inhibitor quinoline. The mechanism of adhesion reduction of steel is similar to the previous case.

Polymer primers form a large group of inhibited coatings. They should provide the three following requirements: (1) elevated adhesion to a metal substrate, (2) protection of the substrate against corrosion, and (3) reliable adhesion to the outer layer of the coating. Polymer compositions have been elaborated in which corrosion inhibitors provide named functions. A new generation of corrosion inhibitors are used today as anticorrosive components of polymer primers [42], namely, triazoles, benzatriazoles, imidazoles, tetrazoles and their blends in combination with substances serving as ion sources to improve coating adhesion, including molybdate, tungstate, niobate, vanadate. They are impregnated into film-forming substances based on water-soluble ethylene oxide polymers, copolymers of ethylene oxide and propylene, polyethylene glycol, polypropylene glycol, polyvinyl alcohol and their blends. Similarly, adhesion is improved if zinc phosphates, benzoates or molybdates [43] are employed as components of the primer coating. An inhibited primer [44] based on an epoxy resin contains a hardener and corrosion inhibitor in the form of cyclohexylamine chromate, which does not impair adhesion. Phosphating compositions consisting of rust converters with corrosion inhibitors represent a new type of primers for steel ware [45].

Brangs & Heinrich (Germany) suggests using a sticky LDPE film, trade name Branofol, to preserve metal ware during storage and transportation. This film is fitted with an adhesive layer containing a volatile corrosion inhibitor. It has been proved an efficient means of protecting metal ware during carriage by sea [46].

A direction in designing inhibited polymer coatings using new principles has been proposed in [47]. A layer of corrosion inhibitor (μ-hexamethylene imine nitrobenzoate) applied onto a steel substrate is polymerized in a low-temperature plasma. The high protecting properties achieved by this coating are attributed to the formation of corrosion-stable Fe nitrides and oxides at the interface with the substrate. In the inhibitor layer new nitrogen- and oxygen-containing organic compounds are formed, displaying elevated ability to suppress corrosion of the steel substrate and strengthen its adhesion to the coating. The improved barrier properties of the coating are explained by the formation of a high-strength hydrophobic film during the plasma-induced polymerization of the coating (Fig. 3.18).

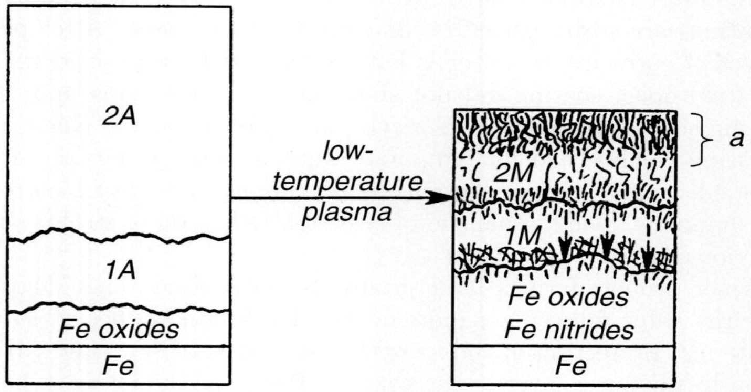

Fig. 3.18. A scheme of plasma-induced modification of the coating based on μ-nitrobenzoate of hexamethylene imine on a steel substrate [47]: (1A) adsorption layer non-flushed by water; (2A) outer flushed layer; (1M) modified non-flushed layer; (2M) plasma-polymerized layer, a – area of elevated density. The arrows show sewing of the modified layer to the substrate

Proceeding from the above, it is possible to distinguish three groups of inhibited polymer coatings (Fig. 3.19). Uniform distribution of the inhibitor in a film-forming material (*a*) is inexpedient in principle. This variant is admissible in the case that the inhibitor does not affect the adhesion and barrier properties of the coating, is not too expensive and the coating is sufficiently thin. A layer of inhibitor can be located as shown in (*b*) within the polymer coating bulk. Consequently, impregnation of the inhibitor leaves adhesion of the coating intact. Hence, it is recommended to use inhibitors that are soluble in the corrosive medium diffusing through the coating. On reaching the inhibited layer, the medium becomes a carrier of the inhibitor particles and transports them to the substrate. The inhibitor transforms the substrate surface layer thus protecting it against corrosion. If the inhibitor will not lower, or elevates, the coating adhesion, it is better to place it on the substrate surface (*c*). The surface layer of the substrate modified by the inhibitor is overlaid with a topcoat functioning as a barrier towards the corrosive medium.

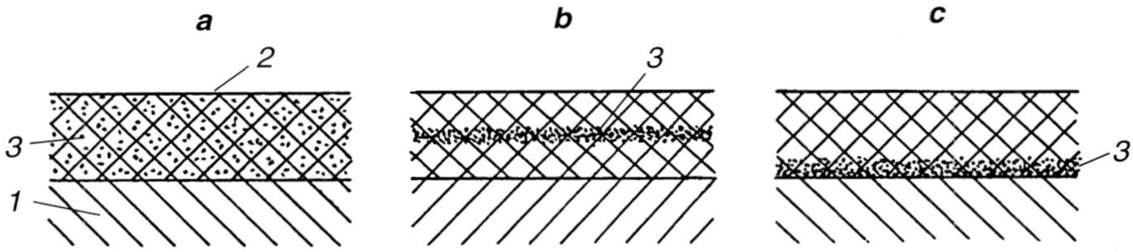

Fig. 3.19. Structure of inhibited polymer coatings: (**a**) the inhibitor is evenly distributed within the film-forming layer, (**b**) the inhibitor layer is localized within the coating volume, (**c**) the inhibitor is concentrated on the substrate surface. (1) metal substrate, (2) polymer coatings, (3) an inhibitor particle

The Above-considered types of polymer coatings follow different mechanisms of protecting metal against corrosion. Being a part of a multilayered structure, the coatings form a complex antirust system for protecting the metal substrate. An example of such combined coating is illustrated in Fig. 3.20. The upper electret layer 5 consists of an electrically polarized polymer functioning as a water-impermeable barrier. This is why the coating is poorly wetted by aqua electrolytes and retards ion diffusion from the environment to the substrate 1. The next layer 4 is filled with water-soluble inhibitor particles. The mechanism of this layer is similar to the coating shown in Fig. 3.19b. Beneath it a layer of electroconductive polymer 3, which serves as a primer, is located, i.e. forms a passivating layer 2 on the substrate surface. In addition, the organic metal layer limits the access of oxidants and the flow of electrons from the metal to oxidant.

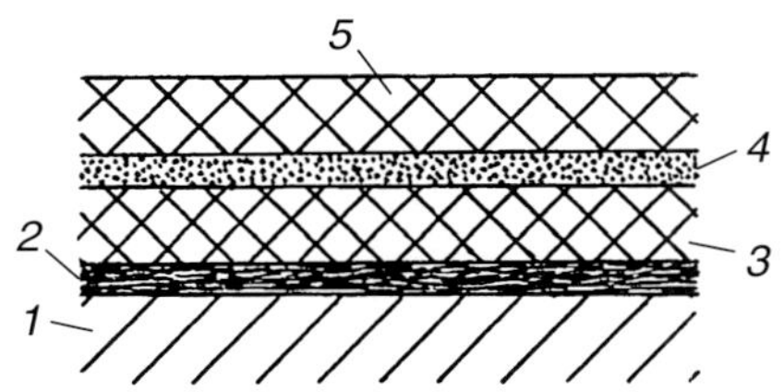

Fig. 3.20. A scheme for a combined polymer antirust coating. (1) metal substrate, (2) passivating layer, (3) a layer of organic metal, (4) a layer containing corrosion inhibitors, (5) electret polymer layer

The above-cited data are proof of the feasibility of developing numerous designs of combined coatings with corrosion inhibitors, using which all mechanisms of anticorrosive protection of metals can in fact be realized. In combination, these designs present a promising base for the creation of smart materials responding to modern requirements of anticorrosion science and technology.

3.3 Coating Techniques

The principal stage of application of anticorrosion polymer coatings on metal ware consists of the uniform deposition of polymer dispersion onto the substrate. Further stages assist in the physical-chemical interaction of the polymer and metal, formation of the polymer film and its cohesion to the article being protected. A combination of procedures for the application of polymer anticorrosion coatings containing CI, illustrated in Fig. 3.21, can form the basis for the methodology of manufacturing such coatings. A brief description including their relations with technological parameters and formulation of CI as coating components is presented below. Primary attention is given to powder coatings and their production technology, which changes on the introduction of CI.

Paint coats and coatings from polymer suspensions are applied to metal ware from liquid compositions containing volatile matter. The compositions are applied to the substrate by brushing, dipping, pouring, pulverizing, electrostatic spraying, or electrophoresis. Use of CI in paint or suspension coatings does not change the technique of coating formation in any way. The inhibitors are mostly dissolved in volatile (aqua or organic) components of the liquid composite material. Depending on the solubility of the inhibitor in water and its effect on the barrier polymer layer formation and coating adhesion, either single-layer or multilayered coatings can be formed. When single-layer coatings are applied, the CI is spread evenly in the liquid composite material. Contact CI for metals are, as a rule, introduced into the primer layer composition applied onto phosphate or chromate sub-layers formed on the steel substrate. Water-soluble inhibitors are impregnated into the film-forming component of the coating that covers the primer layer, as shown in Fig. 3.19.

Powder coatings are formed of powdery compositions based on synthetic film-forming polymers and technological additives in solid and liquid phases. The methods of powder coating application analyzed from the viewpoint of monitoring CI concentration on a given coating area are presented below.

3.3.1 Deposition of Powder Coatings on Heated Ware

When a polymer powder material is brought into contact with metal ware heated above the polymer melting point, the processes of deposition, holding powder particles on the ware surface and formation of a solid polymer layer can be joined together.

A diagram of a polymer inhibited coat formation on a heated article is presented in Fig. 3.22. As soon as polymer particles come into contact with the article surface, they stick to it, warm up and melt. The first stage of this process, consisting of the formation of a monolayer of particles on the article surface, is completed in 1–2 seconds. It is expedient to incorporate contact inhibitor particles into the powder composition so as not to impair

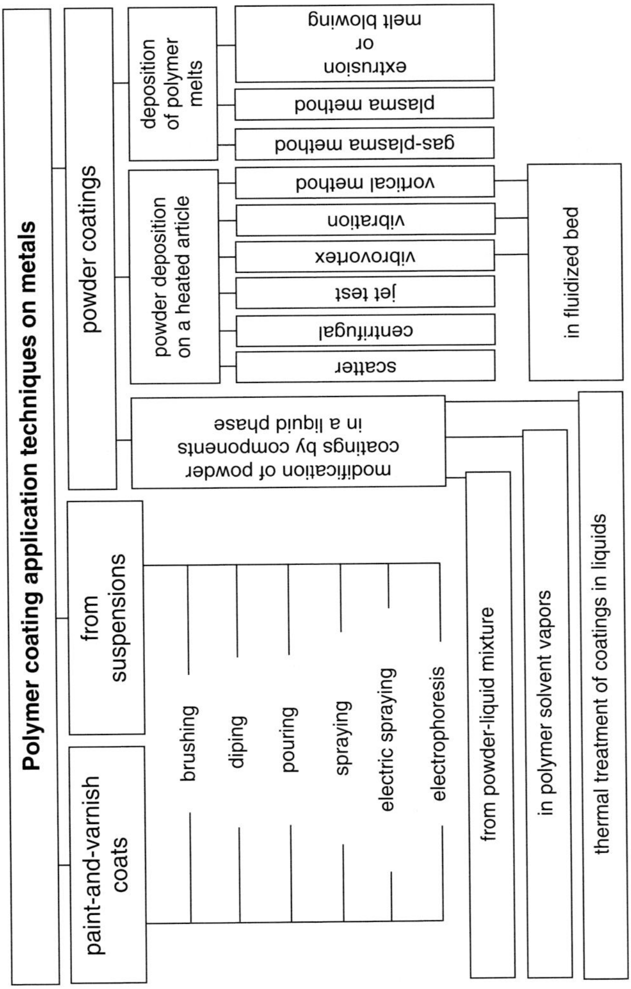

Fig. 3.21. Polymer anticorrosion coating application techniques on metals

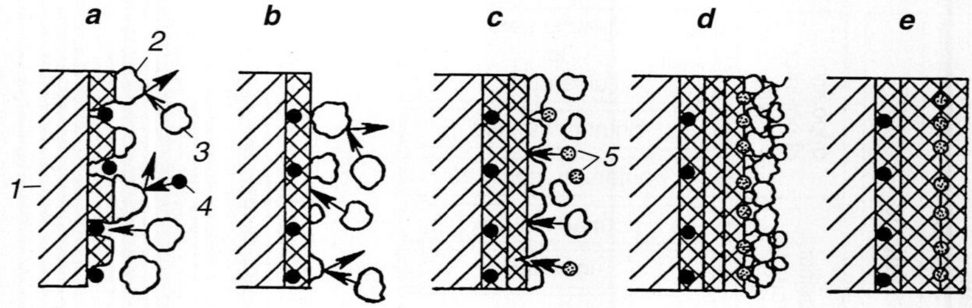

Fig. 3.22. Scheme of formation a polymer coating containing CI on a heated article: (**a**) formation of the first layer, (**b–d**) surfacing stages of powder particles, (**e**) the coating upon melting. (1) article, 2 and (3) polymer particles during melting and in the gas-powder stream, (4) and (5) particles of contact and water-soluble CI

polymer adhesion to metal. As incoming particles come into contact with the molten monolayer, the coating thickness increases. At a certain moment of this stage it is necessary to enrich the powder composition with water-soluble corrosion inhibitors. The heat energy accumulated by the article is used on melting of the polymer particles and is partially dissipated in the ambient space. Increase in coating thickness is accompanied by exponential increase of its heat resistance, which inevitably leads to a moment when the particles stop sticking to the coating. The coating outer layer, which should contain only polymer particles, remains solid. Final coating formation is commonly reached during additional heating of the article. The described procedure makes it possible to improve the coating quality and simplify the formation of coatings of a preset thickness.

Depending on the method of bringing polymer particles into contact with the article we discriminate between coating application in a fluidized bed, blasting with a gas-polymer mixture and bringing into contact with the powder in a loosely poured or compacted state.

Coating application in a *fluidized bed* can be carried out by a vortex, vibrovortical or vibration method. These methods are based on powder fluidizing, i.e. conversion of the powder layer into a pseudo-liquid (boiling bed) under the effect of vibration or up-flow gas streams sufficient to keep the powder particles in a suspended state.

The vortical method of coating deposition consists of immersion of the heated article in the fluidized bed of the polymer powder formed by blowing a gas flow through the porous bottom of a vessel containing the powder (Fig. 3.23). Under certain parameters of the gas flow, the immovable layer of powder particles poured into the vessel to the level H_0 expands up to the level H. The particles acquire mobility and transfer into the airborne state. The criterion for this change is a constant value of the gas pressure differential in the powder bed volume [52].

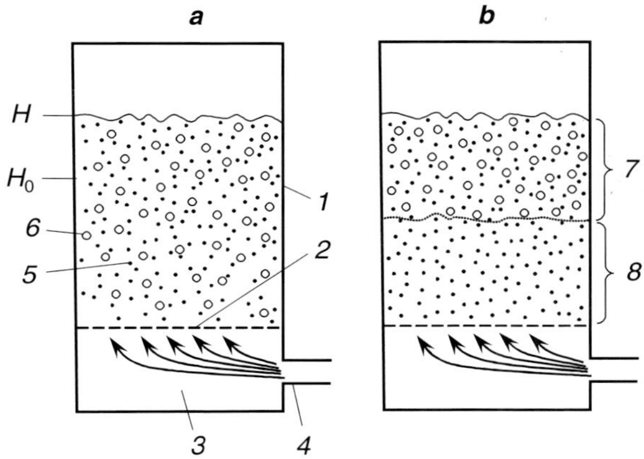

Fig. 3.23. Vortical state of the fluidized bed: (**a**) normal, (**b**) stratification in fractions. (1) vessel (2) porous bottom, (3) inflation chamber, (4) gas socket, (5) polymer powder, (6) CI particles, (7, 8) powder layers enriched with and without CI

It is shown in Fig. 3.24, that the vortical fluidized bed exists for definite ratios between the pressure differential and the gas flow velocity [48]. At small differential and velocity values (AB) powder particles are motionless. The gas flow velocity v_1 corresponds to the onset of transfer to the fluidized bed state (B) and is called the first critical velocity. On reaching the second critical velocity, v_2, powder particles are carried away from the vessel by the gas flow (C).

Depending on the powder properties and fluidizing conditions, different states of the fluidized bed can be realized. As can be seen from Fig. 3.23*b*, a layer of powdery particle mixture may exfoliate in fractions differing in density and particle size (separation).

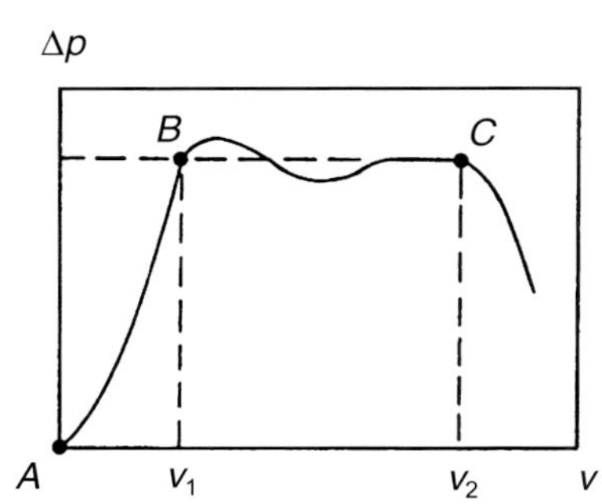

Fig. 3.24. Pressure differential in the powder layer experiencing vortical fluidizing as a function of the gas flow velocity

It is expedient to form single-layer inhibited coatings in a fluidized bed consisting of a homogeneous mixture of the polymer and inhibitor (Fig. 3.23*a*). The double-layer coatings with inhibited primer layer are commonly formed in a separated layer whose scheme is presented in Fig. 3.23*b*. A heated article is first dipped in the upper layer enriched with the inhibitor and then into the lower layer with content dominating by the polymer binder.

The vibration method of coating deposition is based on fluidizing the polymer powder via mechanical oscillations. This method was invented in the USSR in the early 1960s [48]. Vibration is imposed on either (i) a powder bath; (ii) bath bottom, or (iii) the article being covered.

The fluidizing mechanism behaves in response to oscillations as follows. In the case of low oscillation intensity, powder particles covering the vibrator surface become compacted. With increasing intensity the particles start to separate from the surface and become mobile, thus expanding the powder layer. The amplitude of oscillations and acceleration of the particles attenuate exponentially with distance from the vibrator surface. There is a condition of vibration-induced fluidizing of polydispersed polymer dispersions [52]

$$A\omega^2 \geq g \ ,$$

where A is the oscillation amplitude of the area covered by the powder; ω is the oscillation angular velocity; and g is the acceleration of gravity. Oscillation parameters optimum for polymer powders lay within the 5–100 Hz frequency range with accelerations of 3–4 g. We may observe compaction zones and zones of density maximum, loosening and expansion of the powder layer in the dependencies of the fluidized layer height on the oscillation intensity.

Vibration-induced fluidizing is accompanied, as in the vortex method, by the separation of the powder material by particle size and mass. Therefore, application of inhibited layers of the coating can be regulated using only one fluidizing bath incorporating a polymer powder mixture with CI. With this aim, vibrations of different frequencies are imposed in turn on the powder and article in a bath separated into sections by, e.g., baffles.

The vibrovortex method of fluidizing includes a joint action of the gas flow and vibration on the powder material.

The boiling bed surface fluidized by the vibrovortex method is almost devoid of craters or splashes. This is because fine powder particles are mainly blown out of the bath in this case [48]. When the vibration method is used, coarser particles ascend to the boiling bed surface. The vibrovortex effect on the powder causes these processes to compete. A powder layer placed in a cylinder expands depending on the gas flow velocity through this layer for a constant vibration regime (Fig. 3.25) [52]. The height H_0 of the expanding layer is also governed by this regime, all other conditions being equal; its maximum height being H_{max}. Zone I shows a portion of the linear dependence of the expansion on the gas flow velocity. Zone II stretches from the border of zone I to the largest dilatation. As the flow velocity reaches that of zone III, powder particles are carried off from the cylindrical chamber.

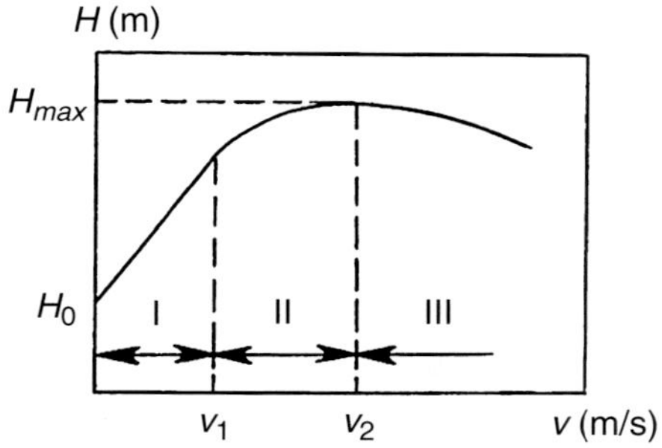

Fig. 3.25. Dependence of the height H of the powder layer fluidized by a vibrovortex method on the gas flow velocity v under a constant vibration regime

It is expedient to apply single-layer coatings of powder polymer mixtures with CI under regimes corresponding to the bounds of zones I and II. When it is necessary to create a CI concentration gradient across the coating thickness, it is applied at the end of zone II where the fluidized bed is most separated.

Jet methods of coating application presume blasting of the heated article by a jet of gas-powder mixture. This technology is convenient for covering items of large size or length, or their separate parts, e.g. in difficult-to-access cavities, hollows and slots. The jet method is typically used to coat inner surfaces of pipes and vessels heated in ovens or by induction currents.

A scheme for applying a polymer coating by the jet method onto a long-length article (wire, rod, tube) is presented in Fig. 3.26 [53]. A circular pipeline channel 1 is connected to a fan blower 2. The spraying zone, located in the pipeline discontinuity, is in the form of the conjugated nozzle 3 and diffuser 4. One of the sections of the pipeline has weepholes 5. As the blower is actuated, the particles in the pipeline are entrained by the gas flow and are transported though. In the spraying zone the air is sucked by the gas issuing from nozzle 3 into the diffuser to reduce powder losses. Air excess is removed through the weepholes 5.

Article 6 passes through inductor 7 of an HFC generator where it is heated above the polymer melting point and then passes through the spraying zone. Powder particles stick to the article or rebound from its surface, get to the rarefaction zone and are drawn into the channel of pipeline 1. The installation is loaded with the powder from bin 8. The ready coating is post-fused in the second inductor 9.

In principle, CI can be introduced into the gas flow in two ways:

- by an additional apparatus generating a gas flow of powder CI;
- via regulated extraction of the polymer-inhibitor mixture of the required composition from the fluidized bed generator.

An example of the second procedure is shown in Fig. 3.27.

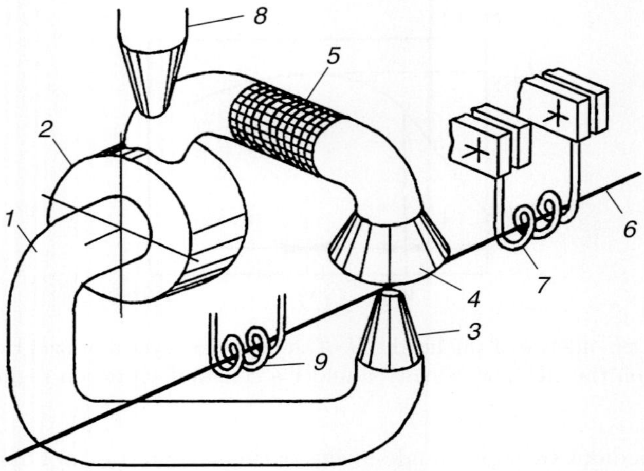

Fig. 3.26. Diagram of jet-spraying polymer coating of a wire. (1) circular pipeline, (2) blower, (3) nozzle, (4) diffuser, (5) filter, (6) wire, (7, 9) inductors (8) bin

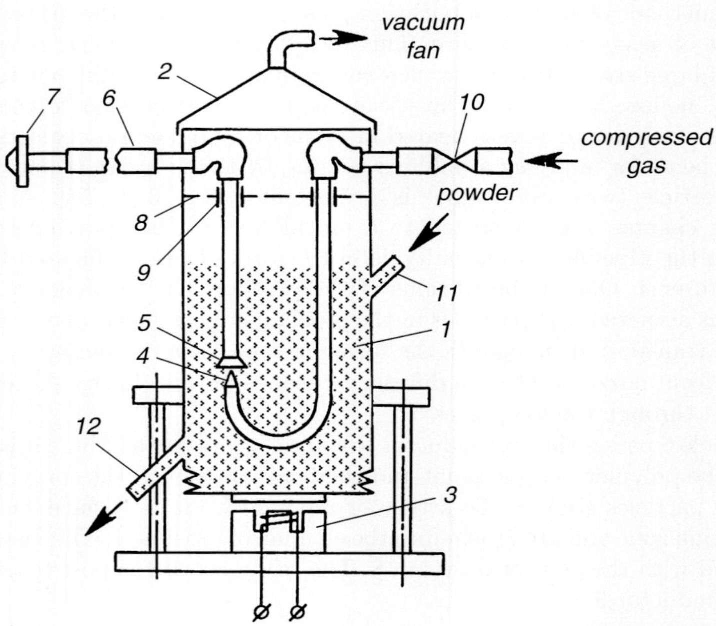

Fig. 3.27. A device for jet spraying of regulated composition powders [54]. (1) tank, (2) hood, (3) vibrator, (4) nozzle, (5) diffuser, (6) pipeline, (7) sprayer, (8) gas-permeable baffle, (9) guide, (10) regulator, (11) feeder, (12) channel

The polymer + CI powder composition is supplied through feeder 11 into tank 1. The spring-supported bottom of the tank perceives oscillations transmitted from vibrator 3 during which the powder composition becomes fluidized and separated. By moving the conjugated nozzle 4 and diffuser 5 along guides 9 it is possible to regulate their position relative to the fluidized bed height to select the required composition of CI concentration.

The gas flow is allowed into the nozzle through the actuator valve 10 and the composition of set content is sieved through diffuser 5. Then the gas-powder mixture is transported along the pipeline 6 to sprayer 7. The feeding rate of fresh powder charge through feeder 11 is such as to keep the volume of the fluidized powder in the tank constant. The excess gas is filtered through sieve 8 and removed from the tank.

The centrifugal method is employed to apply coatings on inner surfaces of the products like a body of revolution, e.g. pipes, bushes and rings. A powder polymer composition found inside a pocket of a rotating heated article is fused in contact with it and spreads over its inner surface under the action of centrifugal forces. The coatings achieved are rather smooth, although it is hard to regulate CI distribution within the coating during application. It is expedient to apply an inhibited primer for polymer coatings formed by centrifugal methods by jet devices. In addition, it is convenient to apply single-layer coatings with evenly distributed CI in the polymer binder by centrifugal methods.

Pouring, more precisely *vibropouring*, is used to coat sheet or roll metal materials. The powder is poured through a sieve onto a preliminary heated article and is melted in contact with it to form a coating. Distribution of CI in the coating is regulated during powder pouring by changing the vibration regimes of the substrate.

3.3.2 Deposition of Polymer Melts

For this group of coating application techniques it is typical to employ powerful heat sources to combine article heating with the polymer melting stage. This process of coating formation from polymer particles in the viscous-flow state has its own specific features.

Since the air carrying molten polymer particles from the melt generator to the substrate is a poor heat insulator, the transported particles cool down quickly. Figure 3.28 illustrates the catastrophic heat energy loss by PE melt particles sprayed in atmosphere by a melt-blowing device [52].

Another group of problems arising in coating deposition from polymer melts is connected with their low thermal stability. The onset temperature of thermally oxidative destruction of structural thermoplastics is much lower than the temperature of the gas flows transporting molten particles to the article. The particles undergo ablation during transportation, which means that a part of the substance is carried away from the surface layer by the

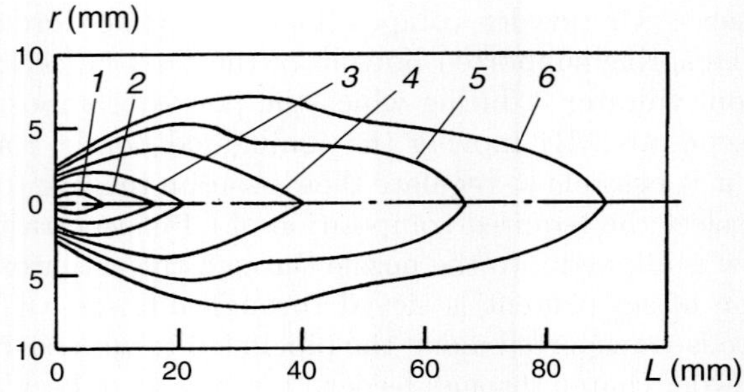

Fig. 3.28. Temperature distribution in a stationary flow of HDPE particles sprayed in atmosphere (T = 20°C): (1) 300°C, (2) 250°C, (3) 200°C, (4) 150°C, (5) 100°C, (6) 80°C

hot gas flow. Fortunately, the ablation processes cannot manage to propagate into the bulk of the particle during such a short duration in the high-temperature gas flow. Therefore, the properties of the polymer material are largely preserved in the coating in spite of the thermal, thermally oxidative and mechanical-chemical destruction processes. Factors that are critical for the quality of these coatings are the optimization of the particle size using the criteria of time and temperature regimes of deposition as well as the fractional homogeneity of particles.

CI are, as a rule, the most unstable components of polymer coatings in respect to elevated temperatures. Their degree of degradation and loss in the course of ablation are strongly dependent on the zone of the high-temperature gas flow into which they are impregnated. Different schemes of arranging the substrate, heat source, gas-polymer and gas-inhibitor flows are shown in Fig. 3.29. Heat-resistant inhibitors can be subjected to thermal treatment jointly with the polymer particles, in which case the inhibitor batcher should be placed in position 5. To alleviate heat effects on the inhibitor it is useful to impregnate it in the gas flow zone where the temperature is lower (position

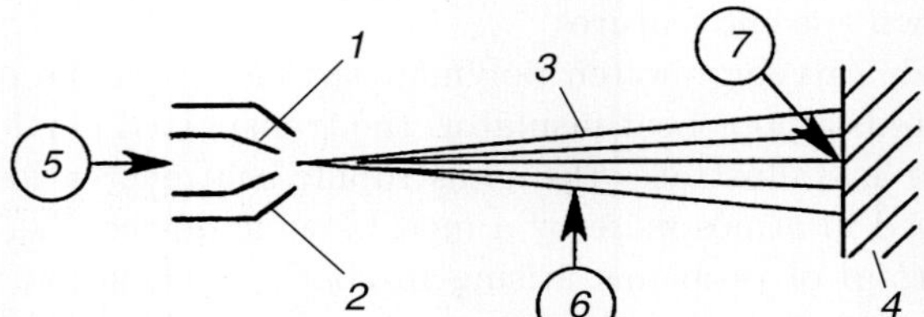

Fig. 3.29. Schemes of CI introduction into the high-temperature gas-polymer flow during coating formation. (1) gas-polymer flow generator, (2) heater, (3) gas-polymer flow, (4) article, (5–7) variants of locating batchers for CI particles

6). Unstable inhibitors can be introduced into the polymer coating in the instance of its formation on the substrate (position 7).

The gas-flame procedure of coating deposition consists of letting the gas-polymer flow pass through a gas jet flame that simultaneously heats the article. Under the action of the high-temperature flow, polymer particles transfer into the viscous-flow state, collide with the article surface and coalesce, thus forming a coating.

A typical design of a gun for the gas-flame coating application is illustrated in Fig. 3.30. A fuel gas, e.g. butane-propane mixture, is fed through regulator 1 and channel 2 into an annular cavity 3 of the gun where it is mixed with the air supplied through channel 5. The fuel mixture comes out through annular slot 6 and burns, forming a flame in the form of a dilative cone. The flame heats the article to be coated to $T = T_{\mathrm{m}} + (50\text{–}60)°\mathrm{C}$, where T_{m} is the polymer melting point. By opening regulator 7 the airflow is directed through tube 8 and hole 9 into a toroidal mixer 10 over the tangent to its surface. Under the action of centrifugal forces the polymer powder begins to circulate. As a result, air in funnel 11 of the mixer rarefies and sucks in new portions of the powder from bin 12. The gas-powder mixture circulating in the mixer is carried away through hole 13 over tube 14 into nozzle 15. The flow of mixture issuing from the nozzle collides with the article surface

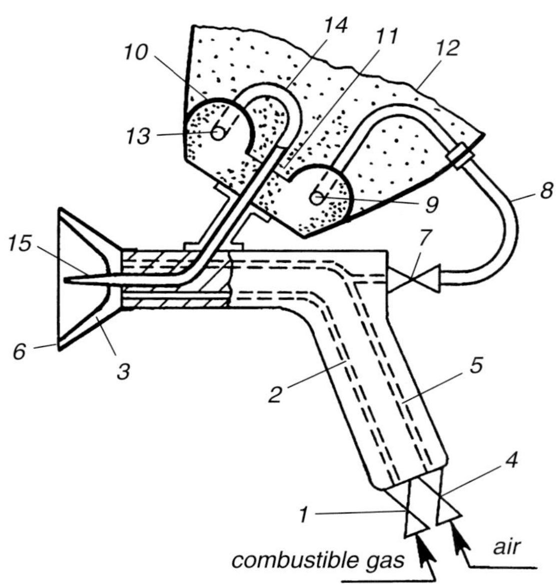

Fig. 3.30. A device for coating deposition [55]. (1, 4) and (7) gas regulators, 2 and (5) channels, (3) cavity, (6) annular slot, (8) and (14) tubes, (9) and (13) holes, (10) mixer, (11) funnel, (12) bin, (15) nozzle

restricted by the flame tongue. Loose polymer particles burn in the flame, thus lowering the dustiness of the working zone.

The gun is useful in depositing homogeneous coatings from a mixture of a polymer powder with CI [55] without direct contact between the flame and powder. It is compact and easy to handle, can be used in field conditions and does not necessitate any special equipment for preparation of the mixture.

The plasma method of deposition includes the following stages:

- a jet of some inert gas (e.g. argon) is heated in an arc discharge to 3000–6000°C, ionizes partially and transfers into the plasma state;
- the polymer powder is introduced into the plasma flow where powder particles are heated and transferred in fractions of a second into the viscous-flow state;
- the particles are transported in the plasma flow to the article surface, collide with it and merge, forming the coating.

The plasma method can be used to form coatings from refractory polymers (e.g. fluoron) on large-size goods (vessels, wagons, airplanes, etc.). The specifics of the deposition of inhibited coatings are connected with keeping CI away from direct contact with the high-temperature plasma. In Fig. 3.31 one can see the design of a device for plasma application of anticorrosion coatings. It consists of a metallized protecting sub-layer incorporating a primary polymer film, and an outer polymer layer.

The device is made in the form of two successive conjugated vortical tubes. A vortical tube is a facility realizing Ranque's effect, which consists of the torsion of a flow of a viscous compressible liquid or gas [57]. The facility operates as follows. Voltage is supplied to wire electrodes 1 and 2 made of electrically negative metal where arc discharge 3 is excited. Particles of the molten metal are sprayed by a gas flow issuing from nozzle 4 and are directed onto the article surface. Gas-powder mixtures of the polymer and CI are fed from sources 9 and 10 to sockets 5 and 6 of vortical tubes 7 and 8. Simultaneously, the arc between non-consumable graphite electrodes (not shown) is extinguished. The primary layer and outer layer of the coating are applied with controlled concentration of the polymer powder and CI. The outer layer acquires the structure shown in Fig. 3.31b where the protecting layer 12 consisting of electrically negative metal particles adheres to the article surface 11, inside which a primary polymer layer 13 is mechanically incorporated and enriched with CI, and to which the primary layer is adhesively bonded with an outer polymer layer 14.

The extrusion method or melt blowing consists of spraying of fibers squeezed from the extruder spinneret by a jet of compressed gas and their deposition on the article surface. The layout chart of the process is presented in Fig. 3.32.

Polymer granulate 1 is processed by a worm extruder 2. The polymer mass in a viscous-flow state is squeezed through the spinneret and is sprayed in the

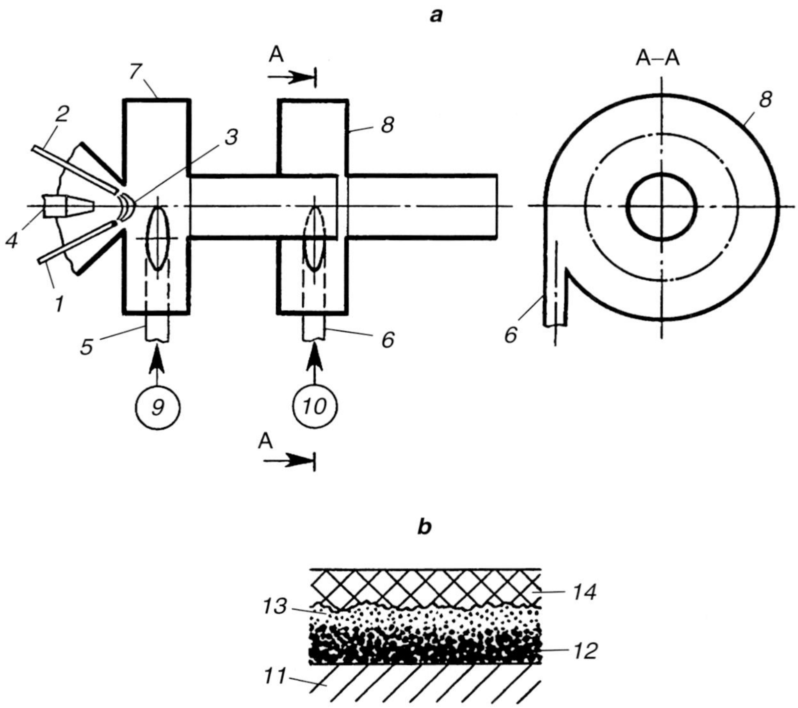

Fig. 3.31. Scheme of a device (a) for plasma application of polymer coatings [56] and a structure (b) of metal-polymer inhibited coating. (1) and (2) electrodes, (3) arc discharge, (4) nozzle, 5 and (6) sockets, (7) and (8) vortical tubes, (9, 10) gas-powder mixture generators, (11) article, (12) protecting metal-plated layer of the coating, (13) prime polymer-inhibited layer, (14) outer polymer layer of the coating

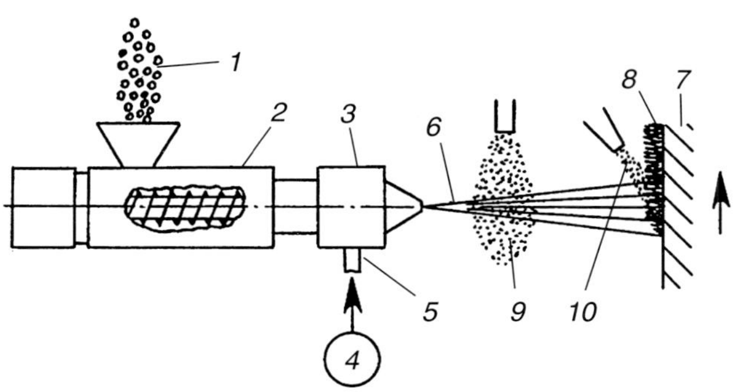

Fig. 3.32. Layout chart of coating deposition by extrusion. (1) polymer granulate, (2) extruder, (3) spray head, (4) gas jet generator, (5) socket, (6) gas-polymer flow, (7) article, (8) coating, (9) aerosol, (10) powder flow

form of fibers by a special head 3 with a jet of gas fed from source 4 through socket 5. The fibers are subjected to orientation stretching, are transported in the gas-polymer flow 6 to the article surface 7 and are deposited as a fibrous mass 8. The described method makes it possible to impregnate CI in the coating composition in both liquid and solid phase at any stage of the process, namely: i) into granulate 1 poured into the extruder batcher; ii) into the gas jet supplied to socket 5 from generator 4 of the gas-inhibitor mixture; iii) into the gaseous medium where fibers are dispersed as a flow of aerosol, or iv) onto the article surface via spraying inhibitor powder 10 during coating formation.

Melt blowing is extensively used in the manufacture of filtering materials [58] but is not fit for application of anticorrosion coatings. The main drawback is the large amount of energy spent melting through the fibrous layer on the article surface without which the pore-free polymer barrier against corrosive media would not be generated. Very often, IR radiators, particularly incandescent quartz lamps, are used as additional heat sources [59]. Their efficiency has been proved for the application of melt-blown coatings onto sheet ware with low heat capacity.

3.3.3 Electrostatic Deposition of Powder Coatings

The electrostatic deposition of polymer powder coatings includes the following stages:

(1) electrostatic charging of powder particles
(2) their transfer to the substrate in the external electric field
(3) deposition and retention of the particles on the metal ground substrate by forces of electrostatic attraction
(4) fusing of the deposited layer

The dielectric polymer particles can be electrostatically charged in the outer electric field, in the gas discharge, following the contact electrification mechanism, or through triboelectrization. The presence of CI, whose electric conductivity differs from that of the polymer, in the powder composition considerably affects these mechanisms.

The formation of a layer of electrically deposited particles on the surface of a cold article includes a number of factors. The charged dielectric particle contacting the substrate experiences: (a) the forces of molecular interactions on the sites of actual contacts; (b) the forces of electrostatic attraction of the particle to the substrate, and (c) Coulomb forces of interactions with neighboring particles. The formation mechanism of electrically deposited layer can be presented as a two-phase process during which a monolayer of particles whose interactions can be neglected is generated followed by a build-up of this layer.

The thickness and density of electrically deposited layers are a function of the dielectric properties of the powder composition and the process parameters. Common to all electrostatic deposition techniques of polymer powders is the existence of some critical thickness of the deposited layer. As soon as this is reached, no particles are deposited any more because of the Coulomb repulsion of the approaching unipolar charged particles by the layer. An effect called "inverse corona" appears in the layer, which means that voltage breakdown occurs in it as a result of raised field intensity, which ionizes the air between the particles, and a flow of ions is generated from the substrate with a charge opposite to that of the particles being deposited. This abruptly slows down the deposition and leads to uneven thickness of the deposited layer, formation of numerous craters and through pores.

The charge value of the electrically deposited layer of dielectric particles diminishes exponentially in time. The charge neutralizes the ions adsorbed from the atmosphere and free charge carriers leak down to the ground substrate thanks to the conductivity of the powder particles. The charge relaxation kinetics is a highly important parameter since it is necessary to keep the deposited layer of powder particles intact until fusing. At $T \sim 20°C$ and normal air humidity this layer of polymer powder does not crumble for a sufficiently long time. Therefore, relaxation of electrostatic charges does not intrude much into the coating deposition process. Impregnation of CI into the powder composition may, however, violate the stability of the deposited layer. So, the applicability criterion of an inhibitor during electrostatic deposition of coatings is the electroconductivity of the polymer powder composition containing a given amount of the inhibitor. This value should not surpass the threshold at which it loses stability.

In Fig. 3.33, comparative information is presented on the adhesive strength of polymer powder coatings formed on aluminum substrates by the electrostatic method under an external field of 2.5–3 kV/cm (the charge density in the layer was $q/m \sim 3 \cdot 10^{-4}$ C/kg) and thermal treatment of a similar thickness poured layer of uncharged particles [52].

It is evident that the strength of the coatings of electrically charged particles is higher within the whole temperature range of formation compared to uncharged ones. With extended time of thermal treatment this difference becomes less pronounced because of the growing effect of thermally oxidative destruction of macromolecules and accompanying processes. An idea has been put forward [52] that this increase in adhesive strength is stimulated by either the action of microcharges in contacting sites of the particles or the effect of the electrically deposited layer on adsorption and desorption processes at the coating-substrate interface.

Fusing of electrically deposited layer is accompanied by relaxation of the electrostatic charge. Note that charge relaxation in a dielectric coating never reaches completion. A part of the charge remains in the polymer layer and induces electrical polarization of the coating. This polarization effects in a

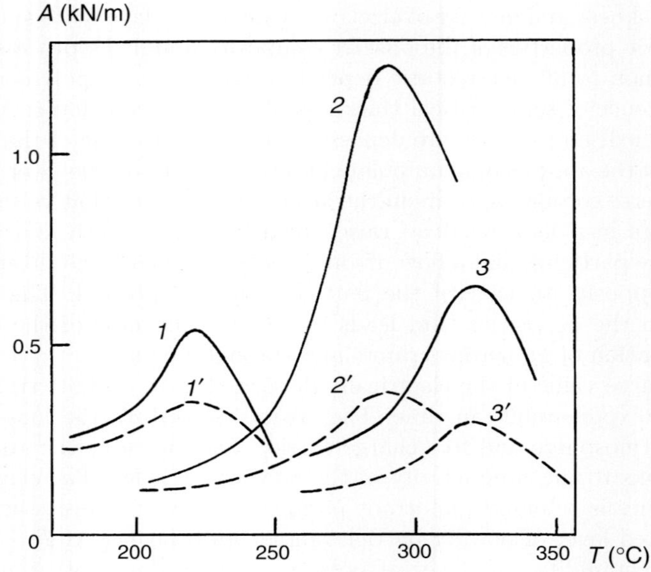

Fig. 3.33. Specific breaking load at delamination of HDPE coating (1, 1′), pentaplast coating (2, 2′) and that of fluoroplast-3 (3, 3′) versus their formation temperature. The coatings have been deposited by electrostatic method (1–3) and pouring (1′–3′)

definite way the performance of the coating. Fusing of thermoplastic powder coatings in an electric field of 20–50 kV/cm and its vector direction from the coating to the substrate significantly improves the resistance of the coatings to electrolytes [60].

Electrostatic methods of polymer coating deposition are traditionally subdivided into those procedures relying on charging of the fluidized bed and the electrojet processes. Their peculiarities connected with application of inhibited coatings are discussed below.

Electrostatic deposition of coatings in the fluidized bed is realized by the vibration, vortical or vibrovortex methods together with the electrical charging of suspended particles.

Powder particles may be charged by the contact electrification mechanism, when they acquire their charge on contact with the electrodes, or by the ion absorption mechanism, when air is ionized under the gas discharge effect followed by charge transfer from ions to powder particles.

An example of electrostatic deposition of coatings in the fluidized bed is illustrated in Fig. 3.34 [61]. Part 1 is transported on the conveyor belt 2 over apparatus 3 for vibrovortex deposition. The deposition bath contains powder composition 4 below which the bath is blocked by baffle 5, which is permeable to the powder particles. The baffle is connected to a high-voltage source 6 and

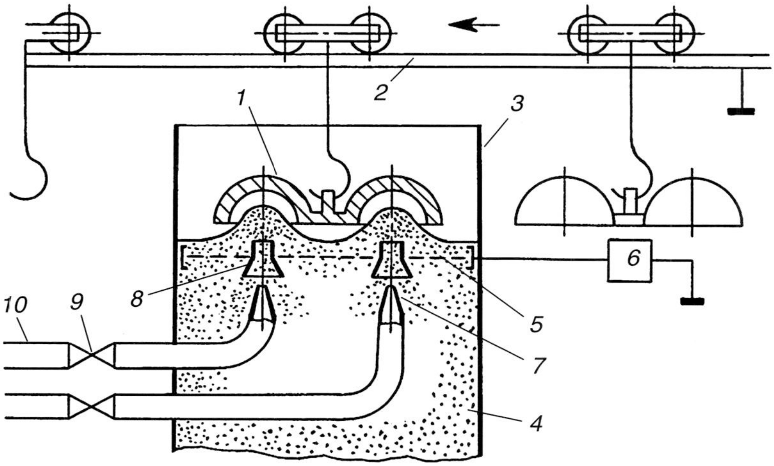

Fig. 3.34. A setup for electrostatic deposition of coatings in the fluidized bed: (1) part, (2) conveyor, (3) apparatus of vibrovortex fluidizing, (4) powder composition, (5) baffle, (6) high-voltage source, (7) nozzle, (8) diffuser, (9) regulator, (10) compressed gas main

is fitted with injectors in the form of nozzle 7 conjugated with diffuser 8. When part 1 passes the deposition position, the systems of fluidizing and electrical charging of the powder are activated. A gas jet from main 10 is fed through regulators 9 into nozzle 7 and lifts suspended particles over the horizontal surface of the fluidized bed, imparting it with a shape congruous to the part surface. This shape is controlled by regulators 9 and by displacing baffle 5 fitted with diffusers 8 along the bath height. As soon as the deposition is finished, apparatus 3 is grounded and the residual charge is picked-up from its metal units. The ready part is lifted over the bath and transported by conveyor 2 for further fusing in a thermostat (not shown), the next part occupies its position and the deposition cycle is repeated.

The setup [61] is used to apply coatings from a mixture of polymer and inhibitor powders. Initially, as the part is located at a considerable distance from the fluidized bed surface, mainly CI particles, which are finer and less dense than the polymer particles, deposit onto it. As a result, a primer for the future coating is formed, after which the part is sunk lower over the fluidized bed surface to enrich it with the polymer particles. After fusing the coating obtains a structure shown in Fig. 3.19c.

Electric conductivity exerts a perceptible effect on crumbling of both polymer and inhibitor particles in the fluidized bed. It should be underlined that electrostatic deposition of inhibitor powders, related mainly to semiconductors, differs a little from that of thermoplastics. In the case that the specific volume resistance is $\rho_v < 10^6 \Omega \cdot$ m, there is a strong probability of crumbling of the deposited CI particles. Inhibitor powders showing $\rho_v = 10^6 - 10^{13} \Omega\cdot$ m are optimum for electrical deposition, those with $\rho_v > 10^{13} \Omega\cdot$ m are just as good for deposition on metal articles as polymer powders but are prone to the generation of inverse corona. Typical dielectrics displaying $\rho_v = 10^{14} - 10^{18} \Omega\cdot$ m are fit for electrical deposition only after atomizing to a high dispersion degree.

Jet electrostatic deposition of polymer coatings includes charging of powder particles transported by a gas stream to a ground metal article, their deposition on its surface and fusing.

There are two methods of charging powder particles that are most applicable in industry, namely contact and ion procedures. In the former case, a charge is imparted to a particle on contact with a metal electrode connected to a high-voltage source. In the case of ion charging, the high-voltage source is connected to corona electrodes, between which the electric charge is generated, creating a flow of ions. The ions, in their turn, precipitate onto the powder particles and charge them. Typical designs of the heads for charging polymer particles are shown in Fig. 3.35.

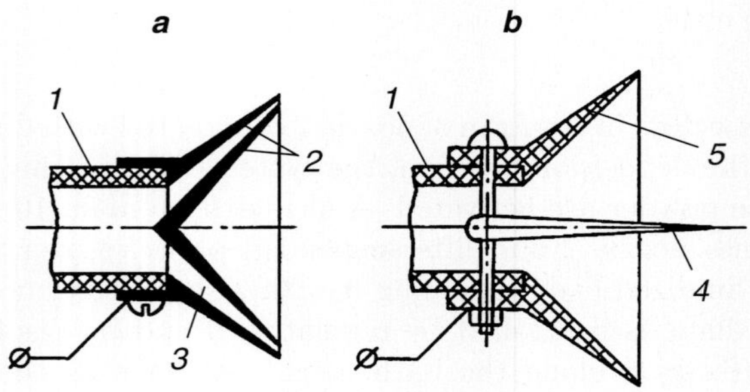

Fig. 3.35. A head for jet-electrostatic deposition of polymer powders charged by contact (**a**) or ion (**b**) mechanism. (1) tube, (2) metal mouthpiece, (3) helical channel, (4) metal needle, (5) nozzle

The powder in the head for contact charging (*a*) is carried by the airflow along the helical channels to a sharp edge of the charging mouthpiece connected to a high-voltage source. The centrifugal force generated by the helical gas-powder stream makes the particles contact the mouthpiece surface and charges them. The strongest charge is acquired by the particles neighboring the mouthpiece edge where the voltage is maximal. The polymer powder is forced into the ion charging head (*b*) through a coronary layer of the electric discharge generated at the head outlet between the metal needle and the nozzle. To regulate the concentration of the polymer and inhibitor in the electrically deposited layer one should control the powder mixture composition issuing from tube 1. Distribution of powder mixture particles in this layer depends on their size, specific mass and electric conductivity.

Such coatings are commonly deposited in a chamber through which a conveyor with articles to be coated is passed. With this aim, high-voltage sources (50–120 kV at 100–250 μA current strength) like rotor electric generators or transformers with rectifiers are used. The specifics of polymer powders as processing materials consist in their high electrifiability when being transported along the channels for deposition. In a number of cases thanks to this feature it is possible to avoid external electric sources for powder charging.

Triboelectric deposition of polymer coatings is a modification of jet technology using the electrostatic field generated by triboelectrization of polymer powders during a closed-cycle circulation.

Triboelectricity is the phenomenon of generating electric charges through friction. Both elements of the friction pair are electrified by opposite charges of equal value. The positive charge acquires a denser body, as in the case of liquids – the substance which has a stronger surface tension. A particle sliding over a body surface from which it is made electrifies negatively. Since ancient times people have drawn triboelectric series, i.e. the series of materials electrified by friction where each preceding material is more electrically positive than the following one. The most well known today are Rose's triboelectric series and those of Henry [62]. Dielectrics, to which most polymer materials belong, are arranged in the triboelectric series in decreasing order of their hardness. External effects, such as films, heating by friction and others, violate the sequence of the triboelectric series.

The technological characteristics of devices for jet spraying in which triboelectrization of polymer powders are employed are close in their characteristics to setups having external high-voltage sources (Fig. 3.35). Some examples of such devices are presented below.

A device for triboelectrostatic application of polymer inhibited coatings [63] is illustrated in Fig. 3.36.

Polymer powder poured in bin 1 is fed through hole 2 into the annular diffuser 3. Using regulator 4 on handle 5 the compressed gas is supplied through hose 6 into channel 7 and is fed through hole 8 into the diffuser cavity. A rarified area appears near the hole into which the gas from the diffuser zone with the maximum cross-sectional area is sucked. A rectilinear section of channel 7 terminating by hole 8 forms, together with the diffuser walls, an injector that excites circulation of the gas stream. Gas excess is removed from the diffuser via the filtration insert 9, which is impermeable to powder particles. The gas stream carries the powder in the diffuser into a continuous circulation. Powder particles acquire a triboelectric charge that is sufficient for electrostatic deposition on the ground metal article. By opening throttle 10 the charged powder is fed into channel 11 and is transported by the gas stream through the spraying mouthpiece to the article surface. The spraying intensity is regulated by pressing the cock of the cam mechanism 13. Gas filtered through inserts 8 is taken over channels 14 through holes 15 to cavity 16. It issues through the annular gap 17 and the charged particles are directed

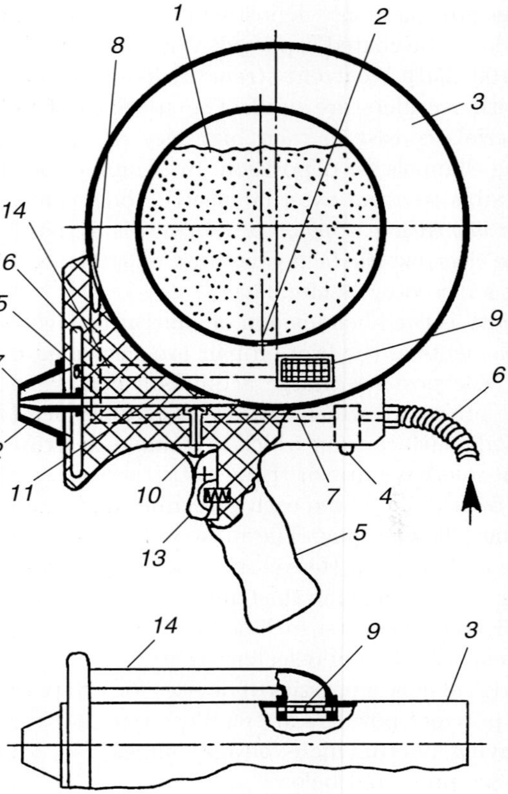

Fig. 3.36. A device for triboelectrostatic application of polymer powder compositions. (1) powder bin; (2, 8 and 15) holes, (3) diffuser, (4) regulator, (5) handle, (6) hose; (7, 11) and (14) channels, (9) filtering insert, (10) throttle, (12) mouthpiece, (13) cam mechanism, (16) cavity, (17) annular gap

on the article. This device [63] is used to apply coatings of polymer powder mixtures with CI in which the inhibitor distribution within the deposited layer is dependent upon its capability to be electrified by friction and its particle size.

An analogous design of a triboelectric generator in the form of a closed diffuser where the polymer powder circulates has been used in an apparatus for coating application [64]. High voltage (tens of kV) is supplied from the generator on the metal bottom of a vibrovortical apparatus for depositing coatings using a fluidized bed in an electrostatic field.

Triboelectrization of polymer powders can also be realized in devices containing a charging unit in the form of a vortical tube. The apparatus for the coating application [65] is an analogue to the device presented in Fig. 3.36. In this device instead of circulation inside the closed diffuser the powder is triboelectrized on sliding within a helical stream along a cylindrical socket of the swirl tube. When leaving the socket, the charged powder particles are attracted by electrostatic forces to the earthed metal article. The potential of this device is almost the same as of the apparatus presented in Fig. 3.36.

This potential can be expanded significantly by the use of an installation (Fig. 3.37) for the application of polymer powder compositions [66]. The gas-polymer powder mixture is fed from the source into helix 2 of the vortical tube and is transported in a spiral stream 3 to socket 4. Inhibitor-1 is supplied from source 5 to a gap between socket 4 and cylindrical female mouthpiece 6 fixed on helix 2 face of the vortical tube and moves in the gap in a spiral stream 7. Inhibitor-2 fed from source 8 is sprayed in flow 9 through nozzle 10 installed along the swirl tube axis.

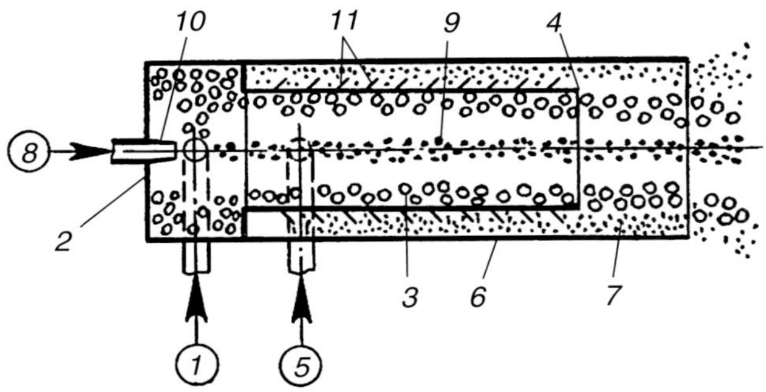

Fig. 3.37. A device for deposition of powder polymer compositions [66]. (1) gas-polymer mixture source; (2) helix; 3 and (7) spiral streams of polymer and inhibitor (1); (4) socket; 5 and (8) sources of gas flows of inhibitor-1 and inhibitor-2; (6) mouthpiece; (9) gas stream carrying inhibitor-2 particles; (10) nozzle; (11) electrodes

The design of this device presupposes regulation of triboelectric charges and electrostatic combining of powder components in the coating. Socket 4, along which the polymer powder is transported in stream 3, is fitted with electrodes 11 on its outer surface. The electrodes are imparted a charge opposite to the tribocharge of the polymer powder. Particles of inhibitor 1 acquire their own tribocharge when running in stream 7 and, on contact with the electrodes, the particles charge similarly to them. Consequently, the concentration of the inhibitor-1 in the coating can be finely adjusted. Thanks to the electrostatic attraction of the polymer and inhibitor, the losses of the inhibitor are cut to a minimum, while electrostatic repulsion makes conditions for competition between particles on deposition on the earthed article. The particles of inhibitor-2 are electrically neutral and are polarized in the fields of the moving charged particles and the deposited powder layer.

3.3.4 Liquid-Induced Modification of Powder Coatings

Many CI are liquid-phase. The application technology of inhibited coatings depends on the character of their interaction with polymer materials.

As has been specified earlier, for the modification of polymer goods it is often expedient to use CI that are thermodynamically compatible with the polymer base, i.e. dissolving or plasticizing the polymer. Such additives commonly decrease the temperature and time of the polymer coating melting. The CI content in the film polymer materials is usually negligible (about 5%), so the inhibitors cannot damage the aggregate state of these coatings. To make allowance for the high film-forming temperature of powder coatings, impregnated inhibitors should have low vapor pressure and be resistant to degradation on coating fusing.

Polymer powders absorb well, as a rule, liquid CI. Most crystalline polymers are insoluble and cannot be plasticized at room temperature. Therefore, adsorption of a liquid on powder particles results in their coagulation and noticeable reduction of powder friability. Impregnation of encapsulated CI eliminates, to some extent, this drawback but raises the price of the coatings and limits their application range [67].

The presence of liquid and plasticizing CI in the coatings improves a number of aspects of their performance. Firstly, their presence reduces residual stresses, which are the main reason of untimely cracking and exfoliation from the substrate. In addition, the elasticity modulus and hardness of the coatings diminish, while the elasticity and relative elongation increase. Moreover, the inhibited coatings display much higher adhesion than non-inhibited ones.

Coating deposition from powder-liquid blends is simplest procedure for obtaining inhibited polymer coatings. It is restricted in application by the following effects: (i) polymer particles wetted by the inhibitor stick together and reduce friability of the powder; (ii) the negative effect of most of the liquid CI on the adhesion of the polymer coating to the metal substrate. These restrictions can be overcome by choosing appropriate CI, which should be, firstly, effective enough even at insignificant concentrations and, secondly, not affect coating adhesion or improve it.

A preservative coating composition based on PE and oil solution of CI [68] presents a condensation product of alkenyl-succinic anhydride and urea, and has the following peculiarity. Mineral oil considerably reduces the adhesion of PE coating to metals, but in combination with the aforementioned CI shows satisfactory adhesive strength. The strength value measured via exfoliation of the coating from aluminum foil is about 1 N/cm. The corrosion rate of the steel substrate under this coating in aqua solutions of acids diminishes by a factor of 2 to 3 compared to the initial PE coating.

A method of obtaining anticorrosion coatings containing a liquid inhibitor reducing adhesion of the coating to the metal substrate has been patented [69]. To improve adhesion, a primary layer of the initial thermoplastic polymer is first applied onto the substrate under technological regimes optimum for reaching perfect adhesion values. After this, an interlayer is formed from a mixture of the same thermoplastic and a compatible technological liquid (esters of organic acids, epoxidized compounds, mineral oil) in concentration sufficient for gel formation. Afterwards, the technological liquid is extracted and micropores of the gel polymer matrix are filled with the inhibiting liquid. With this aim, any CI can be used, even those that are thermodynamically incompatible with the initial polymer. The interlayer treated in this way is overlaid with an upper barrier layer of the same thermoplastic or some other polymer. The three-layered coating displays higher adhesive strength and elevated resistance thanks to the presence of the inhibited layer neighboring the metal substrate.

Dosed treatment of a powder layer deposited on the substrate with a CI aerosol turned out to be highly convenient for the introduction of liquid inhibitors into polymer coatings. Polymer particles can be treated by a liquid CI directly during jet deposition (Fig. 3.32) or during electrostatic powder application. In the latter case, the properties of the electrically deposited layer are altered.

An original method of protecting metal ware against high-temperature corrosion is based on the use of powder-liquid composition [70]. A primary layer of aqua dispersion of hydrocellulose containing activated lignin and tartaric acid is applied to the surface of a steel article. The acid enters into a reaction with iron oxides on the article surface and forms a passivating film. The primary layer is then dried and a suspension of colloidal graphite in a liquid carrier (synthetic liquids or petroleum oils) is applied over it. After this, the coating is roasted above the liquid-carrier flash temperature. As a result, a passivating composite coating is formed, adding high fatigue strength to the steel ware at $T \sim 300°$C.

Coating deposition in solvent vapors is a low-temperature technique during which polymer powder particles coalesce on the substrate. Its development dates back to the 1970 s and includes the following stages [48, 50, 51]:

(1) preparation of the article surface (cleaning, degreasing);
(2) application (commonly by electric deposition) of thermoplastic powder particles on the article surface;
(3) endurance of the article with applied powder layer in vapors of the polymer solvent (swelling) at 75–115°C during 2–3 min;
(4) drying of the coating at 75–150°C for 10–30 min.

There are two modifications to the described method. The first one presupposes deposition of the coating in CI vapor, which is at the same time a solvent of the coating material. To this class of materials belong some CI, and grades IFHAN and VNH-L. According to the second method swelling of the polymer is exercised in the mixture of two vapors: that of CI and polymer solvent compatible with it. Analysis of Fig. 2.35 with kinetic curves of CI evaporation has proven that many CI can be used within this technology.

Note that most organic solvents of polymer materials (hydrocarbons, alcohols, ketones, esters) turn out to be good solvents of commercial CI too [49].

The deposition technique of powder coatings in solvent vapors is convenient for the introduction of CI into coating compositions. Upon swelling and coalescence of powder particles, the inhibitor evaporates from the polymer matrix at a rate an order or two slower than the initial one. It is expedient to use low-boiling-point polymer solvents to force the solvent to evaporate largely during the drying of the coating formed by the second variant. Molecules of the inhibitor encapsulated in the polymer matrix are retained for several years and furnish prolonged protection of the article from corrosion. The drying stage is not obligatory to obtain coatings of the first variant since the CI remaining in the coating plasticizes it.

Thermal treatment of coatings in liquids that are thermodynamically compatible with the polymer material of the coating allows impregnation of liquid-phase CI without impairing its adhesion. This original technology can be realized only in polymer-solvent systems as it is based on the formation of the gel structure in the polymer layer making up just a part of the coating thickness [71]. Its mechanism is as follows.

The coatings are thermally treated in the liquids, which are either CI compatible thermodynamically with the coating material or are suspensions or solutions of CI in the polymer solvent. During heating (below the polymer melting point) and endurance of the coating in CI it diffuses into the coating material. When cooled down the colloidal solute converts into a gel whose polymer matrix contains a liquid in its micropores. The concentration of this liquid in a local volume of the gel is conditioned by that of the colloidal solution. The resultant system is characterized by a gradient of deformation and strength characteristics of the coating across thickness. The polymer layer contacting the substrate is not in this case affected by the liquid, so the properties of the coating–article adhesive joint remain intact.

The discussed coatings have been employed to protect cooling systems made of aluminum alloys of IC engines of large trucks. As a result of machining of face parts of the system, foundry crust that is resistant to corrosion has been removed. This has led to intensive crevice corrosion on flanges of the parts under the action of antifreezes and finally to failure of the cooling system. This problem has been solved by the use of polymer coatings subjected to thermal treatment in inhibiting liquids [72]. The coatings are fitted with a system of communicating micropores whose size diminishes towards the substrate. The pores are filled with CI that is highly effective in respect to non-ferrous metals, namely a mixture of potassium and tri-ethanolamine salts of unsaturated fatty acids of tall oil. These coatings protect not only the articles onto which they are applied but conjugated metal parts in contact with the coating as well against corrosion. Under the effect of mechanical stresses, syneresis and evaporation the CI isolates from the coating in gaseous and liquid phases and fills contact gaps between mated members thus protecting both.

The above analysis of coating deposition methods visualizes that CI impregnated in coating compositions requires specific technological operations. These technologies, equipment and methodology have rightfully occupied a place in rust protection means of metal ware.

3.4 Adhesion

Adhesion, or the bond between the contacting surfaces of the coating and substrate, contributes much to the rustproofing and other service characteristics of polymer coats on metals. Impregnation of CI in polymer film-like compositions imposes a contradictory effect on polymer coating adhesion to metal ware [1,48–51]. Owing to this, as well as to the fact that novel developments in the field of protection against corrosion are referred to as a closed subject in a number of countries, this topic has not been reflected comprehensively even in resumptive monographs on polymer adhesion [73,74].

In experiments on the adhesion of a powder, HDPE modified by oil-soluble contact CI containing salts of organic amines or sulfur acids with either carboxyl or hydroxyl groups has been used. The composition of HDPE + CI has been applied to a preliminary cleaned and degreased 100−μm-thick aluminum foil at a temperature of $T = 155 \pm 5°C$ and a pressure of $p = 7-10$ MPa. The splices have then been subjected to thermal treatment in an oven at 200°C for one hour, which corresponded to the maximum strength of the adhesive joints. The strength has been estimated by delamination tests under a constant deformation rate. Not less than 20 samples were tested in each point to show 10–30% variation factor at a confidence interval of 0.95.

As experimental results have proved (Fig. 3.38), the strength of the splices under study varies depending on the CI content. The adhesive strength maximum is seen to correspond to ~1 wt% of CI content. The greatest strength increase was recorded for the composition containing SIM (the condensation product of alkenyl succinic anhydride and urea). This is the composition that added satisfactory impact resistance, elasticity and adhesion to preservative removable coats (see Sect. 3.3.4) [68]. The compositions containing Ukramin CI have shown less increase in adhesion. Exactly this inhibitor has been used in rustproof coatings [72] in the cooling systems of ICE.

Itis believed that variations in the adhesive strength of coatings under study are induced by plastification of the HDPE by oil-soluble CI. Plastification of polymers is known [75] to change drastically their physical-mechanical characteristics, including toughness, elasticity modulus, strength and hardness. This phenomenon, called antiplastification, has been analyzed in a model experiment using the example of the HDPE–mineral oil system (Fig. 3.39).

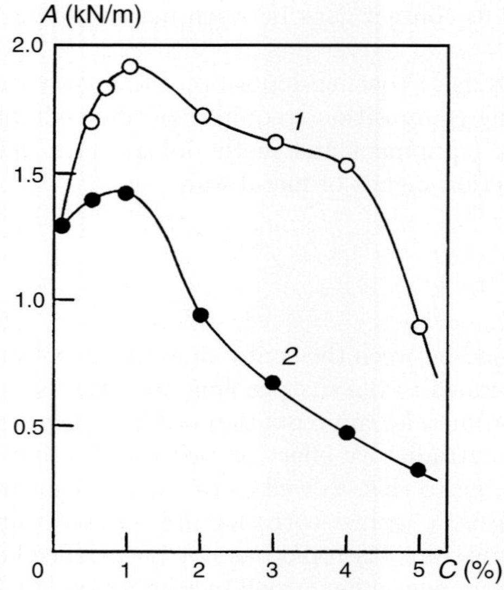

Fig. 3.38. Delamination force of the polymer coating – Al foil adhesive joints versus oil-soluble CI content in HDPE: (1) SIM; (2) Ukramin

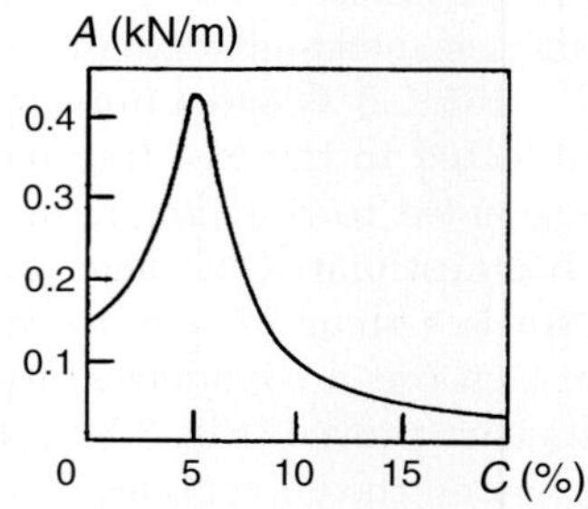

Fig. 3.39. Delamination force of the polymer coating from Cu foil as a function of mineral oil content in the coating

Variations in the adhesive strength in response to oil content can be explained as follows. Initially, adhesion increases, evidently due to the growing mobility of macromolecules, which strengthens the polymer film and contributes to adsorption of macromolecules on the substrate [76]. Further increase of oil content in the coating intensifies the competing process, during which syneresis spurs oil isolation at the coating – substrate interface where it functions as an antiadhesive. The highest adhesion value of the coating to a Cu foil has been recorded at 5% of mineral oil content in HDPE. No chemical interactions were observed in the system considered [77].

Based on this, it can be presumed that the experimental increase in the adhesion strength of inhibited coatings (Fig. 3.36) is also an effect of antiplastification processes in the system. Noncoincidence of the inhibitor concentrations, corresponding to the coating adhesive strength maximums and volume strength of the inhibited plastic [78], is apparently due to the following reasons. Squeezing of the oil component onto the coating–substrate interface makes the largest contribution to the coating adhesion. Actual oil concentration on this interface may surpass substantially the rated one found from the condition of a uniform oil distribution within the composite bulk.

Based on the concentration dependence of adhesion (similar to that presented in Fig. 3.38) the formulation of an antirust coating has been calculated [79]. It is composed of polyethylene and a sulfuric 8-oxyquinalin CI. At 5 wt% content of this CI adhesion of the coating to a steel substrate shows a maximum adhesion $A = 0.4\,\mathrm{kN/m}$ corresponding to the antiplastification peak of polyethylene by CI. It is peculiar to this composition that, after endurance in a 50% aqua solution of sulfuric acid for ten days, adhesion of the initial polyethylene coating drops almost to zero, while adhesion of the inhibited ones, though sure to reduce, remains at a level (0.1–0.2 kN/m) sufficient for normal operation.

The presented data is a confirmation that it is expedient to use oil-soluble CI as components of polymer coatings. If their concentration is optimal, the strength of adhesive joints coating metal ware can improve by 10–20%. The inhibited coatings are more resistant than those made of initial polymers. The crevice corrosion rate of the metal substrate recorded by the polarization resistance method presented in [80] lowers on transfer from the initial polyethylene coating to an inhibited one by three to ten times [81].

Fillers of polymer coatings in the form of fuming oxides (FO) turn out to be a specific type of metal CI. Fuming represents extraction of volatile components (chiefly Zn and Pb, also Sn, In, Cd and others) from a molten slag by a hydrocarbon mixture blown at 1200–1250°C. A by-product of this process is dust particles of nonferrous metal oxides differing in value of standard electrochemical potentials. In Fig. 3.40 information on the properties of thermoplastic coatings filled with FO are presented [13]. When the filler content is optimal (*a*) adhesion of the coating is seen to rise three to five times. Simultaneously, the corrosion rate of the steel substrate estimated by the polarization resistance method [80] reduces essentially (*b*) on exposure to acidic media. Of interest is the fact that the FO concentrations in the coatings corresponding to maximums on the adhesion curves and to minimums on the corrosion curves coincide for each type of polymer binder.

Thus, FO are acting as adhesion promoters and corrosion inhibitors, whose mechanism can be explained as follows. In the course of thermal molding of powder polymer coatings filled by FO the oxides enter into physical-chemical interactions with the polymer melt. As a result, the oxides are reduced in the surface layer of the particles to metals. This fact is confirmed by

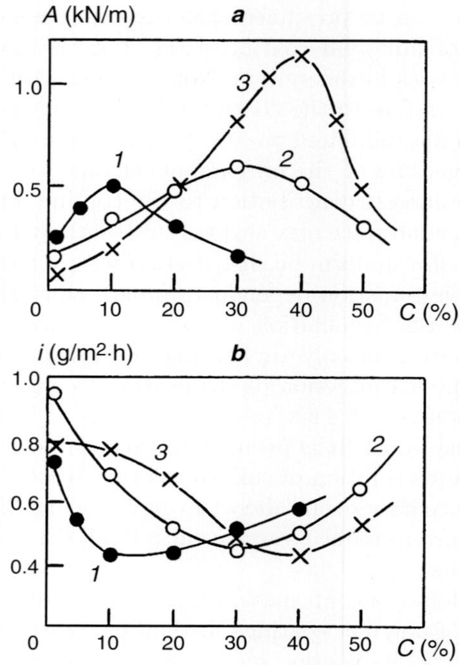

Fig. 3.40. Delamination force of the coating from Al foil (**a**) and steel corrosion rate under the coatings (**b**) as dependent on the filling degree of the coatings by FO in different polymers: (1) PE, (2) PVB, (3) pentaplast

the decreasing ohmic resistance of the coatings during formation and readings from X-ray structural analyses [13]. With prolonged thermal treatment of the coatings, the intensity of the peaks characterizing the oxide crystalline structure diminishes and that of the peaks of the non-oxidized metals augments. This results in the generation of such microchains as "metal 1 – polymer – metal 2" carrying a current in the coating in a viscous-flow state. The microchains function between the filler particles as well as between an individual particle and the metal substrate. As can be seen, the processes of electrochemical interactions between components and electric polarization of the polymer binder take place in the course of the coating formation, which assists in improving the adhesive strength of the coating. Moreover, fuming oxide particles reduced to metals make their own contribution to protection of the steel substrate against corrosion.

In addition, filling of polymer coatings with fuming oxides raises their electric conductivity, thus promoting protection of commodities from radiation and radiointerference. Formulations of the coatings described by extreme portions of the curves in Fig. 3.40 have been patented [82].

However, the introduction of CI into coatings is not always favorable for their adhesion to substrates. Besides, compensation of adhesion lost due to CI presence via electric polarization during solidification of the molten polymer layer is not always feasible. It has been proposed to fill coatings with electrically polarized dielectric particles as a preventing measure against impairing adhesion of inhibited coatings [13]. A composition has been patented [83] for coatings based on a thermoplastic filled with epoxy particles carrying a polarizing charge with a surface density of $\sigma_e = 5 - 100\,nC/cm^2$. During such coating formation the polymer binder becomes polarized in the electric field of the epoxy particles. Thus, there arises an electric component to the adhesion, which ensures a 1.5 times increase in coating adhesion compared to analogous coatings filled with non-electret particles (1.0–1.5 kN/m up to 2.5 kN/m). The corrosion rate of carbon steel covered by the coatings described lowers by a factor of three in acidic media (0.1 N solution of HCl) compared to coatings of non-filled polymers.

As the analysis has shown, impregnation of CI in polymer coating compositions significantly improves their antirust characteristics leaving sufficiently high adhesive and other service characteristics of the coatings.

3.5 The Range of Inhibited Coatings

Inhibited polymer coatings may provide protection against corrosion of metal ware on a permanent or temporary base.

The coatings for permanent protection are to perform these functions during their scheduled service life. In the case that the technological lifespan of a product surpasses that of the permanent coating, it can be repaired by applying new layers.

Coatings for temporary protection are called preservative or rustproof. They are to ensure anticorrosion protection of products in the course of production (inter-operational period), storage and transportation.

Depending on the film-forming base and coating application technology we discriminate between inhibited paint and powder coats, as well as coatings formed by adhesive tapes.

3.5.1 Paint-and-Varnish Coats

Rustproof paint-and-varnish coats are unable to fully hamper permeation of water, oxygen or ions that, on reaching the metal substrate, enter into electrochemical interactions with it. Besides, because of the diffusion of electrolytes, temperature difference, solar radiation, and so on, the barrier properties of the coatings deteriorate with time. Impregnation of CI into paint compositions may raise the protective ability of such coatings. Passivating ions of CI are either adsorbed during coating formation on the substrate or transferred to it during operation when electrolytes split water-soluble CI ions from macromolecules (by hydrolysis or a dissociation mechanism) and transport them to the substrate thus hindering corrosion.

Certain difficulties are connected with the realization of the above idea. Being surfactants or oxidizers in their essence, CI are able to interact with film-forming substances and lose their protective properties, forming undesirable side products that are hostile to the metal substrate, can initiate gelation and so on. This is why, for optimum matching of each CI-paint composition pair, the choice is made on a case-study basis, taking account of technical and economic criteria and is usually at the level of know-how. For modification of paint coats water and oil-soluble CI are commonly employed.

Water-soluble CI significantly improve the protective ability of oil, alkyd, alkyd-styrene and alkyd-nitrocellulose coatings (Fig. 3.41), for example, 3 wt% of water-soluble potassium chromate guanidine. CI has been introduced into the lacquer base, after which a 20-μm-thick single-layer coat has been applied on a plate of low-carbon steel (St 2). The time until the appearance of corrosion traces under the coating has been recorded on exposure to atmosphere at 100% humidity and a temperature of $T = 18$–$25°C$. The results presented in Fig. 3.38 prove that the introduction of this CI has perceptibly improved the anticorrosion properties of lacquer coatings [1].

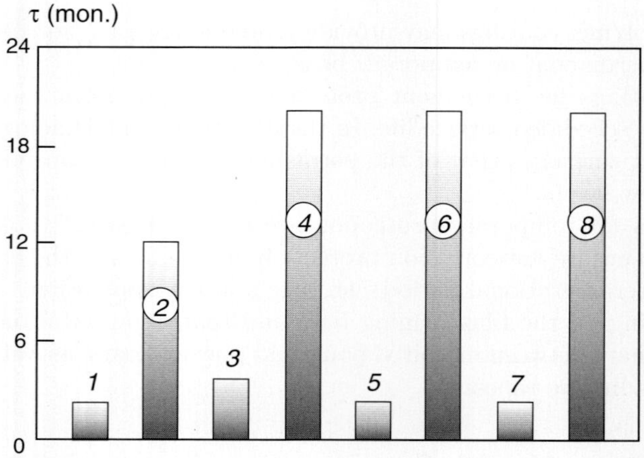

Fig. 3.41. Time until formation of corrosion traces on the steel coated by: (1) natural drying oil; (3, 5) and (7) alkyd, alkyd-styrene and alkyd-nitrocellulose lacquers; (2, 4, 6) and (8) the same modified by potassium chromate guanidine

The electrode potential of the steel plate coated by the inhibited lacquer is seen to shift to the positive side by 300–500 mV (Fig. 3.42). This strong shift in the potential, accompanied by the corrosion rate reduction, is evidence that potassium chromate guanidine hampers anodic reactions of metal ionization.

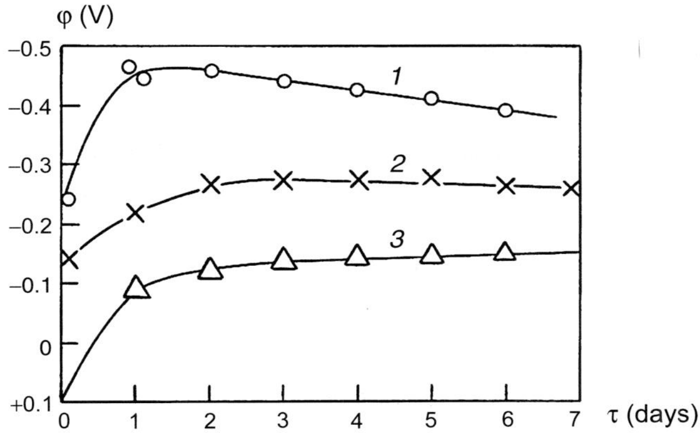

Fig. 3.42. Kinetic dependence of electrode potential of steel samples protected by coatings in 0.1 N solution of Na_2SO_4: (1) without a coating, (2) alkyd lacquer coating, (3) coating of alkyd lacquer modified by potassium chromate guanidine [1]

The comparison of IR absorption spectra of the films of natural drying oil and that modified by potassium chromate guanidine has shown that the amount of hexavalent chromium (800–900 cm^{-1}) diminishes in the inhibited film. This gives grounds for the supposition that chromium forms complex compounds with carbonyl, carboxyl and oxide groups of the drying oil. Apart from this, the intensity of the absorption bands corresponding to the NH_2 group oscillations (1600 and 3100 cm^{-1}) attenuated and new absorption bands (1580 and 1670 cm^{-1}) appeared, proving the formation of amide groups [1]. These results have lead us to the conclusion that the products formed on desiccation of the drying oil have reacted with the guanidine and chromate group of the inhibitor making the coating harder and more resistant to UV radiation.

An important characteristic of coatings containing water-soluble CI is washing out of the inhibitors. Coatings of alkyd-nitrocellulose lacquer are preferable in this respect to those of alkyd lacquer. About 70% of the chromate ions are washed out of the alkyd coatings per day, in contrast to the alkyd-nitrocellulose coatings, in which the chromate ions remain practically unwashed during the first five days and then are washed out but very weakly.

It has been established by express tests that the introduction of chromate inhibitors at amounts below 3 wt% amounts in alkyd paint compositions does not improve their anticorrosion properties as compared to non-inhibited coatings.

The rustproof characteristics of lacquer coatings can be improved by the application of synergetic combinations of CI as film-forming components. Thus, introduction of a synergetic mixture of chromate and guanidine phosphate extends by more than an order of magnitude the time to formation of corrosive damage on the steel surface under the coating in aqua salt solutes. A promising coating composition has been proposed and patented that contains chromates, phosphanates and molybdates of alkaline metals (Na_2HPO_4, Na_3PO_4, $NaNO_2$, K_2CrO_4) as CI and phosphonate of 2–phosphono–butane–1, 2, 4–tricarboxylic acid, nitrio-tris-(methylenephosphonic acid), hydroxyethylidene diphosphonic acid [84]. The described regularities are, on the whole, true for other paint binders modified by water-soluble CI.

Oil-soluble CI exhibit their specificity in paint coats in a different way than water-soluble ones. A paradoxical conclusion has been derived in [1] that the rustproofing ability of paint coats modified by oil-soluble CI is attributed to the passivating properties not of the inhibitor itself but a new passivating agent formed as a result of interactions between the inhibitor and the film-forming agent. This fact corroborates investigation results on the kinetics of anodic metal reactions in aqua extractions of coatings. For example, steel in an extraction of coatings based on the inhibited lacquer does not experience corrosion for a year thanks to strong inhibition of the anodic reaction. The electrode can be polarized to 1.5–2.0 V even under very low current densities. In aqua extractions of initial lacquers and oil-soluble CI the metal turns out to be unprotected against corrosion. This conclusion on the formation of a new passivating agent is also supported by combined investigations by IR spectrometry, X-ray phase analysis and others.

As a result of modifications of paint coats by oil-soluble CI there appeared numerous compositions for preserving metal ware during shipment and storage (see Sect. 3.5.4).

The review of inhibited paint coatings would be incomplete without mentioning coatings with CI that inhibit biological corrosion. Microbiologically influenced corrosion is the deterioration of a metal by a corrosion process that occurs either directly or indirectly as a result of the activity of living organisms [85].

Antifouling coatings hamper the formation of bacterial mucous films on machinery operating in seawater. These films commonly consist of bacteria, microscopic forms of primary and diatomaceous algae. The rate of biomass growth in such films reaches $100\,g/(m^2{\cdot}day)$. Fouling leads to catastrophic impairment of service characteristics of marine machinery, as a result of which vessel velocity reduces by 5 to 10%, marine and oceanological instruments fall out, and wave loads on hydroengineering installations increase many times [86]. Further increase of concentration above 10% does not in fact vary the properties.

Biological failure of coatings is in part attributed to mold fungi and bacteria, for which polymer materials serve as a nutrient medium. Albumen and protein-containing film-forming substances are subject to molding and decay more than all other synthetic polymers. Pigments and fillers improve resistances of coatings to microorganism attacks in a number of cases.

Antifouling coatings contain toxins in their polymer film-forming base that inhibit fungi and bacteria growth and fulfill some other functions, including

- activity at combination with a film-forming agent
- stability under external effects
- safety for man and environment

These toxic components of the coatings can be copper oxides (I), mercury, tin-bis(tributyl)oxide, barium metaborate, anilide of salicylic acid, phenols and their derivatives, compounds of aminoformic acids, etc. [87]. In the area of metal ion toxins copper protoxide, lead and tin derivatives have recently come into extensive use.

Proceeding from the application of specific matter in inhibited coatings we can single out trends for the future:

(i) synergetic mixtures of toxins containing metal ions [88] and toxins based on organic compounds [89];
(ii) nontraditional polymer film-forming agents themselves exerting a toxic effect on microorganisms

A typical example of using nontraditional film-forming matter can be antifouling coatings based on binders and obtained by either

- copolymerization of monomers of different chemical nature and a vinyl monomer [90]
- modification of silicone rubber by named copolymers [91]
- copolymerization of ethylene and monomers containing ammonia salts [92]
- polyisocyanate modification by organofluoric compounds [93]
- modification of thermoplastic and thermosetting binders by heterocyclic amino compounds, carbocyclic vinyl etheroketones, polychlorinated phosphorous compounds [94]

Acute attention has been paid in recent decades to environmental safety of antifouling coatings. Among coatings meeting this requirement one can name a fluoropolymer-based coat Compound-X modified by copper oxides (Star Brite Corp., USA). It is able to inhibit metabolism of marine acorns and other organisms generating fouling selectively but is nontoxic for higher organisms. Along with this goes a coating produced by Rust Ollum Corp (The Netherlands) in which a film-forming nontoxic fungicide called trichloromethyl thiophthalamide is introduced. Such coatings are intended for protection against biological corrosion of walls, ceilings and equipment in industrial enterprises.

Toxins are extracted via diffusion in the coating or partial dissolving of the film-forming agent. Antifouling coats do not furnish protection against electrochemical corrosion as they are intended to dissolve and liberate bactericides into the seawater. They are most often used as a biological barrier on the surface of anticorrosion coats applied onto underwater parts of vessels and marine structures.

Antimicrobial polymer coatings are intended for modification of packaging materials to inhibit spoilage of foodstuffs (bakery, confectionery, dairy, meat, fish, fruit and vegetables) [95], for which they are filled with antimicrobial components [96] such as:

- chemical preservatives (sorbic, benzoic, n-nitrobenzoic acids and their salts and esters)
- antibiotics (nisin, tetracycline and their derivatives)
- metal ions with bactericidal effects, e.g. silver ions

Formation and protective mechanisms of antimicrobial polymer coatings are in principal the same as in inhibited paint-and-varnish coats [97].

3.5.2 Powder Coatings

Powder coatings based on polymers and CI represent a long-term rustproof means for metal ware operating in hostile environments.

Their range depends on the preparation procedure of powder compositions, where the first stage is manufacture of polymer powders. In fact any polymer can be powdered mechanically. Classification of polymers by the criterion of grindability proceeds from a combination of characteristics [98]: degree of heterogeneity, relative elongation on rupture, impact viscosity, and specific grinding work. Regarding energy consumption on grinding, polymer materials can be arranged in the following series:

Short – brittle (oligomer thermosets)	< Resistant to short– brittle fracture (thermoplasts)	< Viscoelastic (polyamides, polyurethanes, polyfluoro– olefins)	< Elastoplastic (elastomers)

The grinding rate of polymers depends on the presence of the second component in the material being decomposed. For example, grinding of HDPE in a disintigrator can be improved by introduction of TiO_2 and some other powder inhibitors. Sometimes an opposite result is possible, that is, grinding of one polymer is retarded by the presence of another. It should be underlined that grinding improves at lower temperatures. Under their brittle temperature elastic polymers break like glass. The brittle temperature of oligomers is close to the T_g of the rigid-chain polymers, which is above $-40°C$; that of elastomers and some crystalline polymers is between $-6°C$ and $-120°C$.

Powders of soluble polymers (predominantly crystalline), or their blends can also be obtained by precipitation of solutions during cooling below their crystallization point. In industry this method, often called chemical, is employed for processing HDPE and polyamides.

Powder polymer-inhibitor blends can be produced by:

(1) dry mixing of powder compositions
(2) blending powder polymers with liquid CI
(3) blending molten components followed by granulation and decomposition
(4) homogenization of components in solution followed by isolation of solid particles via spray drying, crystallization or precipitation using special agents or nonsolvents.

In this regard, the dynamic strength of steel samples covered by powder inhibited coats has been examined during operation in sulfuric acid. The simplest method of estimating corrosive damage of metals turned out to be through changes in the sample mass, although electrochemical dissolution of metals in electrolytes seldom leads to refusal of the product. A more reliable sign is considered to be impairment of mechanical strength due to either local corrosion or corrosion-induced fatigue of the metal.

The tests performed were as follows. Samples in the form of 3-mm-diameter, 150-mm-long rods were covered by powder HDPE and compositions on its base. Modifiers of HDPE were mineral oil and MSDA-1 inhibitor, representing a mixture of dicyclohexylamine salts and synthetic fatty acids. Upon exposure in a 0.2 N solution of H_2SO_4 the samples underwent bending tests in accordance with the relevant standard [99] bearing in mind that dynamic strength is a more sensitive characteristic of corrosive damage than the limiting short-term strength. The dynamic strength was estimated by the number of bends until fracture. The data presented below are the mean values of 10–15 replicate tests.

The initial values of the dynamic strength of the coated samples are seen (Fig. 3.43) to be higher than those without the coatings. Apparently, polymer coatings hamper access of hostile media to the metal and contribute to

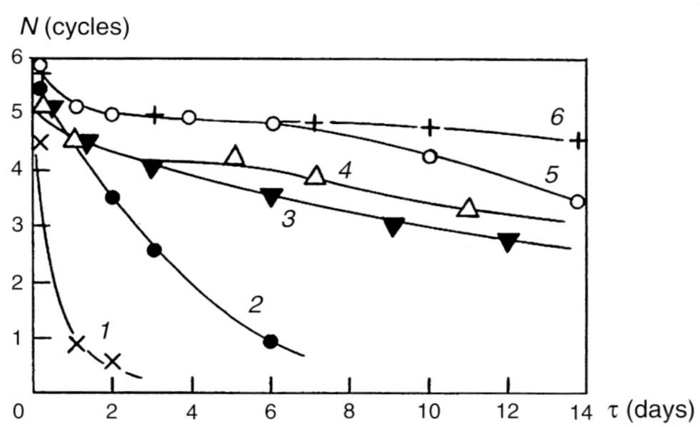

Fig. 3.43. Number of bends preceding fracture of samples versus exposition time in acid for different coating composition: (1) without a coating; (2) HDPE coating; (3) and (4) HDPE-based coat containing CI (5 and 10 wt%); (5) coating modified by oil (20%) + CI (10%)

the endurance of the samples. Molecules of gas adsorbed on the surfaces of freshly formed cracks in solids [100] are known to exert a wedging pressure that hinders fracture healing. Barrier properties of the coatings studied have shown little difference, so this weak dependence of the initial strength of the samples on coating composition may serve as an indirect confirmation of this hypothesis.

The cyclic strength of unprotected samples drops abruptly in the harsh medium (curve 1). The number of loading cycles preceding failure of the samples with PE coatings (curve 2) reduces by not more than 35%. The samples with inhibited coatings turned out to be even less well protected (curves 3–5). They are characterized by three sections on curve $N(\tau)$.

The first section presents noticeable reduction of N in the first two days of exposure, which is probably connected with degradation of the barrier properties of the coatings.

The second section is characterized by negligible changes in N and it extends with increasing CI content in the coatings (curves 3 and 4) and at similar concentration of CI (curves 4 and 5). The CI are isolating by a syneresis mechanism, which is most intensive when the liquid phase content in the polymer material reaches 35–40 wt% and above [101]. Therefore, the samples having the coating of the fifth composition are characterized by the longest second section. Being a plasticizer of HDPE and the CI solvent, oil has an additional function as a carrier that transports CI from the polymer matrix to the coating-metal interface. This is why samples 5 containing CI + Oil solute have a longer second section at similar CI concentration (10%) in coatings 4 and 5.

The third section of the curves corresponds to a linear decrease in dynamic strength of the samples versus exposition time in a hostile medium. The inclination angle of the curves in this region is roughly similar, which is largely due to the specific properties of the inhibitor. These features help to hamper corrosion of metal ware significantly using the described polymer inhibited coatings that combine the functions of barrier and electrochemically active elements.

Among the drawbacks of polymer powder coatings as CI carriers is the elevated temperature of their formation able to induce degradation of the chemical structure of the inhibitor. It is therefore inexpedient to use easily degradable CI in the coatings repaired by gas-flame PE powder spraying [102]. To protect rollers and cylinders of equipment a combined coating can be used [103] as an alternative to the inhibited coating. It includes a protecting gas-flame layer of electrically negative metal and an outer fluoroplastic layer. The outer layer serves as a barrier to media and hampers sticking of processed products to the roller. The coating is characterized by high protective capabilities at the level of the best samples of inhibited coats, but is rather costly. A modification of this inhibited coating [104], which has excellent water repellency, resistance to corrosion and snow sticking, weatherability and lubricity has been patented. The coating is based on acrylic, polyester, urethane and epoxy resins and contains up to 3 wt% fluorographite powder as a modifying agent.

The methods of depositing CI layers on metal ware, including organic hydrogenated methylbenzotriazole [105] or inorganics containing an amorphous phase [106] are often used as a preliminary operation prior to application of the barrier polymer layer isolating inhibitors from atmosphere. A convenient procedure for the deposition of inhibited powder coats is spraying of a polymer melt together with a crystal-nucleating agent [107], for which powder CI can be employed.

3.5.3 Coats Based on Adhesive Tapes

According to generally accepted international standards, oil and gas pipelines are protected against corrosion by specific coats formed by winding tapes with a sticky layer [108–110]. In contrast to hot-applied coatings, the technique of insulating pipes by tapes does not necessitate any special equipment, or any perceptible energy consumption on fusing of the polymer film. The cold-lining process presupposes application of insulation in both industrial conditions and on the site of pipe installation in trenches and is easily automated. Adhesive tapes produced by world-known companies Polyken (USA), SPI-Batiniol (France), Nitto (Japan), Long Products, Ltd. (GB) are extensively used in insulating pipelines. They offer either polyolefine or PVC tapes whose adhesive layer is based on either rubber or rubber-resin adhesive compositions. A large-scale manufacturer of adhesive tapes (grade Polylen) in Russia is Pipeinsulation Co. with affiliated ventures in Novokuibyshevsk and Syzran.

A standard technique of covering pipes with adhesive tapes (in industrial conditions) or wrapping the installed pipeline (e.g. during field repair) consists of the following stages:

1. Application of the primer coat from a liquid quick-drying primer. The primer should moisten the pipe surface well and fill its irregularities, forming a water-impermeable layer. The primer usually contains contact CI and bactericidal additives.
2. Formation of a barrier later via winding a PVC or PE tape with a glued layer containing CI or hydrophobic agents.
3. Winding of the outer layer protecting the anticorrosion insulation against mechanical damage during transportation and installation of the pipeline. This layer is made of a sufficiently thick PE film (0.5 mm), which is the adhesive coat carrier.

The insulation is formed with the help of special equipment that wraps the pipeline in the tapes without folds with about 15–25 mm overlap. A scheme for formation of the protecting coating is illustrated in Fig. 3.44. Application

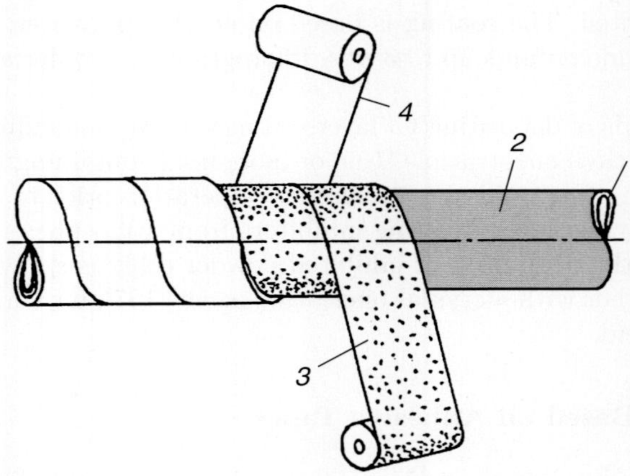

Fig. 3.44. Coating a pipeline by adhesive tapes: (1) steel pipe, (2) primer coat, (3) barrier coating, (4) protecting layer

of such coatings during pipeline maintenance repair is often complicated by the necessity to elevate sections of the pipeline on supports. A pipeline is recovered using jacks or air bags, which is more efficient than excavation.

The methodology of impregnating CI (mainly contact types) in primers is analogous to modification of paint coats.

CI representing a mixture of cyclohexamine salts and synthetic fatty acids or heteroalkylated amines [111], or an ester based on alkenyl-succinic acid and ethylene glycol with acidic number 80–250 mg KOH/g are introduced into the barrier layer composition of the adhesive [112]. Very often, substances able to retard corrosion and simultaneously adding hydrophobic properties to the adhesive serve as its components, e.g. heterocyclic polymers with tertiary nitrogen [113], or water-repellents, e.g. methylsilane triazole and phenylsilane triazole condensation products [114]. Some original technologies for direct impregnation of the adhesive layer with CI during adhesive tape production have been developed. It has been proposed to cool the glue layer on the tape-carrier heated to the viscous-flow state in a gaseous technological medium saturated with inhibitor vapors [115]. CI are not commonly introduced into the adhesive composition of the outer insulation layer.

Adhesive tapes are also extensively employed for protection of pipes with curvilinear and corrugated surfaces used in chemical industry, gas and oil field developments. The problem of lining convoluted pipes is solved [116] by rolling the tube over a roller base wrapped in a polymer tape (Fig. 3.45*a*). Roller axes are arranged at a pitch equal to the distance between crimps of the pipe. For this purpose, pinch mechanisms can also be used [117] made as a sprocket and having a roller on each beam (Fig. 3.45*b*).

In Fig. 3.46, the design of the Polyken Foam Insulation System is shown, furnishing wide opportunities for introduction of CI into the pipe insulation.

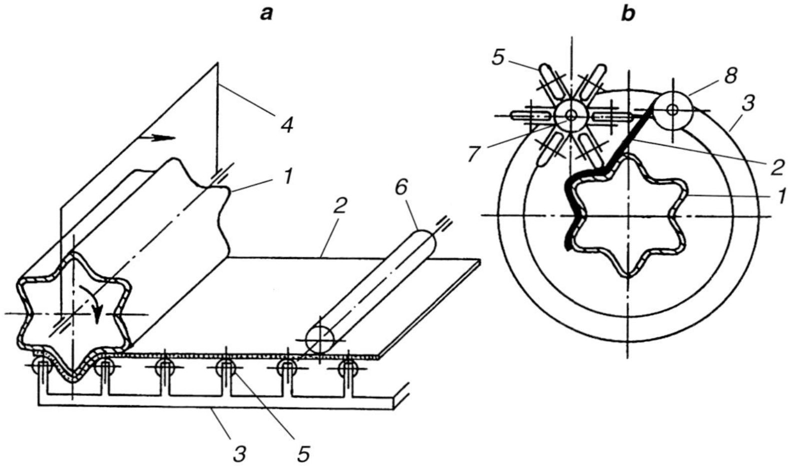

Fig. 3.45. Diagrams of devices for lining convoluted pipes: (**a**) with a roller base, (**b**) with a sprocket pinch mechanism; (1) pipe, (2) adhesive tape, (3) base, (4) drive, (5) roller, (6) shaft, (7) sprocket, (8) spindle

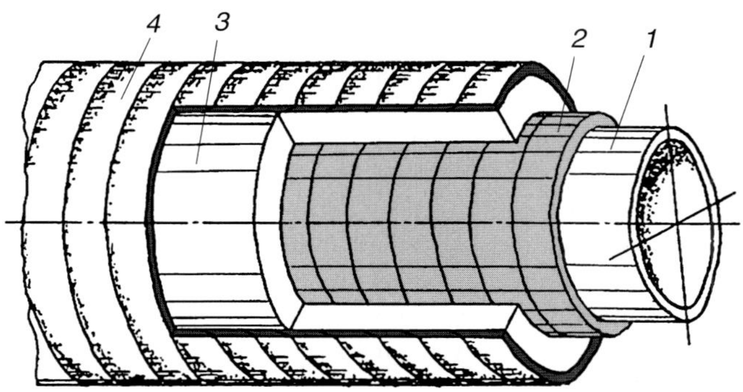

Fig. 3.46. Structure of the Polyken Foam Insulation System: (1) steel pipe, (2) anticorrosion coating, (3) sprayed PU insulation, (4) protective coating

First, a primer coat is applied onto a cleaned and prepared exterior pipe surface. Anticorrosion coatings are then wound onto the pipe in the form of adhesive PE tapes. After this, polyurethane foam is sprayed to form an insulating layer. The resin and activator are mixed at the spray nozzle head and sprayed onto the PE-wrapped pipe. The protective coating is made of adhesive–backed PE outerwrap bonded to PU foam insulation. This coating makes the entire system impervious to water and chemically resistant. The coating also improves the strength of the PU foam system. Once in the ground, Polyken Foam Insulation System efficiently insulates the product in the pipeline.

It is convenient to introduce CI into the primer or adhesive layer of the tape from which the anticorrosion coating is formed or into the PU insulation. In the latter case the PU mass is foamed by a gas saturated with CI.

3.5.4 Preservative Coats

For a long time, man has been fighting corrosion, attempting to avert corrosive damage of metal materials and commodities during production, shipment and shelf life. However modern and completely equipped a machine may be, it will inevitably fail because of the devastating effect of corrosion in case of improper storage. This is why mechanical engineering feels a strong need for coatings to cover the surface tightly, hamper diffusion of the ambient media and inhibit corrosion within a required period. As this function terminates, the coatings are slit and peeled off or washed off in solvents. Preservative coats and coatings for in-process protection commonly include polymer film-forming agents, CI, fungicides, antiseptics and process additives in their composition. The application of such coatings is regulated by special standards [118, 119]. Polymer coatings for temporary protection of metal ware are subdivided into those which are melt-applied, as well as paint and petroleum coats.

Melt-sprayed coatings have ethyl cellulose, acetyl cellulose or cellulose acetobutyrate as a film-forming base. Mineral oil, which dissolves a number of CI of ferrous and nonferrous metals and is perfectly compatible with the majority of cellulose plasticizers, is often introduced into the molten composition to lower adhesion of the coating to the goods being protected.

Such compositions are molten in special baths fitted with exhaust ventilation and a mixer, and heated to the optimum for cellulose-containing blends (160–180°C). The articles to be coated are immersed in the melt and are then cooled down, forming a solidifying film 1–2 mm thick on the surface. This film presents a diffusive barrier to corrosive matter and protects to some extent from mechanical damage. The polymer film is found in the state of a gel, thus the liquid phase is separated by the syneresis mechanism. When liberating at the coating-article interface the liquid phase reduces adhesion practically to zero thanks to its oil base coating. CI dissolved in the oil are liberated onto the article surface and adsorb on the surface thus realizing the inhibiting mechanism of anticorrosion protection. Depending on the efficiency of the CI, the terms of storing steel ware may be from five to ten years indoors and up to three years in a shed.

Coatings removed from articles can be remolten and reused for preservation.

Paint-and-varnish coatings for temporal protection of metal ware are subdivided into wipe-off and detachable.

Wipe-off coats are not numerous. Coating compositions based on neutralized synthetic fatty acids modified by a mixture of water- and oil-soluble CI are diluted by white spirit to the required viscosity. They are either sprayed, brushed or applied by dipping. These coatings are most often used to protect against corrosion at elevated humidity and are washed off by organic solvents or hot aqueous solution of soda.

Wipe-off coats based on alkyd-styrene lacquer and oil-soluble contact CI form a bright goldish film on the surface protecting both ferrous and nonferrous metal ware from corrosion during storage in unheated depots.

Detachable or removable paint coats are applied from compositions on the base of polymers in organic solvents or aqueous latexes.

Coatings of the first group are able to protect metal ware against the effects of industrial atmosphere containing sulfur dioxide and nitrogen oxides. Polyvinyl chloride and vinyl chloride copolymers as well as chlorine, and fluorine derivatives of polyethylene constitute their film-forming base. Their compositions also contain oil-soluble CI, pigments, plasticizers and antiblock agents. The coatings are air-sprayed by applying multiple layers with intervening drying. Being 200 to 300 μm thick, the coatings are able to preserve products of ferrous and nonferrous metals for several years (up to ten). However information on the duration of the anticorrosion protection of goods with paint coats in different storage conditions is very contradictory [1].

Detachable coats based on latexes are much cheaper. Their compositions are based on water dispersions or emulsions of the following polymers:

- copolymers of acrylic acid and its derivatives
- butadiene-styrene and other rubbers
- copolymers of polyvinyl alcohol and partially saponified polyvinyl acetate with vinyl monomers

CI are a requisite component of these compositions since their water may initiate corrosion of metal substrates that are already at the stage of film forming. Various water-soluble inhibitors can be used in such coatings, including organic and inorganic chromic acid salts, sodium or ammonium benzoates, sodium nitrite, boric acid salts and others [50,51]. As latexes represent unstable colloidal systems, they are impregnated with surfactants whose functions are fulfilled in part by CI. Such coatings are air-sprayed in two or three layers.

Petroleum inhibiting preservative coats cannot be strictly related to plastics. They should be mentioned here since these coats contain high-molecular substances as a protecting and film-forming component. Besides this class of inhibited materials is progressing rapidly, showing a strong future in combating corrosion in machines and equipment.

Coatings of this type are based on film-forming inhibited petroleum compositions. In English-speaking countries these coatings are called "Fluids", in Russia "PINS" (abbreviation). They are used as a temporal anticorrosion means and consist of high-molecular film-forming oil products containing CI and solvents. When applied to metal ware the coating is formed upon evaporation of the solvents. Their properties are close to those of paint coats and wax films. These fluids belong to the class of wipe-off coats, so can be removed by mineral thinners.

The fluids are subdivided into the following groups of coats [120]. The first group is intended for long-term preservation of metal ware stored in the open air, as well as for stationary metal structures and machinery. They consist of bitumen-rubber-filled or bitumen-polymer-wax, or polymer-wax-filled compositions, along with bentonite, silica gel, commercial carbon, metal oxides as fillers as well as oil-soluble CI and organic solvents as requisite components. The fluids of grades Tectyl 506 (Valvoline Oil Co., USA), SACI 200 (Witco Chemical,USA), Rustline AG 191 (Kalubroil, Italy) form up to 200-μm-thick solid protective films on metal surfaces.

Another group of compositions of this type is used to protect metal friction joints during transportation and preservation. They represent compositions thickened by polymers of soap, silica gel and soap-hydrocarbon lubricants impregnated with anti-frictional fillers (molybdenum or graphite disulfide), oil-soluble CI, anti-seizure additives and surfactants. To this type of Fluids belong such compositions as Tenox Film 30 (Sophes Co., France), Koral 530 (Hungary), SACI 500 and Shell Ensis Fluid 256 (Shell, GB).

Substances for corrosion inhibition in difficult-to-access places on metal ware are a special group of Fluids. A pioneer in creation of such Fluids is a Swedish automobile association "Motormenniens Rykosferbund". Its soap-polymer-wax compositions are distinguished by high content of oil-soluble CI, surfactants and hydrocarbon solvents. The compositions are produced as aerosols and are extensively employed in automobile, railroad, aviation and agricultural spheres. To name but a few among a broad spectrum of products, e.g. Tectyl 309 ML, Castrol Rustilo 160 (Castrol Industrie GmbH,Germany), and Bygging Professional 120.

There are also compositions that display high tip leakage temperature and elevated thixotropy, including Protex Car Dit (Milloil, Italy), Profet V 800 (Kalubroil, Italy), and Tectyl 309 AG.

A separate group of Fluids is intended for protection of spare parts and semis during in-process storage. They embrace mineral and synthetic oils with polymer and soap-polymer thickeners with high oil-soluble CI content. Water-based compositions in aerosols have gained wide popularity recently, including such Fluids as Rustline AE 170 and Protex W/52 on a water base, Contractor SSN (Kluber Lubrication AG, Germany) based on an organic chlorine solvent and Contractor A-40, which is a concentrate to be diluted before use.

It should be emphasized that inhibited polymer coatings are a separate class of polymer coats. The necessity of CI impregnation into their polymer base, the requirement of preserving coating adhesion, the demand of inhibitor delivery to the metal substrate, etc. have stipulated their original compositions and specific technological means of manufacture. Note that inhibited coatings are more expensive than usual polymer ones but this is quickly compensated thanks to their high service characteristics and perfect corrosion suppression features.

In recent decades a novel type of coatings based on organic metals that constitutes an alternative to inhibited coatings has appeared. However, their apparent advantages in terms of ecological safety do not look so beneficial after taking into account the technological difficulties of their processing, high cost, production organization of synthesizing conducting polymers and so on. Coatings based on stock-produced polymers and CI have been a traditional product of the chemical industry for a long time and show, for the time being, far more advantages supported by a long experience. The question of what type of coatings will become most efficient in the near future in technical and economic respects is still disputable. Only time will show, although it is already evident today that a wide range of polymer inhibited coatings are required to fight corrosion. Each type of these coatings should occupy its own place in the arsenal of rustproofing measures; the largerthe arsenal, the less damage will be inflicted by corrosion.

References

1. I.L. Rosenfeld and F.I. Rubinstein. *Antirust Primers and Inhibited Paint Coats.* Moscow, Khimia, 1980.
2. N.I. Egorenkov, D.G. Lin and V.A. Belyi. Effect of metals on melt oxidation of polyethylene. *J. Polym. Sci.: Polym. Chem.*, 1975, Vol. 13, pp. 1493–1498.
3. A.S. Mikhnevich and L.S. Pinchuk. Potentiostatic investigation of protective properties of polyethylene coatings on carbon steel, *Protection of Metals*, 1972, Vol. 8, No. 1, pp. 126–127.
4. A.S. Khanna. *Introduction to High Temperature Oxidation and Corrosion.* NY, ASM Int., 2002.
5. L.S. Pinchuk, A.S. Mikhnevich and L.S. Koretskaya. On the protective ability of pentonic coatings in electrolytes, *Protection of Metals*, 1973, Vol. 9, No. 1, pp. 120–121.
6. J.J. Senkevich. Degradation of an alkyd polymer coating characterized by AC impedance. *J. Mater. Sci.*, 2000, Vol. 35, No. 6, pp. 1359–1364.
7. L. Nielsen. The penetration of filled polymers. *J. Macromol. Sci.,* 1967, Vol. 1A, pp. 929–942.
8. Yu.S. Lipatov. *Physical Chemistry of Filled Polymers.* Moscow, Khimiya, 1977.
9. L.S. Pinchuk. Investigations of capillary liquid permeation into metal-polymer joints. *Trans. Belarus AS, ser. Phys.-Tech. Sci*, 1978, No. 2, pp. 119–122.
10. L.S. Pinchuk and A.S. Mikhnevich. On the substrate effect upon formation of spherulitic thermoplastic structure from melt. *Proc. Belarus AS*, 1976, Vol. 20, No. 2, pp. 137–140.
11. V.A. Goldade and L.S. Pinchuk. Protective ability of polymer coats against liquid media. *Trans. Belarus AS, Ser. Phys.-Tech. Sci.*, 1982, No. 2, pp. 50–54.
12. H.H. Uhlig and R.W. Revie. *Corrosion and Corrosion Control.* 3-rd ed. NY, Chichester, Brisbane, Toronto, Singapore, John Wiley & Sons, 1984.

13. V.N. Kestelman, L.S. Pinchuk and V.A. Goldade. *Electrets in Engineering: Fundamentals and Applications.* Boston /Dordrecht/ London, Kluwer Academic, 2000.
14. V.G. Plevachuk, I.M. Vertyachikh, VA. Goldade and L.S. Pinchuk. Effect of the charge of a polymer electret on the spreading of liquids. *Vysokomolecul. Soedinenia,* 1995, Vol. A37, No. 10, pp. 1071–1074.
15. I.M. Vertyachikh, V.A. Goldade, A.S. Neverov and L.S. Pinchuk. The effect of electric field of a polymer electret on sorption of organic solvent vapors, *Vysokomolecul. Soedinenia,* 1982, Vol. B24, No. 9, pp. 683–686.
16. Y.I. Voronezhtsev. Effect of electric polarization of polymer coatings on their diffusive penetrability, *Trans. Belarus AS, Ser. Phys.-Tech. Sci.,* 1986, No. 4, pp. 42–44.
17. A.A. Okhlopkova, Pinchuk, VA. Goldade and V.N. Kestelman. On the mechanism of thermoplastics modification by ultradispersed inorganic fillers. *Intern. J. Polymeric Mater.,* 2000, Vol. 46, pp. 11–25.
18. B. Wessling and J. Posdorfer. Corrosion prevention with an organic metal (Polyanilin): corrosion test results. *Electrochimica Acta,* 1999, Vol. 44, pp. 2139–2147.
19. T.P. McAndrew. Corrosion prevention with electrically conductive polymers. *Trends in Polym. Sci.,* 1997, Vol. 5, No. 1, pp. 7–12.
20. F. Garnier. Les polymeres conducteurs. *La Recherche,* 1987, No. 193, pp. 1306–1312.
21. R. Racicot, R. Brown and S.C Yang. Corrosion protection of aluminium alloys by double-strand Polyaniline. *Synt. Metals,* 1997, Vol. 85, No. 1–3, pp. 1263–1264.
22. K. Jdla, A. Talo, H. Niemi, et al. An XPS and AFM study of polypyrrole coatings on mild steel. *Surf. & Interface Anal.,* 1997, Vol. 25, No.11, pp. 837–854.
23. A. Yfantis, I. Paloumpa, D. Schmeiber and D. Yfantis. Novel corrosion-resistant films for Mg alloys. *Surf. and Coat. Techn.,* 2002, No. 151–152, pp. 400–404.
24. N.V. Krstaijic, B.N. Grgur, S.M. Jovanovic and M.V. Vojnovic. Corrosion protection of mild steel by polypyrrole coatings in acids sulfate solutions. *Electrochem. Acta,* 1997, Vol. 42, No. 11, pp. 1685–1691.
25. L.C. Lei, M. Wu, H.M. Shu, et al. Preparation and EIS study on polymide/polyaniline blend film for corrosion protection. *Polym. Adv. Technol.,* 2001, Vol. 12, No. 11–12, pp. 720–723.
26. L.M. Liu, and K. Levon. Undoped polyaniline – surfactant complex for corrosion prevention. *J. Appl. Polym. Sci.,* 1999, Vol. 73, No. 14, pp. 2849–2856.
27. C.K. Tan and D.J. Blackwood. Corrosion protection by multilayered conducting polymer coatings. *Corr. Sci.,* 2003, Vol. 45, pp. 545–557.
28. Beschichtungs system auf basis eines organischen metalls. *Galvanotechnik,* 1997, Vol. 88, No. 5, p. 1568.
29. Kluthe und Ormecon mit qemeinsamen vertrieb. *Galvanotechnik,* 1998, Vol. 89, No. 12, pp. 4060–4061.
30. G.M. Spinks, A.J. Dominis, G.G. Wallace and D.E. Tallman. Electroactive conducting polymers for corrosion control. Part 2. Ferrous metals. *J. Solid State Electrochem.,* 2002, Vol. 6, No. 2, pp. 85–100.
31. P. Li, T.C. Tan and J.Y. Lee. Corrosion protection of mild steel by electroactive polyanilin coatings. *Synth. Metals,* 1997, Vol. 88, pp. 237–242.

32. N. Ahmad and A.G. Mac Diarmid. Inhibition of corrosion of steels with the exploitation of condudting polymers. *Synth. Metals*, 1996, Vol. 78, No. 2, pp. 103–110.
33. S. Wencheng and O.I. Jude. Electrodeposition mechanism, adhesion and corrosion performance of polypyrrole and poly(N–methylpyrrole) coatings on steel substrates. *Synth. Metals*, 2000, Vol. 114, pp. 225–234.
34. I. Krylova. Painting by electrodeposition on the eve of the 21-st century. *Progr. Org. Coat.*, 2001, Vol. 42, pp. 119–131.
35. S. Roul, S.K. Dhawan and R. Chandra. Compensated sulphonated polyaniline – correlation of processability and crystalline structure. *Synth. Metals*, 2001, Vol. 124, No. 2–3, pp. 295–299.
36. B. Del Amo, R. Romagnoli and V.F. Vetere. Steel corrosion protector by means of alkyd paints pigmented with calcium acid phosphate. *Ind. and Eng. Chem. Res.*, 1999, Vol. 38, No. 6, pp. 2310–2314.
37. U.S. Patent 3833513, C 23 F 11/10. Gels inhibiting corrosion electric contacts. J. Fath, 1974.
38. Russian Patent Application 93053613/04, C 09 D 195/00, 5/08. *A composition for anticorrosion protection of metals.* Yu.I. Blokhin, E.E. Nifantyev, V.A. Mityagin, et al, 1996.
39. Peelable coating protects flanges, *Chem. End.* (U. S.), 1998, Vol. 105, No. 13, p. 23.
40. Re–usable corrosion barrier coating, *Anti-Corros. Met. and Mater.*, 1999, Vol. 46, No. 2, pp. 128–129.
41. Russian Patent 2034885, C 09 D 127/06, 5/20. *A film coating composition for temporary anticorrosion protection.* Yu.G. Skrypnik, N.V. Vasilyeva and V.V. Popov, 1995.
42. U.S. Patent 6146701, B 05 D 31/00. *Process for improving the adhesion of polymeric materials to metal surfaces.* D. Ferrier, 2000.
43. U.S. Patent 5100942, C 08 L 75/00. *Corrosion resistant acrylic coatings.* Ch.R. Hegedus, D.I. Hirst, W.I. Green and A. T. Eng, 1992.
44. Russian Patent Application 93010524/04, C 09 D 5/12. *Primer-inhibitor for anticorrosion protection of aluminum alloys.* M.V. Kirsanova, L.G. Shodet, G.M. Tseitlin and S.V. Vedyakin, 1996.
45. Yu.I. Kuznetsov. Corrosion inhibitors in conversion coatings. Part III. *Protection of Metals*, 2001, Vol. 37, No. 2, pp. 119–125.
46. Transparent schutz. Korrosionsschutz fur transport und lagerung. *Production*, 1997, Vol. 36, No. 40, p. 15.
47. A.M. Dorfman, A.M. Lyakhovich, V.I. Povstugar and S.G. Bystrov. Plasma modification of a protective coating formed by hexamethylenimine m-nitrobenzoate on iron. *Protection of Metals,* 2000, Vol. 36, No. 3, pp. 291–297.
48. A.D. Yakovlev, V.F. Zdor and V.I. Kaplan. *Powder Polymer Materials and Coatings.* 2nd edn., Leningrad, Khimiya, 1979.
49. M. Ash. and I. Ash. *Handbook of Paint and Coating Raw Materials.* 2nd edn. NY, Synapse Information Resources Inc., 2003.
50. Z.W. Wicks, F. N. Jones and S.P. Pappas. *Organic Coatings. Science and Technology.* Vol. 2: *Applications, Properties, and Performance.* Chichester, Wiley, 1994.
51. J.N. Undady, ed. *Coatings and Coatings Processes for Metals.* NY, ASM Publ., 1998.

52. V.A. Dovgyalo and O.R. Yurkevich. *Composite Materials and Coatings Based on Dispersed Polymers. Technological processes.* Minsk, Nauka, 1992.
53. U.S.S.R. Patent 466913, B 05 B 1/00. *Device for coating application from powder materials.* L.S. Pinchuk, A.S. Mikhnevich, O.G. Kokhanovskyi and L.S. Koretskaja, 1975.
54. U.S.S.R. Patent 383634, B 44 D 1/095. *Device for application of polymer coatings.* V.A. Belyi, A.S. Mikhnevich and L.S. Pinchuk, 1973.
55. U.S.S.R. Patent 761017, B 05 B 7/20. *Device for application of polymer coatings.* K.M. Gilman, V.V. Kirpa, F.S. Mamedov, L.S. Pinchuk and M.A. Poluyanov, 1980.
56. U.S.S.R. Patent 730382, B 05 C 19/00, B 05 B 7/18. *A Device for coating deposition.* V.A. Goldade, A.S. Neverov, L.S. Pinchuk, et al., 1980.
57. U.S. Patent 1952281, B 05 B 5/08. *Method and apparatus for obtaining from fluid under pressure two currents of fluids at different temperatures.* G.J. Ranque, 1934.
58. L.S. Pinchuk, V.A. Goldade, A.V. Makarevich, V.N. Kestelman. *Melt Blowing. Equipment, Technology, and Polymer Fibrous Materials.* Berlin Heidelberg New York, Springer-Verlag, 2002.
59. U.S.S.R. Patent 510273, B 05 B 7/20. *Device for coating application from polymer melts.* V.A. Belyi, A.S. Mikhnevich, L.S. Pinchuk, et al., 1976.
60. U.S.S.R. Patent 587674, B 05 D 1/04. *Method for coating deposition on metal substrates.* V.A. Belyi, V.A. Goldade and L.S. Pinchuk, 1977.
61. U.S.S.R. Patent 509296, B 05 B 5/08, B 05 D 1/22. *Installation for coating application from polymer powders.* V.M. Katsnelson, A.S. Mikhnevich, L.S. Pinchuk, et al., 1976.
62. D. Dowson. *History of Tribology.* Bury St. Edmunds, Mechanical Engineering Publ., 1997.
63. U.S.S.R. Patent 388793, B 05 B 5/02. *Device for application of polymer coatings.* V.A. Belyi, A.S. Mikhnevich and L.S. Pinchuk, 1973.
64. U.S.S.R. Patent 403446, B 05 B 5/02. *Apparatus for coating deposition.* V.A. Belyi, A.S. Mikhnevich and L.S. Pinchuk, 1973.
65. U.S.S.R. Patent 601053, B 05 B 5/02. *Device for application of polymer powder coatings.* L.S. Pinchuk, V.A. Belyi and V.A. Goldade, 1978.
66. U.S.S.R. Patent 533401, B 05 B 5/08, B 05 B 19/02. *Device for application of polymer powder compositions.* V.A. Belyi, V.A. Goldade, A.S. Neverov and L.S. Pinchuk, 1976.
67. U.S.S.R. Patent 1236743, B 09 D 3/10, F 16 J 15/20. *Sealing coating (its modifications).* K.M. Gilman, L.S. Pinchuk, V.A. Goldade, et al., 1983.
68. D.D.R. Patent 263418, C 09 D 5/08, 5/20, 3/727. *Masse fur eine korrosionsschutzschicht.* V.A. Goldade, Ya. M. Zolotovitskyi, A.S. Neverov, L.S. Pinchuk, 1989.
69. U.S.S.R. Patent 531554, B 05 D 1/38, 7/14. *A method of coating application.* V.A. Belyi, L.S. Koretskaya, A.S. Nevero, and L.S. Pinchuk, 1976.
70. U.S.S.R. Patent 1208676, C 23 C 8/00. *A method of protecting metal surfaces from high-temperature corrosion.* K.M. Gilman, V.A. Goldade, L.S. Pinchuk, et al., 1985.
71. U.S.S.R. Patent 571078, C 08 J 7/08. *A method of thermal treatment of thermoplastic coatings.* L.S. Pinchuk, K.M. Gilman, V.A. Goldade, et al., 1977.
72. U.S.S.R. Patent 872874, F 16 J 15/00. *Sealing coating.* V.A. Goldade, A.S. Neverov, L.S. Pinchuk and I.T. Solovyev, 1981.

73. A. Pizzi, K. L. Mittal, ed. *Handbook of Adhesive Technology.* NY, Basel, Hong Kong, Marcel Dekker, Inc., 1994.
74. R. J. Hussey and J. Wilson. *Structural Adhesives Directory and Databook.* London, Chapman & Hall, 1996.
75. P.V. Kozlov and S.P. Papkov. *Physico-Chemical Foundations of Polymer Plastification.* Moscow, Khimiya, 1982.
76. V.L. Vakula and L.M. Pritykin. *Polymer Adhesion Basic Physico-Chemical Principles.* Hertfordshire, Simon & Schuster Int. Group, 1991.
77. A.S. Neverov. *Scientific foundations of developing low-modular engineering materials based on thermoplastics and liquids of restricted compatibility.* Dr. Sci. Thesis, Gomel, 1993.
78. U.S.S.R. Patent 761537, C 09 K 3/10, C 08 L 23/06. *Composition for a sealing material.* L.S. Pinchuk, K.M. Gilman, V.A. Goldade, et al., 1980.
79. U.S.S.R. Patent 1374753, C 09 D 5/08, 3/74. *Composition for anticorrosion coatings.* V.A. Goldade, L.S. Pinchuk, O.E. Reinvald, et al., 1986.
80. V.A. Goldade, Ya.M. Zolotovitsky, A.S. Neverov and L.S. Pinchuk. Estimation of protecting capability of inhibited film materials by the method of polarization resistance. *Protection of Metals,* 1982, Vol. 18, No. 6, pp. 946-949.
81. V.A. Goldade, A.S. Neverov, L.S. Pinchuk and G.V. Rechits. On adhesion of inhibited polymer composites to metals. *Trans. Belarus AS, Ser. Phys-Tech. Sci.*, 1984, No. 4, pp. 34–36.
82. U.S.S.R. Patent 666189, C 09 D 5/08, C 08 K 3/22, C 08 L 23/06. *Composition for coatings based on a thermoplastic polymer.* V.A. Belyi, K.M. Gilman, V.A. Goldade, V.V. Kirpa and L.S. Pinchuk, 1979.
83. U.S.S.R. Patent 1358383, C 09 D 5/03, 3/74. *A composition for coatings.* Yu.I. Voronezhtsev, I.M. Vertyachikh, V.A. Goldade, L.S. Pinchuk and V.V. Snezhkov, 1987.
84. ES Patent 8802626, C 23 F 11/18, 11/173, F 25 D 47/00. *Inhibiting corrosion in aquatic systems.* B. Greaves and S.J. Grenside, 1988.
85. E. Heitz, H.-C. Fleming, W. Sand. *Microbially Influenced Corrosion of Materials. Scientific and Engineering Aspects.* Berlin Heidelberg New York, Springer, 1996.
86. Sh.W. Dean, G.H.-D. Delgadillo, J.B. Bushman, eds. *Marine Corrosion in Tropical Environments.* Philadelphia, ASTM Press, 2000.
87. A.D. Yakovlev. *Chemistry and Technology of Paint and Varnish Coatings*: Handbook for Higher Educational Institutions, Leningrad, Khimiya, 1989.
88. U.S. Patent 5071479, C 09 D 5/16. *Biocidic composition.* R. Gruening and B. Ring, 1990.
89. U.S. Patent 5069717, C 09 D 5/14. *Non-fouling compositions containing diphenyl esters and lysozyme, methods of fighting algae and coating compositions containing antifouling additives.* S. E. Sherda, N.J. Willingboro and R.J. Mehta, 1991.
90. JP Patent 5058464, C 09 D 143/04, 151/00, 5/00. *Coating composition not fouling in water.* M. Sigaru and D. Hirosi, 1993.
91. JP Patent 6019070, C 09 D 133/14, 183/02. *Coating composition not fouling under water.* N. Masaesi et al., 1994.
92. JP Patent 6019069, C 09 D 201/02. *Film-forming composition of nonfouling paint.* Ya. Naoki, H. Akio et al., 1994.
93. U.S. Patent 6261695, B 32 B 27/00. *Coating composition, treating method and treated substrate.* K. Kazusuke, 2001.

94. WO Patent 92/20747, C 09 D 5/14, A 01 N 25/34. *Marine compositions with selectively concentrated domains of tin-free organic additives against fouling with microorganisms.* V.D. McGiniss and R.J. Dick, 1992.
95. W.A. Jenkins and J.P. Harrington. *Packaging Foods with Plastics.* Detroit, Technomic Publ. Co., 1991.
96. R. Singhal and P. Kulkarni. Food spoilage, preservation techniques and role of packaging. *Modern Food Packaging,* Mumbai, Indian Inst. of Packaging, 1998, pp. 21–34.
97. V.A. Goldade, L.S. Pinchuk, A.V. Makarevich, et al. New polymeric packaging film for food produkts. *Modern Food Packaging*, Mumbai, Indian Inst. of Packaging, 1998, pp. 234–242.
98. A.D. Yakovlev. *Powder Paints.* Leningrad, Khimiya, 1987.
99. Russian State Standard 1579-93. Wire. Bending test method.
100. V.I. Lihtman, Shchukin and P.A. Rehbinder. *Physico-chemical Mechanics of Materials.* Moscow, Publ. USSR AS, 1962.
101. V.A. Belyi, V.A. Goldade, L.S. Pinchuk, et al. Inhibited polymer materials for protection against corrosion. *Proc. USSR AS*, 1982, Vol. 267, No. 5, pp. 1163–1166.
102. JP Patent claim 2002069604, C 23 C 4/04, B 05 D 7/14, C 23 F 15/00. *Preparing method for lining film for corrosion resistance.* H. Wakabayashi, K. Sonoya, N. Kiga and M. Tamano, 2002.
103. U.S. Patent 5283121, B 29 B 15/00. *Corrosion and abrasion resistant industrial roll coating with non-sticking properties.* B.A. Bordner, 1994.
104. JP Patent 6248223, C 09 D 127/12, B 29 C 67/04, B 29 B 27/18. *Material excellent in water repellency, resistance to corrosion and snow sticking, weather ability and lubricity and its production.* G. Yamauchi, 1994.
105. U.S. Patent 5874026, C 23 F 11/14. *Method of forming corrosion inhibiting films with hydrogenated benzotriazole derivatives.* J. P. Pilsits, J.S. Gill and A.M. Cognetti, 1999.
106. U.S. Patent claim 2001024729, C 23 F 11/06, B 23 B 9/06, 18/00. *Corrosion resistant coatings containing an amorphous phase.* D.M. Price, N.M. McGowan, D.R. Webb, et al., 2001.
107. DE Patent 3605555, B 05 D 3/02, 3/10, 5/12, 7/16, 7/24. *Process for producing a corrosion–protective plastic layer.* R. Ostwald and G. Voit, 1987.
108. I.V. Kuznetsov, V.F. Novoselov, P.I. Tugunov and V.F. Kotov. *Anticorrosion Protection of Pipelines and Reservoirs.* Handbook. Moscow, Nedra, 1992.
109. *Handbook of Valves, Piping and Pipelines.* London, Elsevier, 1986.
110. *Coatings and Coatings Processes for Metals.* Ohio, ASM Int., 1998.
111. Belarus Patent 6083, C 09 J 7/02. *Glue composition for adhesive insulating tapes.* L.S. Pinchuk, I.V. Tsarenko, V.A. Goldade, A.V. Makarevich et al., 2003.
112. U.S.S.R. Patent 1652326, C 09 J 7/02. *Glue for adhesive tapes.* S.V. Bezrukov, Yu.I. Voronezhtsev, V.A. Goldade, L.S. Pinchuk, et al., 1991.
113. EP 471336, C 09 J 8/02. *Polymeric adhesive and anti-corrosion agents.* I. Ahmed, T.W. Johnson, H.L. Hsieh, 1992.
114. U.S.S.R. Patent 1709724, C 09 J 7/02. *Adhesive tape.* A.N. Myasischeva, N.A. Suklyasyan, A.I. Khrakovsky and S.A. Krekoten, 1995.
115. Belarus Patent 6326, C 09 J 7/02. *Method to manufacture adhesive tapes.* L.S. Pinchuk, N.I. Tishkov, V.A. Goldade, A.V. Makarevich, et al., 2004.

116. U.S.S.R. Patent 1609701, B 29 C 63/06. *Facing method of outer surface of pipe.* N.I. Tishkov, E.M. Markov, V.A. Goldade, L.S. Pinchuk, et al., 1990.
117. U.S.S.R. Patent 1620320, B 29 C 63/14. *Device for facing outer surface of pipes by tape insulating material.* E.M. Markov, V.A. Goldade, L.S. Pinchuk, et al., 1991.
118. Russian State Standard 9.014-78. Temporary anticorrosion protection of goods. General requirements.
119. Russian State Standard 9.028-74. In-process anticorrosion protection of blanks, parts and assembly units of metal ware. General requirements.
120. T.I. Bogdanova and Yu.N. Shekhter. *Inhibited Petroleum Compositions for Antirust Protection.* Moscow, Khimiya, 1984.

4 Structural Inhibited Plastics

Structural inhibited plastics have found application in the sealing of movable and stationary metal-polymer joints operating in hostile environment. They are most efficient in weakly and moderately loaded friction joints, plain bearings and seals. Their main role in metal-polymer friction joints is the reduction or suppression of mechanochemical wear.

Despite the fact that the peculiarities of metal wearing in contact with polymers were determined more than 30 years ago [35], the serviceability of metal-polymer friction joints has resisted estimates for a long time. The origin of this corrosion wear mode was considered only for plastics processing equipment, in which metals are in contact with the moving polymer melt [36].

In the present chapter we have attempted to generalize the span of phenomena that accompanies friction and defines the mechanochemical behavior of metal wear in conjunction with polymers, and to expound the methods for reducing corrosive wear of metals via imparting the properties of wear inhibitors to polymers. In this regard, wear mechanisms of friction parts in corrosion-active media, corrosion and migration capabilities of plastic components, chemical activity of polymer tribodestruction products and some other factors stimulating corrosion in metal-polymer systems are analyzed. Special attention is paid to the methods of combating wear by electrical fields and potentials arising on metal contact with polymers as well as corrosion inhibitors (CI) impregnated into the polymer or formed during friction.

4.1 Corrosive Wear of Metals

During operation in liquid and gaseous hostile media (acids, alkali, salt solutes and melts, hot gases, etc.), some parts and friction joints of chemical, oil extracting, mining, metallurgical, food, microbiological and other equipment experience severe wearing. The simultaneous action of the two effects, friction and aggressive medium, leads to the generation of a specific kind of wear called mechanochemical wear.

Mechanochemical wear is an intricate process in which corrosive damage and mechanical wear are interrelated [1,2]. The role of corrosive and mechanical factors in corrosion-induced wear of metals in electrolytes is estimated in different ways. Some authors consider this process as purely corrosive, others as mechanical, the rest consider both factors as equally significant. In accordance with the terminological standard ASTM G40, corrosive wear is wear in which both chemical and electrochemical reactions with the environment are critical. A somewhat wider interpretation that includes a terminological aspect is given in Russian State standard 26764: "Mechanochemical wear is the wear resulted from the mechanical action accompanied by a chemical and/or electrochemical interaction of the material with the environment". We will continue to use the term "corrosive wear" as well as "mechanochemical wear".

Proceeding from the above, corrosive wear is conditioned by a group of effects of different nature. It is actually impossible to differentiate between these effects or to analyze their influence on friction and wear separately as they are in fact inseparable; what is more, it is incorrect to speak about additivity of their interactions.

Among the factors governing corrosive wear, it is justifiable to mention the following [1]:

- chemical composition and microstructure of contacting surfaces
- roughness and porosity of contacting surfaces
- presence or absence of surface cracks and grain boundaries
- surface hardening degree or its stress state (e.g. cold-working in the case of metals or alloys)
- presence of electrical potential difference and current passage between contacting surfaces
- temperature and pressure in the tribosystem
- reactivity of the medium between the contacting surfaces

The science engaged in studying the interactions of tribosurfaces with the chemical environment is called *tribochemistry*. This is a complex field of knowledge as it analyzes not only mechanical, thermal, macro- and microstructural processes on the friction surface but also thermochemistry and reactive kinetics as well as catalytic processes arising during friction and wear. This intricacy is clearly expressed in Heinike's book [3], where he suggests a system of terms reflecting his viewpoint on the place of tribochemistry in tribology. He depicts tribology in the form of a tree with two branches, one of which is triboscience and the other is triboengineering. The scientific branch is subdivided into tribochemistry and tribophysics. Thribochemistry is divided in turn into tribological

- diffusion, sublimation, gas absorption, sorption and adsorption
- oxidation, reduction, corrosion, catalysis and electroplating.

This classification has perhaps gone too far in its usage of complex terms but it vividly visualizes the multitude of chemical phenomena emerging under conditions of contact friction. Besides, the subdivision of tribochemistry into branches should not be understood as to indicate the nonexistence of individual chemical mechanisms in one particular contact. The chemical, electrochemical and mechanical aspects of friction-induced failure of metals in hostile media are set forth below.

4.1.1 Chemical and Oxidative Wear

Chemical reactions on the friction surfaces may in fact appear in any kind of media, although their effect on the friction behavior and wear of the friction pair is often far from apparent. *Chemical wear* in its broadest sense could mean abrasive wear of the tribocomponents in a corrosive liquid [1, 4]. Nevertheless, it would be unfair to relate chemical wear to an individual wear mode since it is just one of the contributing processes exerting a significant effect on other wear modes.

Moderate wear in a normal atmosphere is sometimes described by the term *oxidative wear* [1, 5]. Oxidative wear takes place if an oxide film has been formed on the friction surface. It can be a result of the oxidation of the metal surface by either the air oxygen or lubricant. The film strongly affects friction and wear and plays a dual role. On one hand, it protects the friction surface from severe wearing and seizure, on the other hand, being harder than the oxidative wear products, it evokes abrasive wear.

Chemical and oxidative effects during friction and wear of metals have been studied at length in atmospheric conditions, corrosive liquid media and high temperatures. Nevertheless, it is hard to establish any common patterns in these wear factors. Let us take some examples.

It has been proposed in [6] to make allowance for some critical thickness of the oxide film at which wear changes from the moderate to severer mode. According to Molgaard [7], the mechanical properties of the oxide layer are more important than its thickness. He has shown that the transferred oxide layers may inhibit further oxidation of the base metal by isolating it from the hostile environment until these layers are worn out. This process of the formation and removal of oxide layers can impose a determining effect on the wear rate variation.

The authors of [8] have detected that transition from mild to strong (severe) wear as a result of fracture of the oxide film occurs faster under reciprocating motion compared to unidirectional sliding. The relation between terms "mild wear" and "oxidative wear" was revealed in earlier works in which the mild wear mechanism was attributed to the presence of oxide films protecting the metal surface from failure during friction. The transition mechanism from the mild to severe wear depending on the load and friction velocity was discussed in [9]. The authors attributed variations in the wear rate with friction path to run-in. The dependencies of the friction coefficient on loading and friction path were obtained for different Fe-Cr alloys. It was found that to attain mild wear an alloy containing 3% Cr required twice as much run-in time as that with 1% Cr. This essential difference was explained by a lower rate of high-temperature oxidation of the alloy with 3% Cr during the run-in stage.

In his early classical work Welsh [10] reported the friction coefficient and wear rate variations for the mild-severe-mild wear transition as a function of the load growth (Fig. 4.1). The rise in local temperatures on loading P_2 leads to transformation of the surface layer material into a martensite that forms an oxide layer. This reduces the friction coefficient and abruptly returns the tribosystem into the mild wear mode.

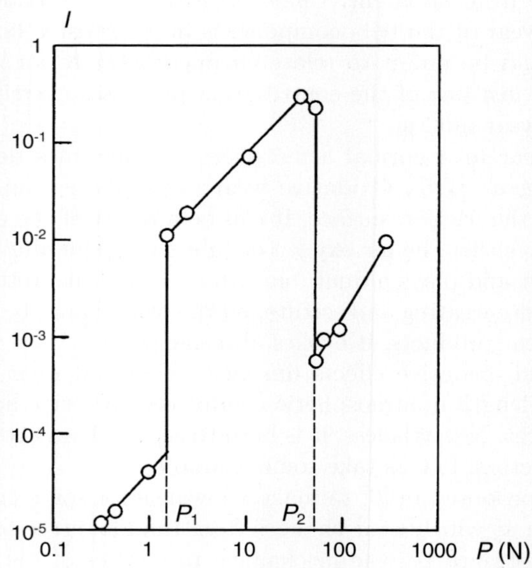

Fig. 4.1. Wear rate of steel with 0.52% carbon content depending on loading. Velocity – 0.1 m/s [10]

These variations of the friction coefficient as a function of loading have led to Quinn's theory of oxidative wear [11, 12]. At the base of the theory lies the idea that transition from one type of friction coefficient to loading dependence to another is a function of the iron oxide type formed on the contact surface during friction. The lowest friction coefficients (mild wear) are characteristic of Fe_3O_4, the highest of FeO. The theory predicts correctly enough the rates of oxidative wear for carbon low-alloy steels.

Heat energy for chemical and oxidative wearing can be gained from three types of sources: 1) frictional heating, 2) the heat generated during exothermal reactions on the frictional surface, 3) external heat sources. Estimates of temperature regimes of stainless steels and alloys during oxidative wear can be found in [13, 14].

It was reported in [13] that Fe-Cr alloys with slight amounts of alloying additions of Ni and Mo initially show a period of severe wearing followed by a transition to mild wear. It has been found that the time to reach a stable state decreases as the temperature rises. The difference between the run-in period of the two alloys with different Ni and Mo content is evident at room temperature but is practically absent at 200°C. The authors have derived a relation for the total wear value (W) as a function of time (τ), transient time (τ_1), severe wear rate (W_s) and mild wear rate (W_m):

$$W = W_s\tau_1 + W_m(\tau - \tau_1) . \tag{4.1}$$

The equation for the full (mean) wear rate as a function of transient time ratio to the full time of friction (τ_1/τ) is obtained by dividing the relation above by τ. Since wear rates equalize on transition to mild wear a conclusion has been made that the apparent wear rate increase of one of the alloys at room temperature takes place at the expense of a longer run-in of the other alloy at the severe wear stage.

The analysis of the facts above drives one towards the idea that heat and chemical phenomena run hand in hand during friction. It is hard to distinguish these phenomena when accusing one of them of lagging behind during transfer.

Numerous investigations have proved that tribooxidation of surfaces strained by mechanical action occurs 100 to 1000 times faster than that of non-strained surfaces [3]. This assertion should not however be generalized to the case of specific surface treatment such as shot blasting (so-called impact pre-hardening), which inhibits corrosion. Since oxidation can actually be accelerated by mechanical friction, the interrelation between the chemical and mechanical effects on the rubbing surfaces is not unambiguous. Different factors that complicate the prediction of wear rates of the rubbing parts in hostile media are presented in Fig. 4.2.

4.1.2 Metal Failure in Electrolytes Under Mechanical Stresses

Below is a brief account of the mechanical aspect contribution into corrosive wear. Failure of materials is treated today as a discrete process of jerky local elementary acts repeated periodically in time [15]. Since metal friction has a pulse behavior and metals wear by the fatigue mechanism [16], so friction and wear in corrosion-active media should be treated as the interrelation of the process with adsorption and corrosive fatigue. Adsorption fatigue usually paves the way for corrosion processes in the materials and precedes stronger effects in the media where the rubbing bodies operate [17].

The mechanical properties of the majority of crystalline materials are conditioned by their dislocation behavior. Consequently, the effect of the medium on the mechanical properties of solid bodies can be induced by interactions with dislocations and imperfections of the surface structure. The conditions at the solid-medium interface are dependent on the free surface energy. It is generally accepted that the strength of solid bodies is a value proportional to their surface energy.

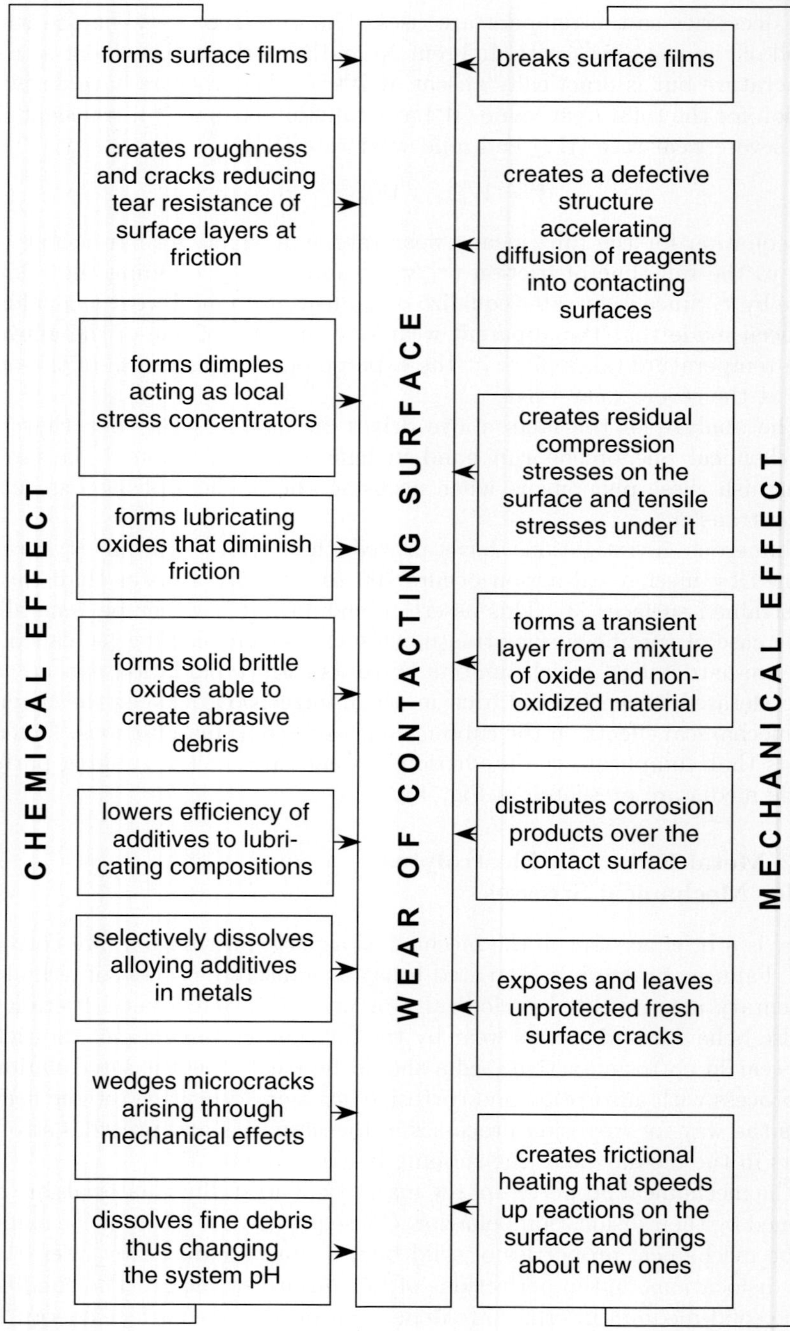

Fig. 4.2. Mutual effects of chemical and mechanical factors at wearing [1]

The effect of the medium on metal deformation commonly begins on reduction of the surface energy during physical adsorption. Afterwards, various secondary phenomena, including chemical, electrical, etc. are superimposed on this process. A certain activation energy, dependent on the process specifics, is necessary for these phenomena to occur. In the case of metal strain and failure in corrosion-active media, mechanical and electrochemical activation are of particular importance [18].

From the multitude of intricate corrosion processes in the presence of mechanical action (friction, erosion, vibration, cavitation, fretting and so on) it is justified to touch upon corrosion types joined under a single failure mode induced by mechanical stresses. These are the stresses that govern the corrosion wear rate of metals during friction. Such processes are usually called corrosion stress-induced cracking in the case that the mechanical action is effective only in one definite direction, or otherwise termed corrosion fatigue in the case that compressive and tensile stresses alternate within cycles. In spite of the differences between the appearance of these corrosion types, they have much in common, e.g. fundamental mechanisms, the causes, and they overlap to a certain degree [19].

Stress corrosion cracking (SCC) represents a process of crack nucleation in metals under the simultaneous action of corrosive medium and tensile (external or internal) stresses. A few typical examples of corrosive media for some metals and alloys are given in Table 4.1.

Table 4.1. Specific media inducing stress-corrosion cracking [19]

Metal	Medium
Low-carbon steel, high-strength steels*	Solutions containing NO_3^-, OH^-, H_2O
Austenitic stainless steels	Solutions containing Cl^-, OH^-, Br^-
α-Brass	NH_3, amines
Titanium alloys (8% Al, 1% Mo, 1% V)	Solutions containing Cl^-, Br^-, I^-
Aluminum alloys	H_2O, NaCl solutions

*Steels with yield point >1240 MPa or Rockwell's hardness $H_R > 40$.

The following properties are characteristic of SCC [20]:

- brittle character of cracks
- crack propagation in a direction perpendicular to the tensile stresses
- formation of intercrystalline, trans-crystalline or mixed types of cracks with branching
- dependence of time when cracking starts on the level of applied external stresses

Several theories of SCC differing in some details but having common general principles have been put forward, two of which are the electrochemical and adsorption theories. The electrochemical theory is based on the recently revealed formation of galvanic elements in the crack apex between the base metal or alloying elements, or additives (pure metals are not subject to SCC) during plastic deformation. The cathodic polarization effectively hampers SCC and stops it if the applied potential becomes lower than a critical value.

The critical potential of corrosion-induced cracking is such that above it the ions breaking the medium are adsorbed on the metal surface and below it the ions are desorbed. It may be more positive than the corrosion potential or more negative. Inhibiting ions unable to evoke cracking compete with the breaking ions for adsorption sites. To attain the surface concentration of adsorbed breaking ions sufficient for cracking a higher potential is necessary. In the case that the breaking ions cannot adsorb under the potential applied (lower than the critical one) cracking stops.

Proceeding from the analysis of existing theories and concepts, the mechanism of SCC can be presented as follows [3]. Microcracks start to form in surface defects sites where stresses are concentrated. As soon as the oxide film ruptures on the crack bottom under mechanical action, a fresh surface is formed that acquires a growing electrochemical potential and exerts a corroding effect within a short time interval (10^{-3}–10^{-1} s), thus widening the crack. Hence, the material becomes more strained and the stress in the crack apex grows quickly, reaching a value exceeding the strength limit of the reaction product (oxide film) and the cycle restarts.

Corrosion fatigue represents metal cracking as a result of the joint action of the corrosive medium and a periodic or variable loading. In contrast to the SCC process, which is specific for a given combination of the material and medium, the corrosion fatigue may arise in any material and any corrosive medium [21]. The material gradually breaks down in local sites leading to the onset and propagation of cracks, variation of properties and finally to total failure. An important feature of this type of failure is the absence of the breaking stress σ_f characteristic of the material behavior in air (Fig. 4.3).

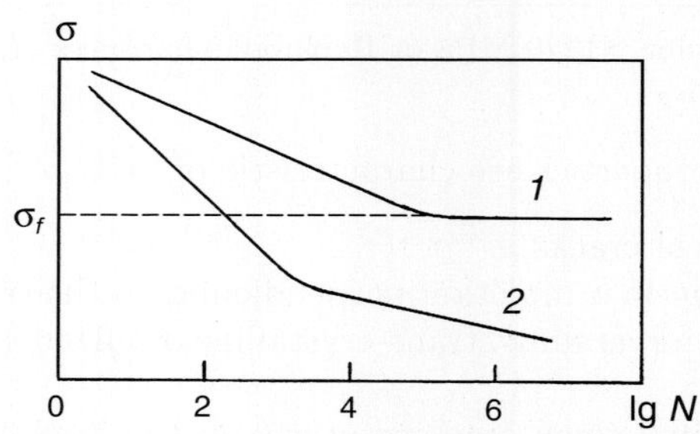

Fig. 4.3. Steel curves σ – lg N in air (1) and corrosive medium (2) at cyclic loading; σ_f – fatigue limit [19]

The effect of corrosion process on fatigue is exhibited mainly in the acceleration of plastic deformation accompanied by the formation of projections and dimples on the metal surface, which assis in crack generation. This is why failure induced by corrosion fatigue is not the result of the additive effect of corrosion and fatigue but always exceeds their sum. The chief mechanisms of fatigue corrosion are basically the same as for SCC [3]: stress concentration on the corrosion pitting bed, electrochemical effect on the plastically deformed crack bottom area, formation of a region where the cathode is less deformed, and finally surface energy decrease via adsorption processes (Rehbinder's effect).

Notice that pure metals resistant to SCC are subject to corrosion fatigue to the same extent as they are to general corrosion. Therefore, the resistance of a material to corrosion fatigue is more dependent on its corrosion stability than on its tensile strength. This is the major difference between SCC and corrosion fatigue.

4.1.3 Electrochemical Processes During Friction

In contrast to the processes commonly considered in electrochemistry, electrochemical processes occur during friction under conditions of moving and deforming discrete contacts of individual microasperities. The participation of electrolytes as a liquid layer in the friction pair leads to potential leaps of φ_{13} and φ_{23} at the metal-solution interface and to the contact potential difference φ_{12} in metal contacts (see Fig. 1.7) [22]. As a result, a short-circuited galvanic microelement appears with a probability of redox reactions on its electrodes.

Taking into account the discreteness of the friction contact, heterogeneity of the friction surfaces, micro-inhomogeneity of the metal structure, local plastic deformations and difference in energy between the states of fresh surface and areas covered with oxide films, the friction pair can be treated as an intricate multielectrode system.

The electrode potential value of a metal during friction can differ greatly from that measured under static conditions [18]. This is attributed to, firstly, mechanical cleaning of the surface of oxide films, secondly, the formation of secondary structures that shift the potential to the positive or negative region, and thirdly, the potential shift being affected by plastic deformation leading to the generation of dislocations in the metal crystalline lattice.

Since friction is a pulse oscillating process, its oscillation is probable along with the potential shift. In this case, it is always possible to isolate the constant and variable constituents, φ, of the friction process. During relative sliding of microasperities, the inter-electrode gap varies periodically, changing the hydrodynamic pressure of the liquid. In the local contacts of microasperities, the surfaces are cleaned of their oxide films. These phenomena result in abrupt oscillations of the potential with a frequency that depends on the sliding velocity, loading and surface roughness.

Oscillations may exert a strong effect on adsorption processes in the frictional contact. Adsorption of particles on the electrode with a certain potential is known [23] to occur at a finite speed. Under low oscillation frequencies the adsorption manages to follow the potential and participate in the variation of the interfacial layer structure. At high frequencies the adsorption mechanism does not work, giving place to electrostatic charging of the layer as a condenser, i.e. the generation of the double electric layer (DEL). A mechanical model of the interfacial DEL has been elaborated by Shepenkov [24]. It follows from the model that, if a periodic mechanical force acts on the double layer from the side of the liquid or electrode, the electrode potential will vary periodically with the same excitation frequency.

Being a fundamental characteristic of the metal-electrolyte interface electric state, the electrode potential φ of the system may impose a perceptible effect on the friction and wear of metals in electrolytes. The effect of polarization on the friction coefficient and surface damage of metals has been detected and comprehensively studied by Bowden and Young [25]. It has been proven that the dependence of the friction coefficient on the electrode potential takes the form of an electrocapillary curve (for platinum, nickel, cadmium, lead and copper). Variations in the friction coefficient on polarization of the friction zone are attributed to the growing wedging effect of DEL between the rubbing surfaces. On immersion of a friction pair into an acidic electrolyte, either hydrogen or oxygen will deposit on the rubbing metal surfaces as the potential difference varies. The friction coefficient shows a maximum forvalues of the potential at which neither oxygen nor hydrogen deposit.

A well-known discovery in the field of friction and wear, selective transfer (ST), is currently interpreted from the standpoint of electrochemistry [24]. Selective transfer involves the transfer of one or more particular constituents of a material to the counterface in preference to others in the material [26,27]. In the course of frication between a bronze-steel pair in glycerin the surface of the copper alloy undergoes selective dissolution due to electrochemical reactions. This is because all components of the alloy show much higher chemical activity compared to copper and serve as anodes. For instance, the standard electrode potentials of the components in a copper alloy with tin, lead and zinc have the following values [19]: +0,337 V (Cu); −0,136 V (Sn); −0,126 V (Pb); −0,763 V (Zn). The selective anodic dissolution of components and their removal with the lubricant from the friction zone bring about a fine copper layer on the bronze surface. Then, copper ions precipitate onto the cathodic areas of steel and form a copper film. The copper layers formed in this way follow the deformation almost completely and ensure a minimal friction coefficient leading to raised wear resistance and durability of the friction joints. The proof of the electrochemical nature of ST is the established relationship between the wear resistance of metals and the stationary potential value of the friction surface [28, 29].

4.1.4 Mechanochemical Kinetics of Metal Wear

Mechanochemical wear is a complex process during which electrochemical solution of metals catalyzes fatigue failure. Friction, in turn, activates electrochemical corrosion. As can be seen, the process of mechanochemical wear of metals in electrolytes has a combined fatigue-electrochemical nature.

A corrosive macroelement arising in corrosion-active media during friction of unlike metals (Fig. 4.4) is characterized by a certain value of the frictional electrode potential φ_{fr}. The jerky character of friction due to the discreteness of the process brings about a variable constituent to the potential.

The main result of interactions between the medium and the metal turns out to be adsorption governed by the steady potential φ_{fr}. Note that the localized adsorption of surfactants is strongly influenced by the potential fluctuations. Adsorption promotes surface energy reduction, which is also affected by the potential or its fluctuations. As a result, surface barriers decrease, alleviating the emergence of dislocations on the surface. According to the heterogeneous electrochemical theory of corrosion, microelements arising in the structure are involved in active dissolution. Such metal solvation on weaker surface areas facilitates fatigue failure of these sites.

Let us examine the chief factors affecting the kinetics of mechanochemical wear of metals in electrolytes.

Corrosion rates and those of mechanochemical wear of stainless steels and alloys widely applied in friction joints of chemical equipment are presented in Table 4.2 [30]. Owing to the formation of passivating protective films on contact with hostile media, these materials display high corrosion resistance. As can be seen from the table, corrosion rates grow during friction by a factor of thousand. Under such conditions, the material wears largely due to corrosion even in a weak solution of sulfuric acid for both sliding friction over a softer material (PE) and under abrasive action (ceramics).

Electrochemical studies have shown [31, 32] that the stationary potential of the metal shifts to the negative side during friction in electrolytes and increases the initial and full passivation currents (Fig. 4.5). It should be underlined that the rate of mechanochemical wear is much lower in the passivation region of the potentials (characterized by i_{fp} currents) during sliding friction (e.g. over PE and graphite) than in the stationary regimes. It has also been found [32] that the maximal wear resistance of a friction pair corresponds to the minimal negative potential value. Hence, total correlation is observed between the wear values and electrode potentials. In addition, variations in the electrode potential during friction can be an indicator of dynamic equilibrium of the formation and failure processes of secondary structures on the friction pair surfaces.

The wear rate of rustproof steels and alloys depends considerably on the ambient medium composition and activity. For instance, when austenite steel (position 2 in Table 4.2) is rubbed in a pair with Al_2O_3 ceramics, its wear rate is higher for more aggressive media. It is important that the friction

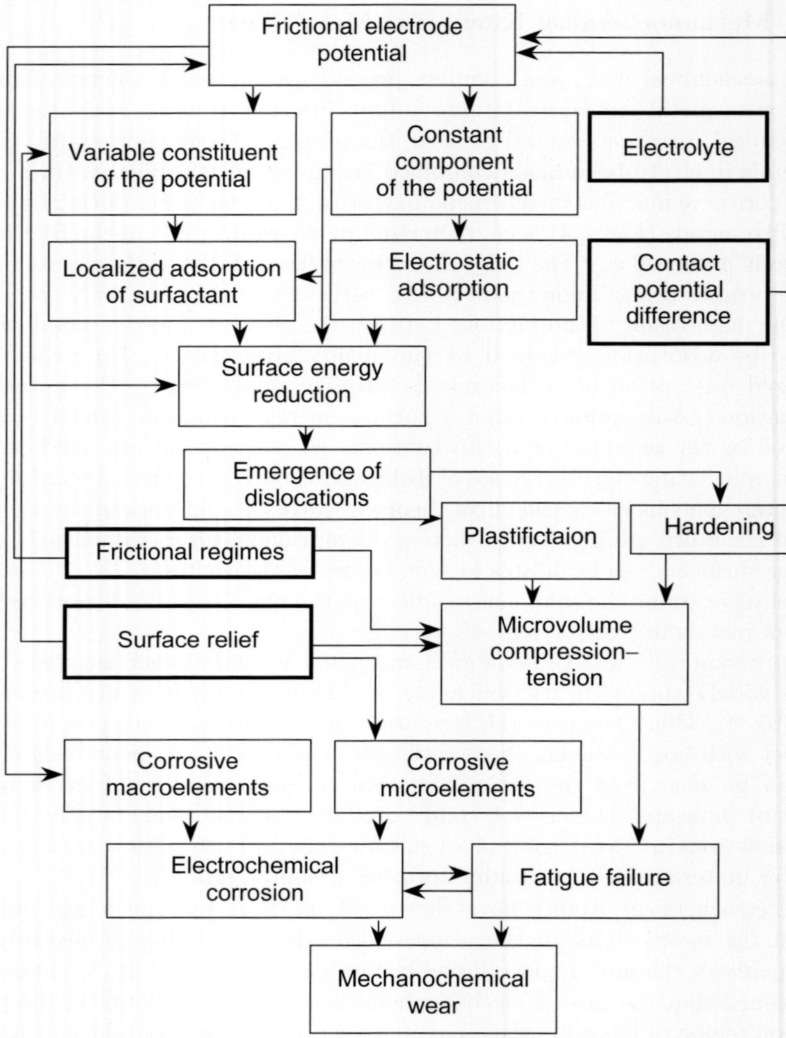

Fig. 4.4. Block diagram of mechanochemical wearing [17]

coefficient of the pair diminishes with increasing hostility of the medium. The picture is different under sliding friction conditions. For example, the mechanochemical wear rate and the friction coefficient of the steel-graphite friction pair decrease just in both the cathodic and passivation regions due to the inhibition of the anodic process of steel solvation.

In addition, the mechanical action intensity affects the mechanochemical wear rate [31]. As the pressure in the steel-graphite friction pair rises from 0.05 to 0.4 MPa, the rate of mechanochemical wear increases from 0.15 to

Table 4.2. Corrosion rates (C) and mechanochemical wear rates (I) of steels and alloys at friction over PE and ceramics in 1 N solution of H_2SO_4 at $p = 0.4$ MPa [30]

No.	Material (Amount of Main Alloying Additives, %)	C, mm/year without friction	Friction over PE			Friction over Ceramics		
			C, mm/year	I, mm/year	C/I, %	C, mm/year	I, mm/year	C/I, %
1	Austenite-ferrite steel (0.08C; 21Cr; 6Ni; 2Mo; 0.3Ti)	0.001	1.80	1.80	100	5.0	6.0	83
2	Austenite steel (0.08C; 18Cr; 10Ni; 0.3Cu; 0.3Ti)	0.001	1.00	1.00	100	6.4	7.3	87
3	Austenite steel (0.1C; 17Cr; 13Ni; 3Mo; 0.4Ti)	0.001	0.80	0.84	95	5.0	6.0	83
4	Nickel + iron-based alloy (0.06C; 23Cr; 28Ni; 0.8Si)	0.0005	0.70	0.70	100	5.0	6.1	82
5	Nickel-based alloy (0.03C; 1Fe; 16Mo; 4W; 15Cr)	0.0005	0.70	0.70	100	5.0	6.1	82

3.2 mm/year, although a further increase in pressure to 1.2 MPa does not significantly affect the wear rate. Consequently, the steel surface becomes activated under friction with graphite at a pressure of 0.4 MPa. This is consistent with the shift of the stationary potential of steel to the negative side and increase in the mechanochemical wear rate. When rubbing against kapron, steel's potential has already shifted to the negative domain at $p = 0.05$ MPa, but the mechanochemical wear rate does not in fact change as the pressure grows.

Hence, the dominant factors affecting the wear rate of metals during friction in electrolytes are the corrosion activity of the medium, the pressure in the friction pair and the properties of the counterbody.

Electrochemical processes that occur on the actual contact spot during friction have been comprehensively studied [30] to reveal the mechanism of mechanochemical wearing. Friction was modeled as an intermittent interaction of a sample with a single surface asperity. The asperity slides over a

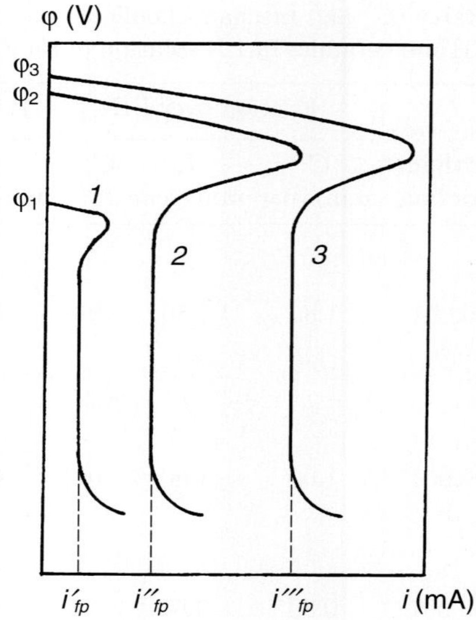

Fig. 4.5. Anodic potentiostatic curves for steel without friction (1), at sliding fiction against PE (2) and under abrasive effect of ceramics (3); i_{fp} – current of full passivation

metal surface, deforms it and breaks the passivating film partially or fully. An intensive process of anodic dilution of the metal takes place on the freshly formed surface. Hence, the mechanism of mechanochemical wear can be represented by two dominating and interrelated processes, namely corrosion of freshly formed surfaces and fatigue failure of the metal on actual contact spots under repeated deformation.

Proceeding from the investigation results, the following major regularities of metal wear under friction in electrolytes were suggested by Lazarev [33] as applied to various friction regimes and contact conditions:

- The wear rate on liquid lubrication by electrolytes is defined only by metal dissolution and intensified friction, and can be varied in response to the corrosion rate under static conditions to a value that characterizes the mechanical failure of the metal.
- Under boundary lubrication in conditions of elastic contact the wear rate is defined by the corrosion processes on freshly formed surfaces in the actual contact spots. In conditions of plastic contact the wear rate is composed of two constituents, that is, the low-cycle fatigue in the actual contact spots and solution on freshly formed surfaces.
- Under microcutting (abrasive effect) metal particles already separate at the beginning of the first cycle, therefore the wear rate in any electrolytes is governed above all by the mechanical factor.

Two interrelated trends can be distinguished in the present day research into electrochemical processes during metal friction: (i) the use of electrochemical parameters to derive information on contact processes and (ii) friction and wear monitoring of metals via the regulation of the electrode potential of tribosystems. The second trend, which presupposes an intrusion into the course of the electrochemical processes on the frictional contact, has been actively developing in the recent two or three decades, especially after the discovery of selective transfer under friction, whose nature is largely electrochemical. The topic of the present chapter is connected exactly with this trend.

It should be emphasized that use of inhibited plastics in metal-polymer friction joints assists in a number of cases in solving the problem of fighting mechanochemical wear through the suppression of corrosion processes in the friction zone by CI liberated by the polymer counterbody. Yet inhibited plastics are not the only means of corrosion inhibition in metal-polymer joints.

4.1.5 Means of Inhibiting Mechanochemical Wear in Polymer-Metal Pairs

The analysis of corrosion factors during polymer-metal pair wear has proved that the main path is electrochemical protection of the metal counterbody, neutralization of corrosion agents formed in the friction zone and suppression of corrosion processes in the polymer-metal contact. These directions are realized by the means illustrated in Fig. 4.6 [37]. They are subdivided into two groups according to the use of special substances or physical fields and power effects.

Wear inhibitors (WI) are substances able to alleviate mechanochemical wear in friction joints [18]. The notion of WI helps to generalize numerous means of fighting corrosion in machines and mechanisms independently of their friction joint design and composition.

WI in gaseous media are related to gas films, reducing media and inert gases. Gas films decrease the actual contact spot of the rubbing bodies thus alleviating tribochemical reactions in the metal-polymer pairs whose products can be corrosion-active matter. Reducing media neutralize corrosion agents formed in the friction zone. Inert gases create an oxidizer-free friction zone and diminish the intensity of corrosion processes.

Liquid WI represent lubricating substances possessing a property of suppressing corrosion. Inhibited lubricants are oils containing corrosion inhibitors. WI absorb on the counterbody surface, force the corrosion-active agents out of the system and interact with active molecules of oils to avert their self-oxidation. Tribologically polymerizable compositions form high-molecular compounds (friction polymers) in the friction zone during interactions of the materials with the medium, which alleviates wear. By absorbing and combining products of tribochemical reactions the polymerized compositions function as WI. Conducting lubricants are able to form electrical

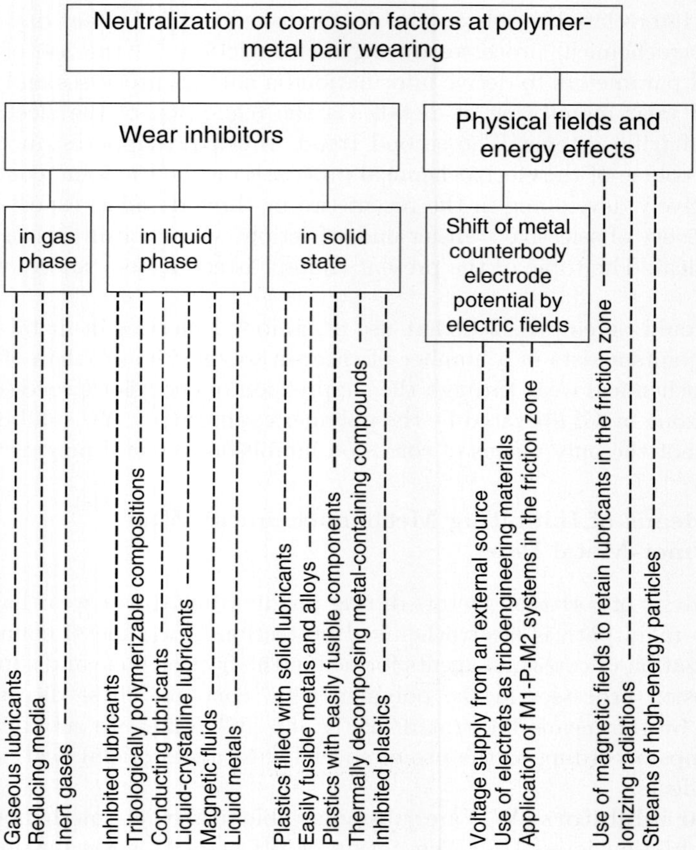

Fig. 4.6. Means of decreasing polymer-metal pair wearing

circuits to remove static charges from the friction zone and to polarize the metal counterbody. Lubricating oils containing liquid-crystalline additives display elevated triboengineering characteristics and assist in removing wear and corrosion products from the friction zone. Oriented structures formed in liquid-crystalline compositions under the electric potential of the frictional bodies can also be used to lower wear of the metal-polymer pairs. Magnetic fluids form a continuous film under the action of magnetic fields that insulates the friction zone from the ambient corrosive medium. Liquid metals generate a diffusive barrier for hostile media and, consequently, can protect the metal.

Solid WI are used as triboengineering structural or lubricating materials. Plastics filled with solid lubricants are included in this class of materials since they lower the heat-induced strain in frictional joints through a reduction of the corrosion activity of the products of tribochemical reactions. Fusible metals and alloys of frictional parts may also ensure a protective effect of the metal counterbody. Plastics with fusible components that absorb frictional heat exert a much lower corrosive effect on metal parts than the original polymers. Being components of lubricating compositions or thermally decomposing plastics, metal-containing compounds form protecting layers, which in addition assist in heat removal from the friction zone. Inhibited plastics are a new type of structural plastics containing CI, which will be examined in detail later.

In its true sense, any component of a tribosystem promoting suppression of unfavorable tribochemical processes in the friction zone, irrespective of whether it was impregnated or formed internally, can be related to WI [34]. WI can be formed both as a result of physical-chemical interactions of polymers and metals, or physical fields and energy effects on a part or friction joint. This can be, e.g. thermochemical or radiation-thermal treatment, ion implantation, superposition of electrical and magnetic fields, shifts of electrode potentials, passage of electrical current, etc.

Physical fields and energy effects are aimed at the regulation of the electrode potential of a metal counterbody and the third body structure and properties.

Shift of a metal counterbody electrode potential with the help of electric fields is a classical procedure used in electrochemistry to reduce mechanochemical wear [18]. This can be achieved by voltage supply from an external electric source, although this is rarely used in triboengineering. A promising approach is the application of electrets as triboengineering materials. In this case, the electret field serves as the source of the polarization of the metal counterbody. The polarization current generated during frictional heating will be able to create metal 1-polymer-metal 2 system (M1-P-M2) using a certain design of the friction joint. The last two methods will be dwelt upon in the next chapter.

Magnetic fields can be effectively applied to retain insulating lubricating layers possessing magnetic properties within the friction zone. These include magnetic fluids and some fusible metals and alloys referred to as corrosion inhibitors.

Ionizing radiations are employed to neutralize corrosion-active macroradicals arising from tribochemical reactions with polymers. γ-radiation is known to result in the volume and surface modification of polymer materials as well as variation in the friction coefficient and wear rate of plastic parts [38].

Streams of high-energy particles effectively inhibit wear of a number of polymer-metal friction pairs in vacuum [39]. The main role of exposure is in the efficient cleaning of frictional surfaces from contaminants, above all of water molecules. Under the action of a stream of particles structural defects in the polymer surface layer are healed and concentration of corrosion-active radicals in the friction zone decreases.

4.2 Electrical Processes in Movable Metal-Polymer Contacts

Among the numerous phenomena occurring during friction between polymers and metals, electrical processes appear to be less studied despite the fact that their aftereffects are far from subordinate. What is more, experimental evidence in this domain is not consistent but is frequently contradictory. The revived interest in electrical phenomena has been driven by their diversity in dynamic contacts of heterogeneous surfaces and by the wide potential for using these phenomena in the diagnostics of tribojoints and monitoring friction and wear processes using electromagnetic effects.

Frictional interactions between a polymer and metal are accompanied by variations in the electrical properties of the contacting surfaces. This involves the formation of surface and volume charges, generation of electrical currents, polarization, emission of electrons, etc. A number of experimental and theoretical works prove the essential contribution of electrical phenomena to metal hydrogenation, tribopolymerization and frictional transfer, electroerosive wear and other processes. In the present chapter we are going to disclose the major electrical phenomena and processes affecting friction and wear of polymers in contact with metals.

4.2.1 Electrization and Polarization of Polymers During Friction

Electrical charges arise in response to contact interactions of two bodies, be it friction, impact, mechanical treatment or another effect. Real triboelectrization is known to appear only in the case of asymmetrical charging of homogeneous materials [40]. In the majority of cases electrization of bodies at friction is conditioned by so-called *contact electrization*, whereas friction alone leads only to an increasing number of contact areas. The mechanism of contact electrization is based on the generation of the potential difference between the contacting bodies. For metals this value is dependent upon the work required for the liberation of electrons, for dielectrics it is dependent on electrical states of the surface with a quasi-ion structure and the low ionization energy electron levels [3, 40].

Aside from contact electrization, local accumulation of charges takes place as a result of mechanical separation and formation of a double electrical layer (DEL). Mechanical separation of charges is brought about by exfoliation of adhesive films from the metal or semiconducting samples. DEL may arise in response to chemical interactions of two phases or as a result of selective adsorption of similar ions, e.g. in the presence of oriented dipoles on the contact surface of one of the phases. Electrization can also be induced by the donor-acceptor (DA) interaction, since in agreement with the electrostatic theory of adhesion DEL are formed at the interface of two substances at the expense of DA links and govern the efficiency of adhesive interactions [41].

It has been suggested in [42] that uncompensated charges are formed from *adsorption layers*, due to which either polar molecules or charged particles compensating inner charges accumulate on the solid surface. As for the intrinsic charge of the dielectric, it is compensated by selective adsorption of ions of unlike sign. Therefore, layers adsorbed on the dielectric surface turn out to be charged too. During friction between two bodies, these layers intermix and are redistributed between the rubbing surfaces. Compensation of intrinsic charges is thus violated, leading to triboelectrization.

The quantitative characteristic of polymer triboelectrization is the generation of electric charge whose sign and value are dependent on a number of factors, including polymer structure, air humidity, contact surface state, relative displacement velocity of the bodies, among other. Empirical information on the sign of the charge and mutual electrization of materials is presented by many authors as a triboelectret series arranged in a certain manner [43,44] so that each material in the series is charged positively when in contact with any lower in position material. This rule is not, however, devoid of exceptions.

The electrization level of a material is commonly evaluated by recording the electrical potential, electrical surface charge and/or electrization current. The chief value characterizing the charged state of dielectric surfaces is the *surface charge density*. High values of the charge density is dependent upon the electrical strength and relative permittivity constant of the material. The maximum charge density should not surpass the value corresponding to air breakdown electric intensity $E_b = 3 \cdot 10^6$ V/m:

$$\sigma_{\max} = \varepsilon\varepsilon_0 E_b = 2.6 \cdot 10^{-5} \mathrm{C/m}^2 .$$

Electrical charges generated by static electrization scatter not only over the surface but across the material bulk as well. The direct consequence of friction-induced electrization process in polymers is the emergence of the *electret state.*

G. Sessler [45] defines an electret as a dielectric with a quasi-constant electrical charge. The term quasi-constant means that the relaxation time during which the electret charge declines greatly exceeds the period needed for study of the given electret. Electret properties are connected with dipolar (or actual) polarization and formation of spatial or surface charges. In contrast to actual polarization, which is conditioned by the freezing of dipoles oriented predominantly in one direction, actual charges form the layers of entrapped positive and negative carriers on both sides of the sample surface. The electret can be charged by carriers shifted within the molecular or domain structures in the dielectric volume. In this case, electrization bears the features of actual dipolar polarization. If the charges shift within each domain to their boundaries, this state is called Maxwell-Wagner polarization [46].

According to numerous investigations at the end of the 20th century polymers acquire a *triboelectret state* as a result of intensive electrophysical processes during frictional interactions with a solid counterbody [47]. The formation mechanism of the triboelectret state in polymers can be presented in a general form as follows. Surface and spatial charges generated during electrization polarize the polymer in their field. As a result, the polymer transfers into the electret state whose parameters are dependent upon the injected charges and polarization.

The principal characteristics of the triboelectret state in polymers recorded experimentally are: i) the efficient surface charge density (ESCD) value and ii) the thermally stimulated depolarization (TSD) current spectrum, i.e. the discharge current dependence of the electret on its heating temperature. The analysis of TSD spectra helped to estimate the parameters of the triboelectret state, including the homo- to heterocharge relation in a dielectric, activation energy of the charge relaxation processes, relaxation time and others.

The triboelectret state parameters of polymer materials are largely dependent on friction regimes and the polymer origin. In general, the ESCD value of polymer samples varies widely and has a maximum (Fig. 4.7) as the friction power grows. The tribocharge relaxation time, estimated by the decline in ESCD during prolonged exposure, also increases.

The TSD current spectra of triboelectrets of polar as well as nonpolar polymers show three peaks (Table 4.3). PE peak II is attributed to relaxation of the residual polarization as a result of accumulation and orientation

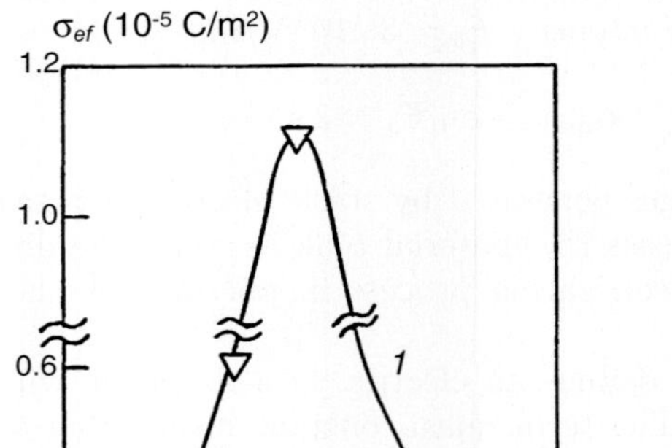

Fig. 4.7. Effect of friction velocity of HDPE over steel in different media on ESCD (σ_{ef}): (1) water; (2) Vaseline oil; (3) 0.1 N solution of KCl; (4) 1 N solution of KCl; (5) glycerin; (6) castor oil [47]

Table 4.3. Characteristics of HDPE and PTFCE triboelectrets under various friction velocities over steel samples [48]

Polymer (Dielectric Permittivity)	Friction Velocity, m/s	Temperature in the Friction Zone, K	ESCD, 10^{-5} C/m^2	Temperature of TSD Current Maximums, K			Activation energy for TSD Current Maximums, eV		
				I	II	III	I	II	III
HDPE	0.5	300	−3.07	351	–	408	1.47	–	–
($\varepsilon = 2.3$)	1.0	317	−3.94	353	400	413	1.65	–	–
	2.0	340	−4.76	358	400	415	1.96	–	–
PTFCE	0.3	298	−1.77	315	391	425	–	1.59	2.64
($\varepsilon = 2.7$)	1.0	320	−3.87	–	394	435	–	1.86	2.91
	2.4	358	−5.86	–	393	451	–	1.51	2.72

of strongly polar groups of thermally oxidative and mechanical destruction products of PE in the polymer surface layers [48] . The analysis of TSD spectra of the studied polymers has underlined that the triboelectret state is greatly dependent on the volume charge, whose high-temperature relaxation is induced by molecular mobility at the amorphous and crystalline interfaces on the crystallite surfaces (α_c-relaxation), large-scale displacement of macromolecular chains during crystallite melting (α'-relaxation) and raised conductivity of the polymer (Maxwell's relaxation).

The electret state created in the polymer essentially affects its wear against metals. The dependence of mass wear rate on ESCD value is close to parabolic (Fig. 4.8).

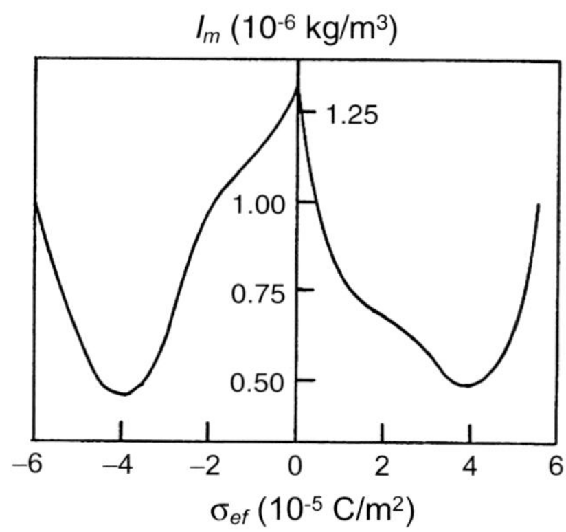

Fig. 4.8. Mass wear rate (I_m) dependence on ESCD (σ_{ef}) value and sign for electret PE coatings. Friction regime: $p = 0.5\,\text{MPa}$; $v = 0.5\,\text{m/s}$ [47]

The presence of dispersed fillers in the polymer material in low amounts may intensify electrization, increase the residual charge and change the friction coefficient. Introduction of the filler in the electret state exerts a still stronger effect on polymer electrization on frictional interaction with metals. Depending on the direction of the field intensity vector formed by the filler particles, the field generated by triboelectrization can be attenuated or intensified. This means that the principle of the electret-triboelectrization superposition is realized [49], which can be used to regulate the parameters of frictional interactions. Thus, by the introduction of the electret filler, e.g. mechanically activated F-3 powder, it is possible to decrease the friction force (Fig. 4.9).

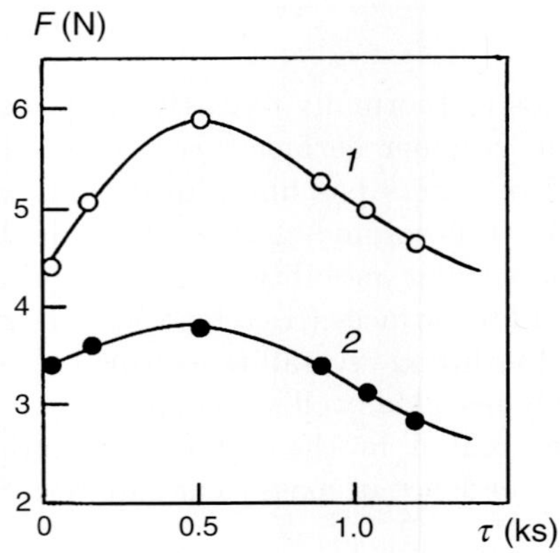

Fig. 4.9. Variation kinetics of friction force in the polymer composite-steel pair. The composite is based on HDPE (75 wt%) and fluoroplastic F-3 powder: (1) non-activated powder; (2) mechanically activated powder

4.2.2 Thermally Stimulated Currents in Metal-Polymer-Metal Systems

Under certain conditions polymers may display the properties characteristic of low-molecular electrolytes. So, it would be justified to anticipate *electrochemical polarization*, i.e. variations of electrode potentials under the action of the current passing through the polymer. This phenomenon is accompanied by changes in the composition of the near-electrode electrolyte layers, meaning the formation of substances different from the electrode and electrolyte. However, few publications contain information on electrochemical polarization of polymers and the data presented are rather contradictory. In [50, 51] the authors consider a metal-polymer-metal system as a galvanic element with a high inner resistance and they attribute voltages generated in the electrodes to their oxidation. In other works [52, 53] the authors do not admit the effect of the electrode nature on polarization processes. The problems of polymer polarization in contact with metals have been elaborated comprehensively in [54, 55] and will be briefly set forth below.

Some 30 years ago the phenomenon of electric current generated in the metal 1-polymer-metal 2 pairs (M1-P-M2) drew the attention of a number of investigators. In 1975 a manuscript by Japanese scientists was published [56] on abnormal thermal currents in PTFE placed between two electrodes of unlike metals. Analogous currents were detected in movable metal-polymer joints [57]. Later on a series of papers was published on this topic [50–52,58]. It was found that the spontaneous currents are in fact thermally stimulated and depend upon the voltage in the open circuit formed between open electrodes. An idea of the electrochemical nature of this voltage has been formulated in [59] and an electrochemical model has been put forward in [50,51] treating the electrode reactions at the polymer-metal interface as a voltage stimulus in the open circuit. As will be shown further, this galvanic effect may noticeably influence the surface properties of the polymer, particularly, its friction and wear processes in contact with metals.

The present authors have studied thermally stimulated currents (TSC) in movable joints of unlike metals separated by a polymer plate. The experimental setup (Fig. 4.10) consisted of an insulated copper disc 5 fitted with a rotary drive and an aluminum sample 2 in contact with the disc with a polymer coating 4. Friction regimes were regulated within rotation velocities 0.1 to 0.5 m/s. The disc 5 and sample 2 were connected to the galvanometer 7 with electrical contact to the disc via the copper needle 8. The frictional heating temperature was recorded by a thermocouple 3 and potentiometer 6. The current recorded under loading regimes with frictional heating to $T = 220°C$

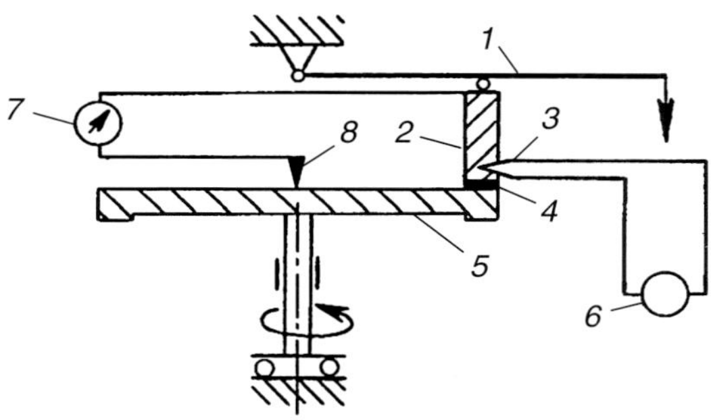

Fig. 4.10. Scheme of experimental setup for studying TSC in movable joints M1-P-M2: (1) lever load unit; (2) aluminum cylinder with a polymer coat on the face; (3) thermocouple; (4) polymer coating; (5) copper disc; (6) potentiometer; (7) galvanometer; (8) copper needle

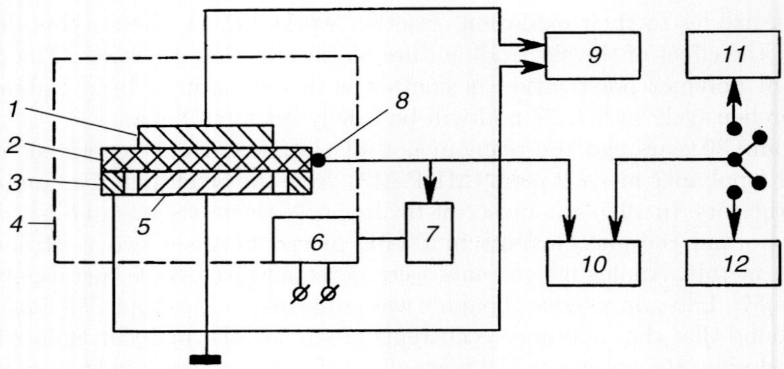

Fig. 4.11. Block-diagram of a device for recording TSC in M1-P-M2 systems: 1, (5) electrodes; (2) polymer sample; (3) collar ring; (4) oven; (6) heater; (7) programmed temperature control; (8) temperature sensor; (9) low current amplifier; (10) plotter; (11) digital voltmeter; (12) self-recording potentiometer

in the sample-coating-disc circuit was directed from the electrically positive copper to the electrically negative aluminum.

To eliminate discrepancies in the parameters of the electrochemical processes in frictional contacts a series of experiments was conducted using M1-P-M2 model systems. A scheme of the experimental setup is shown in Fig. 4.11; it operates as follows. The electrical current is generated in the circuit during heating of a polymer lining placed between short-circuited electrodes of unlike metals. The current direction is dependent on the combination of M1 and M2 metals, i.e. on their relative position in the electrochemical series (see Table 1.1), except for the case of a pair with a noble metal contacting another metal able to form oxides with deviations from the stoichiometric composition. On exchange of the M1 and M2 electrodes, the current direction in the circuit changes. So we can assert that TSC is induced neither by the presence of the temperature gradient in the sample nor by anisotropy of the polymer properties but by the electrode potential difference of the metals. Below (Table 4.4) TSC values of the samples with a PVB lining ($h = 600\,\mu\mathrm{m}$) at $T = 150°\mathrm{C}$ as well as the potential difference of standard electrodes M1 and M2 are presented.

Table 4.4. Some electric characteristics of M1-P-M2 systems

Pairs of Electrodes M1 and M2	Current, μA	Electrode Potential Difference, V
Cu-Al	0.240	2.00
Pb-Al	0.155	1.53
Cu-Pb	0.085	0.40

Let us consider the role of electrode processes at the polymer-metal interface and their effect on current generation in M1-P-M2 systems.

The electrochemical model proposed in [50] describes voltage formation in an open circuit. The authors consider the M1-P-M2 system as a galvanic element whose negative electrode reacts with the ambient moisture and forms metal oxides following the scheme:

$$x\mathrm{M} + y\mathrm{H_2O} \rightarrow \mathrm{M}_x\mathrm{O}_y + 2_y\mathrm{H}^+ + 2ye^- \;. \tag{4.2}$$

As a result, a pair $\mathrm{M} - \mathrm{M}_x\mathrm{O}_y$ is formed on the negative electrode with its own characteristic electrode potential. Liberated protons are injected into the polymer and participate in its conductivity.

If this model is taken as a system where one electrode is not subjected to oxidation within the studied temperature interval (aurum is taken for such a reference electrode in [50]), it can perfectly describe the experimental data. In other cases [60] the consideration of metal oxidation according to scheme (4.2) for both electrodes was suggested. Then, voltage in the open circuit and, consequently, the current will be generated by the potential difference ΔU_{ox} of oxidized electrodes of two metals $\mathrm{M1} - \mathrm{M}_x\mathrm{O}_y$ and $\mathrm{M2} - \mathrm{M}_x\mathrm{O}_y$.

Let us consider the temperature dependencies of TSC in PVB for different pairs of electrodes (Fig. 4.12). The current maximum observed within the 160–170°C temperature range is independent of electrode materials. As for M1-P-M2 systems the current exceeds by more than an order of magnitude that of the system with electrodes of similar metals.

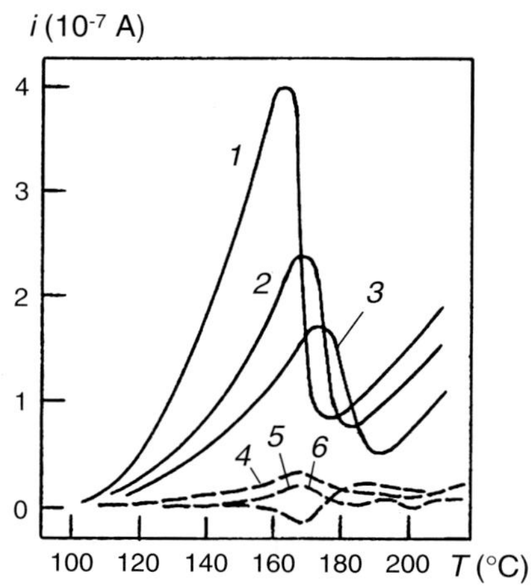

Fig. 4.12. Temperature dependence of TSC in PVB (600 µm thick) for the following pairs of electrodes: Cu–Al (1), Cu–Zn (2), Cu–Pb (3), Al–Al (4), Cu–Cu (5), Pb–Pb (6)

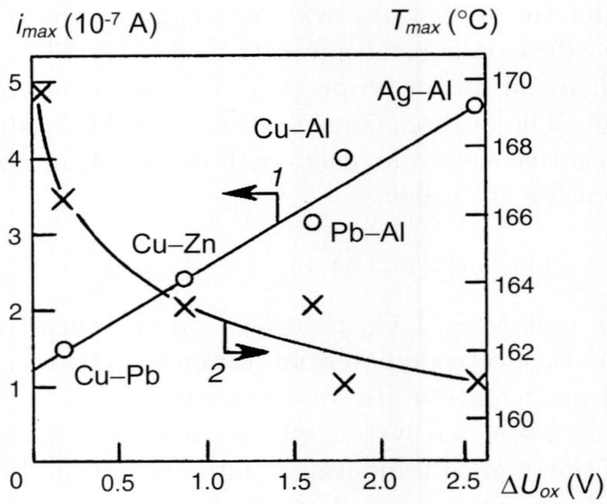

Fig. 4.13. Dependence of maximum current $i_{\max}$ (1) in PVB samples and maximum temperature $T_{\max}$ (2) on electrode potential difference ΔU_{ox} of oxidized metal pairs M1 and M2

In the case of similar metals the current in the M-P-M system is rather weak and the voltage in the open circuit may arise from the inner resistance of the polymer lining. Electrode processes at the polymer-metal interface will mainly keep to scheme (4.2), which is supported by low current values and the weak polarity of the systems with similar electrodes (Fig. 4.12, curves 4–6).

The maximum current $i_{\max}$ shown in Fig. 4.12 is conditioned by the nature of the electrode metals. This value $i_{\max}$ is related to the electrode potential difference of the oxide metal pairs (Fig. 4.13, Table 4.5). It is evident (Fig. 4.13) that extrapolation of line *l* to the ordinate gives zero current at $\Delta U_{\text{ox}} = 0$. This is a qualitative confirmation of the insignificant current in the M-P-M system with electrodes of similar metals. There is, however, a certain deviation of experimental points from the linear dependence *l*, which is explained by the formation of numerous oxides by such metals as, e.g. Ag, Cu, etc., while the ΔU_{ox} values presented in [50] are a result of averaging of the potentials of these oxides.

There exist other explanations for the generation of TSC in M1-P-M2 systems. For instance, the authors of [58] attempted to prove that voltage in the open circuit of the system with a layer of polyvinyl alcohol depended on the difference of work functions of the electrodes. This explanation can hardly be accepted at least because there is always an oxide film on the metal surface that experiences interfacial transformations at the polymer-metal interface. It is evident (Table 4.5) that the work function difference can be correlated neither with the current value in the M1-P-M2 systems nor with its direction.

Table 4.5. Electrode characteristics of metal pairs M1-P-M2 in systems; ΔU – potential difference of standard metal electrodes [19], ΔU_{ox} – potential difference of oxide pair [50], $\Delta\varphi$ – difference of metal work functions

Electrode Pairs	ΔU, V	ΔU_{ox}, V	$\Delta\varphi$, eV
Pt–Al	$1.20 - (-1.66) = 2.86$	–	$6.27 - 4.20 = 2.07$
Ag–Al	$0.08 - (-1.66) = 2.46$	$1.18 - (-1.35) = 2.53$	$4.45 - 4.20 = 0.25$
Cu–Al	$0.34 - (-1.66) = 2.00$	$0.42 - (-1.35) = 1.77$	$4.45 - 4.20 = 0.25$
Pb–Al	$-0.13 - (-1.66) = 1.53$	$0.25 - (-1.35) = 1.60$	$4.00 - 4.20 = -0.20$
Cu–Cu	$1.50 - 0.34 = 1.16$	–	$5.20 - 4.45 = 0.75$
Cu–Zn	$0.34 - (-0.76) = 1.10$	$0.42 - (-0.42) = 0.84$	$4.45 - 4.00 = 0.45$
Cu–Fe	$0.34 - (-0.44) = 0.78$	–	$4.45 - 4.77 = -0.32$
Cu–Ni	$0.34 - (-0.25) = 0.59$	–	$4.45 - 4.75 = -0.30$
Cu–Pb	$0.34 - (-0.13) = 0.47$	$0.42 - 0.25 = 0.17$	$4.45 - 4.00 = 0.45$

For instance, for Pb–Al, Cu–Fe, Cu–Ni pairs the work function difference does not agree with the experimentally determined direction of the current.

The considered electrochemical model has brought us to certain conclusions on the mechanism of polymer polarization in the M1-P-M2 systems. It is evident that definite electrode potentials are established on metals faces in contact with the polymer material. These potential values are affected by the work function of the electrons and metal affinity to the corresponding structural elements of the polymer lining. These potentials are noncoincident with a standard electrochemical series of metals. This is natural, since the properties of polymer materials as electrolytes are not identical to those of salt solutions of the corresponding metals, for which the standard electrode potentials of metals are determined. The studied metals in pairs separated by the PVB lining are arranged in the following series:

Al Zn Ti Pb Fe Ni Ag Pt Au Cu,

which is not in line with the standard electrochemical series [19]

Al Ti Zn Fe Ni Pb Cu Ag Pt Au.

This series is mostly preserved in relation to other thermoplastic polymers as well.

TSC in the M1-P-M2 system results in polarization of the polymer and generation of the electret state in the polymer lining. This new type of polarized state was called a *metal-polymer electret* (MPE) [54, 55] due to a dependence of the electret state of the polymer on the presence of metal inclusions in the polarized dielectrics structure. The value and stability of MPE charge are comparable to thermoelectrets [45], although the polarizing charge is concentrated in the surface layers. The charge density and metal content in polarized samples diminish exponentially from the surface to the MPE depth. The analysis of IR spectra of the polarized PVB and PE samples has brought us to the conclusion of the formation of metal-containing compounds (copper salts of carbonic acids) in the near-electrode layer of the polymer during polarization in contact with the Cu–Al pair. Accumulation of metal-containing compounds takes place in the polymer layer, which may be up to 80 μm thick.

Structural changes in the MPE are induced by two competing processes. Firstly, by the increasing crystallinity under the effect of the polarizing electric field and, secondly, by the formation of the volume charge and permeation of metal atoms into the polymer matrix bulk, which hampers crystallization in the surface layers. Evidently, just these variations in the degree of crystallinity can be a reason for improved physical-mechanical (strength, adhesion) and physical-chemical (sorption, diffusion) characteristics of the materials subjected to polarization in contact with unlike metals.

4.2.3 Metal Corrosion and Wear in Contact with Metal-Polymer Electrets

Let us consider some service characteristics of MPE from the viewpoint of new approaches to application of triboengineering materials in anticorrosion techniques.

Frictional characteristics of MPE were studied within two directions. The first is connected with the effective use of polarization currents in M1-P-M2 systems aimed at improved durability of friction pairs. The second direction is intended to improve wear resistance of polymer materials through their structural modification.

Electrical polarization of friction joints is an effective means of fighting mechanochemical wear. Polarization parameters are chosen so as to transfer the metal into a stable passive state. In other words, the metal potential maintained during operation of the friction joint is different from a standard electrode one. The sources of such polarization can be metal-polymer parts and joints of machines and equipment containing M1-P-M2 pairs.

Model experiments were carried out to study the relation between friction and kinetics of electrochemical processes occurring during polarization of metal-polymer pairs [62]. The experiments employed a pendulum tribometer 1 (Fig. 4.14a) whose advantage is the presence of only one friction pair for examination at a time. The tribometer consists of a pendulum 2, a support 3, and a prism 4 on which the pendulum hangs. Support 3 is made as a vessel containing an electrolyte into which the friction surface in the form of one of the prism faces is immersed. The pendulum is set in motion at a constant initial amplitude. Attenuation of oscillations is recorded in terms of contracting amplitudes of the sinusoidal signal from the inductive pickup 6, into which the bow-shaped core 5 is in turn inserted as the pendulum oscillates.

The friction pair was polarized following a scheme shown in Fig. 4.14b. The working electrode was a metal tray 13 installed on a dielectric support 3. Prism 4 of the pendulum pressed against the bottom of the tray with electrolyte. As an auxiliary electrode, a platinum wire 12 was used. The reference

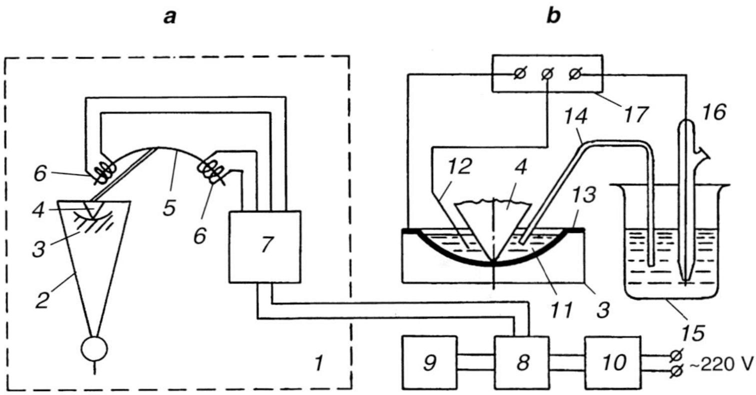

Fig. 4.14. Scheme of a pendulum tribometer (**a**) and electrochemical cell for polarization of the friction zone (**b**): (1) tribometer; (2) pendulum; (3) support; (4) prism; (5) bow-shaped core; (6) inductive pickup; (7) rectifier unit with a generator; (8) amplifier; (9) self-recording potentiometer; (10) voltage stabilizer; (11) electrolyte; (12) platinum electrode; (13) metal tray; (14) electrolytic switch; (15) vessel with electrolyte; (16) reference electrode; (17) potentiostat

silver chloride 16 was connected to a cell via the electrolytic switch 14. Tray 13 was made of copper, steel or aluminum, the prisms 4 were made of PMMA, PTFE, PE and PA–6. For the electrolyte, aqua solutions of NaCl were used. The metal trays were polarized in a potentiostatic regime by varying the load (pendulum mass was varied from 50 to 250 g), during which the polarization current, amplitude and time of pendulum oscillations were recorded.

Characteristic diagrams of polarization current variations during pendulum start and stop are illustrated in Fig. 4.15. A jump in polarization current was observed at the moment when the pendulum was set into motion. This jump is attributed to the disturbance of the film products of electrochemical reaction occurring on the tray surface during motion of the prism. Its value, all other conditions being equal, diminished within the Cu-steel-Al series and corresponds to the position of these metals in the series of the standard electrode potentials [19].

The optimum polarization potential of the metal support, coinciding with the longest time of oscillations (i.e. minimum friction), greatly depends on the material of the prism. Differences in the frictional and electrochemical characteristics of the friction pairs are very unlikely to be governed by the deformation and strength parameters of the prism material (they differ insignificantly in the range chosen) or by mechanical impurities formed on rupture of the oxide film on the metal support surface. One of the factors inducing variations of polarization currents at incipient friction is most likely adsorption of potential-defining ions of the electrolyte on the friction surface of the prism and debris. This factor is also apparently determinative

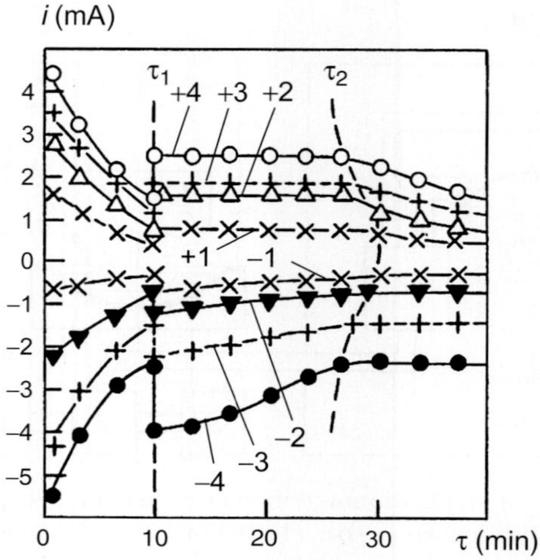

Fig. 4.15. Variation kinetics of polarizing currents of PA-steel friction pair. Numbers on the curves correspond to polarization voltage in volts. Solution of NaCl (10^{-4} mole/l); load – 0.3 MPa; τ_1 – start; τ_2 – stop of the pendulum

for the optimum value of the polarization potential. This assumption agrees with conclusions on the polymer effect on polarization kinetics of metals in electrolyte solutions (see Sect. 1.2.3).

In work [63] the source of the friction zone polarization was an M1-P-M2 sample (the prism and tray were made of steel) formed by two electrodes of aluminum and copper foil separated by a PVB film. It is shown in Fig. 4.16 that the total electrode potential shifts to the anodic region in the absence of polarization and varies only slightly during pendulum oscillation (curve 3). Heating of the metal-polymer sample to $T = (85 \pm 5)$°C leads to the generation of the polarization current of 0.005 mA of some polarity (curves 2 and 4) whose value is consistent with the shift of the total electrode potential. Temperature increase to 105°C results in growth of the polarization current to 0.01 mA and a still greater shift of the electrode potential (curves 1 and 5). The highest velocity of the potential shift in the negative region (cathodic polarization) coincides with the period from the pendulum start till stop, after which the potential stabilizes.

Even insignificant alterations of the polarization potential noticeably affect friction of the support. The cathodic polarization occurring on connection of the aluminum electrode of the M1-P-M2 system prolongs pendulum oscillations compared to anodic polarization. These patterns are affected by the electrical polarization, which changes the rheological properties of the oxide layers and secondary structures formed on the surface of metal parts in the conducting media. The shift of the oxide potential to that of the electrode of a cleaned metal is probably the reason for plastic deformation of the surface

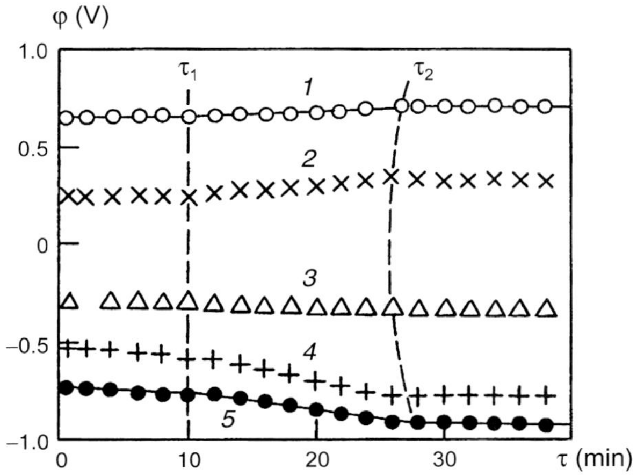

Fig. 4.16. Change of the total electrode potential φ of the support and prism in time t depending on polarization current (mA): (1) 0.01; (2) 0,005; (3) 0; (4) (−0,005); (5) (−0.01); τ_1 and τ_2 – start and stop moments; $p = 0.1\,\text{MPa}$; NaCl solution 10^{-1} mole/l

layers of the rubbing bodies, which is exhibited in variations of the friction coefficient.

Proceeding from the above, current generated in M1-P-M2 systems is able to ensure stable polarization of the friction pair. Polarization takes place when such materials as PA, PPl, PVC, polyolefines and some other thermoplastics are used as the lining material. The polarization sources may appear in seals containing polymer linings, in friction joints whose parts are covered by polymer coats, in reinforced polymer articles and so on. These sources may perceptibly affect friction and wear of metal parts operating in hostile media or found between facings in the electrical contact.

Wear of MPE has been studied by the method of a layer-by-layer removal of the material in planes parallel to the friction surface [64]. The most noticeable changes in the wear rate were observed in the surface layers, where the major portion of the volume charge was concentrated (Fig. 4.17).

Wearing of the surface in contact with aluminum is less than that in contact with copper. The greater residual charge was found to correspond to less wear (curve 1). In all cases, the polarized samples showed less wear than the non-electret ones (curve 3). A general tendency observed was in decreasing wear rate of the electrets with growing TSD charge density.

These changes in MPE friction properties are attributed to the formation of the volume charge, which is accompanied by variations in the degree of crystallinity and formation of a peculiar quasi-structure of metal particles in the polymer matrix. The above-cited information gives grounds for the

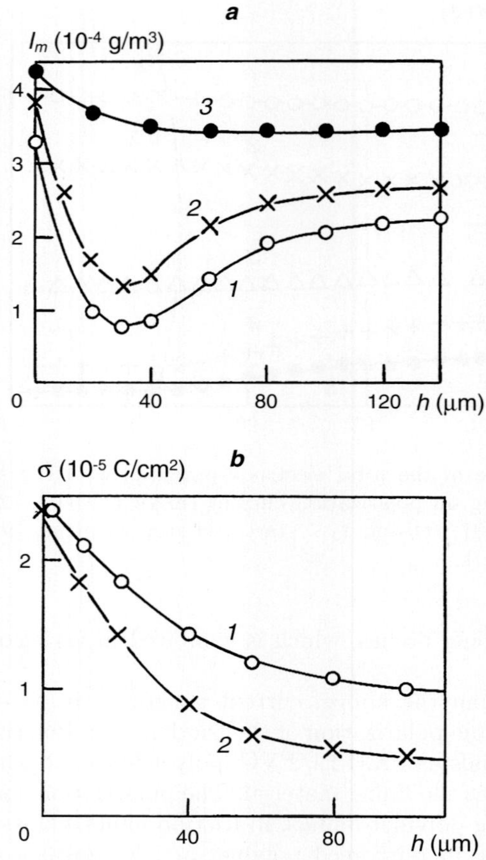

Fig. 4.17. Effect of removed layer thickness from the side contacted aluminum (1) and copper (2) on mass wear rate (**a**) and TSC charge density (**b**); (3) non-electret sample

recommendation of MPE as a means of improving frictional properties of materials and extending the service life of friction joints (see Sect. 4.2.4).

Based on this, it is possible to make the important conclusion that friction and wear of metals can be monitored by maintaining a given optimal potential, defined for each concrete case. As has been mentioned, aside of the potentials of the internal and external polarization sources, this can be regulated using special components introduced into the medium or composition of the rubbing bodies (see Sect. 4.4).

Antirust properties of MPE are most pronounced in protecting and sealing coatings. Polymer coatings intended for rust prevention of metals should possess a combination of physico-chemical properties [65]: electrochemical and insulating, capability to retard diffusion and transfer of agents corrosive to the metal surface, ability to provide passivation protect for metals, adhesive and mechanical properties.

The information presented below is a proof that polarization of polymer coatings is in fact able to regulate all mentioned characteristics.

As has been noted earlier (see Sect. 1.2.1), using outer sources and changing electrode potential of the substrate, it is possible to control the anticorrosion properties of polymer coatings. So, it will be logical to anticipate that anticorrosion properties of polymer coatings will be improved by changing the charge state of the coating material rather than the substrate potential.

The effect of charge distribution in polymer coatings on their protective ability has been estimated judging by the area of the coating peeling in a conducting fluid, variations of the electrode potential of the substrate and strength of the adhesive coating-substrate joint.

It can be seen from Fig. 3.9 that the strength of adhesive bonds of PVB-A1 determined by endurance in the electrolyte by the peeling method is considerably higher in coatings formed in an electric field and whose volume charge field is directed from the coating to the substrate.

Test results of the coatings with a puncture made in a concentrated NaCl solution are presented in Table 4.6. The coatings of group A have been formed on a steel substrate using a classical vibrovortical technique. The coatings of group B have been molten in an electrical field of $E = 20\,\mathrm{kV/cm}$ by connecting the upper electrode to the positive pole of the high-voltage source and the substrate to the earth. The coatings of group B were made as MPE. With this aim, a polymer layer applied to the substrate was overlaid with a copper foil attached to the substrate. After thermal treatment the foil was peeled off.

From the analysis of the results follows that the electret coatings (group B) whose negatively charged surface contacted the metal substrate peel slower and their substrates downgrade to a lesser degree (shift of the electrode potential) compared to the electret coatings of group A. Similar results were

Table 4.6. Peeling area S and change of electrode potential ΔU of the substrate with time

Coating Group	Coating Material	S (cm^2) Depending on Time (days)					ΔU (V) in 5
		1	2	3	4	5	
A	PPl	0.25	0.74	2.03	2.76	3.97	0.24
	PVB	3.03	6.24	10.32	–	–	–
B	PPl	0	0.01	0.02	0.02	0.04	0
	PVB	0.01	0.02	0.04	0.07	0.08	0.02
C	PPl	0	0.01	0.02	0.02	0.03	0
	PVB	0.01	0.02	0.05	0.07	0.08	0.02

shown by MPE-based coatings (group B) formed without application of the external electrical sources.

This suggests that resistance of the hardware in corrosion-active media can be raised substantially using electret coatings with the electret charge.

4.2.4 Wear Inhibiting by Electrical Fields

Properties of engineering materials are known to vary under the action of electromagnetic fields. For instance, wear of metal-polymer friction joints was found to reduce substantially in response to exposure to a flow of charged particles in conditions of electrochemical reactions. Besides, methods of regulating the conductivity of materials and polarization of crystals by electrical fields, as well as by so-called training methods of semi-conducting substances and many others are known in the art.

Let us examine some examples of improved service characteristics of metal-polymer friction joints with the help of electrical fields. A metal-polymer joint (MPJ) is a combination of metal and polymer parts operating in coordination [66]. The durability of an MPJ depends on its design, the properties of constituent materials and the operation conditions, including temperature, pressure, mutual displacement velocity, ambient media, physical fields, radiation etc. During operation, an MPJ undergoes certain changes in its material structure, wearing, etc. that impair its performances and life of the joint as a whole (Fig. 4.18, solid arrows).

A specific feature of MPJ is the presence of electrode elements in its structure, i.e. it contains metals and polymers displaying electrolyte properties, being at the same time a dielectric able to polarize in the electric field.

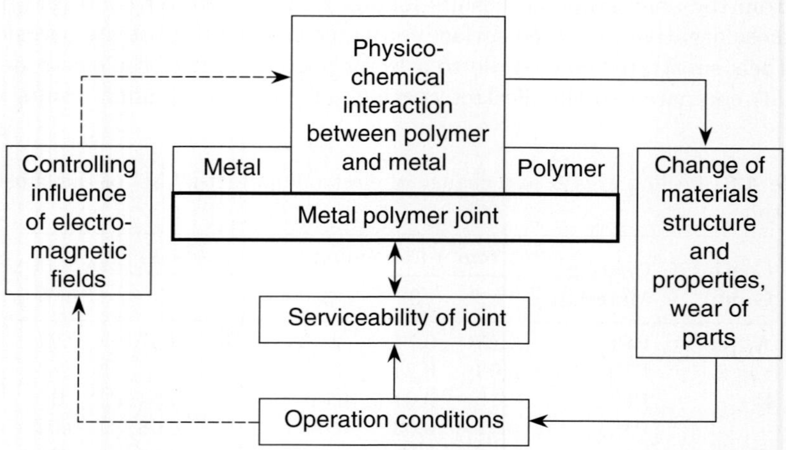

Fig. 4.18. Schematic diagram of monitoring serviceability of a metal-polymer joint during its operation

The diverse influence of service conditions on MPJ durability can be suppressed using electric or magnetic fields. It is possible to modify the structure of polymer materials, improve their strength, adhesion, raise resistance to corrosion, reduce permeability, etc. [55] by electrical polarization. Under the effect of the magnetic field the particles of a magneto-solid filler or structural units of a polymer material become oriented, modifying the structure and service characteristics of polymer composites [66, 67]. In general, the application of the electromagnetic field assists in the regulation of the structure of the surface layers of the contacting parts and, consequently, to monitor interactions between components of the MPJ.

The task of raising the efficiency of the MPJ can be solved when the electromagnetic field is actuated in response to changes in operation conditions, i.e. in the presence of a closed tracking loop over operation conditions with a feedback that switches the electromagnetic field. A control circuit with the electromagnetic field is shown by a dashed line in Fig. 4.18. The result of the controlling action can be either electrization or polarization of the polymer, shift of the electrode potential of the metal, orientation of filler particles or structural units of the polymer material leading to either lubrication, ameliorated permeability of a seal, or other effect. The feedback is exercised via operation conditions and their variations initiate the regulating effect of the electromagnetic field.

Furthermore, electrical polarization of polymer materials in contact with metals may improve the frictional characteristics of the rubbing parts in production and raise efficiency during their operation. Below some examples are presented of engineering and design solutions of realizing this potential.

Let us consider *a device for the production of polymer parts* [68] with different lubricant content that does not impair their deformation or strength characteristics. The forming surface of one of the molding elements of the device is fitted with dielectric and conducting components. The conducting ones are attached to the second molding element, which is made of some other metal. The difference between electrode potentials of the molding elements does not exceed 2–3 V.

One modification of the device is illustrated in Fig. 4.19. A mold for production of a bearing of a thermosetting material consists of a punch 1 and die 2. The steel punch is fixed on the upper plate 9 of the press and is equipped with the heating units 8. The die is fixed through an insulating pad 4 to the lower plate and contains a dielectric insert 3 with pusher 5 whose aluminized face shapes the bearing surface of the bearing. The electrical circuit 7 is shut off by closing the mold, so molding and thermal treatment of the polymer material 6 are accompanied by TSC passage between the pusher 5 and punch 1. The value of the current depends on the electrode potential difference of steel and aluminum ($\Delta U = 1.22\,\mathrm{V}$). As a result, the bearing bottom becomes polarized and acquires the properties of MPE. Further treatment in a self-lubricating liquid makes the bearing sorb the liquid to a less degree than its

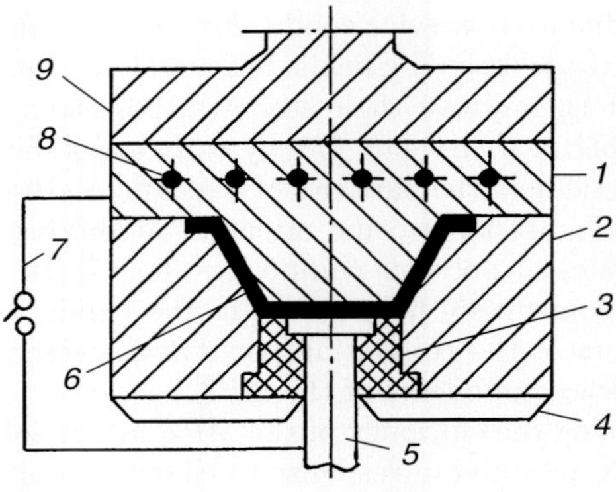

Fig. 4.19. A device for manufacturing polymer articles

conical body since the polarized polymer sorbs low-molecular liquids less than the non-polarized one [61]. Thus, the bottom, which carries the axial load, strengthens and gains resistance to loading, while the body walls modified by the lubricating fluid obtain a property of self-lubrication.

One more method of molding from polymer compositions [69] improves the strength of the products and their antifrictional characteristics by transforming the polymer material into the electret state. The essence of the method consists in the following. An inorganic fibrous filler, e.g. glass or basalt fiber, is exposed to a constant electric field of 10–50 kV/cm at a temperature of 300–500°C for two hours. In the course of the treatment the filler gains the electret charge. After this, the fibers are mixed with the thermoplast and the mixture is molten and fed into the mold. During molding at 170–210°C the melt is subjected to the effect of an electrical field of 10–25 kV/cm. As a result of interaction of the polarizing field and that of the electret filler the fibers become oriented in the binder melt, which also transfers into the electret state. The strength and frictional characteristics of the articles produced by this technique are 25–60% better than those without the field application. The service characteristics of the articles can be regulated by varying the field intensity and filler loading.

The problem of designing reliable and compact sources of electrical polarization of friction joints is still a weak point of the technique. Polarization of the friction zone by extneral voltage sources has not found wide enough application since they complicate the structure of the machines. So, it is important to find some alternative sources of polarization proceeding from design and technological peculiarities of the friction joint.

An example of successful use of the polarized state of polymers for abating mechanochemical wear of friction parts is a *sleeve bearing*, illustrated in Fig. 4.20 [70]. The friction joint consists of a shaft 1 and a bushing consisting of three coaxial bushes installed in the body 5. The middle bush 4 is made of a polymer material, while the bush facing the shaft 2 is made of a metal with more positive electrode potential than that of the outer bush 3. The bushes are grounded electrically to metal cover 6, whereas spacer 7 is placed between the body and the cover. Frictional heating of the metal-polymer bushing during operation generates current in the polymer layer whose direction is determined by the position of the bush metals within the electrode potential series (see Sect. 4.2.2). As can be seen, the metal-polymer bushing functions as an electrical generator that polarizes the friction zone. As the shaft comes into contact with bush 2, the electrode potential shifts in the positive direction. If the shaft is made of a metal or alloy susceptible to passivation under anodic polarization in a conducting liquid encircling, e.g. a steel friction joint, then even at a negligible shift the electrode potential of the shaft will reach the values corresponding to the passive state of the steel. As a result, mechanochemical wearing of the shaft decreases.

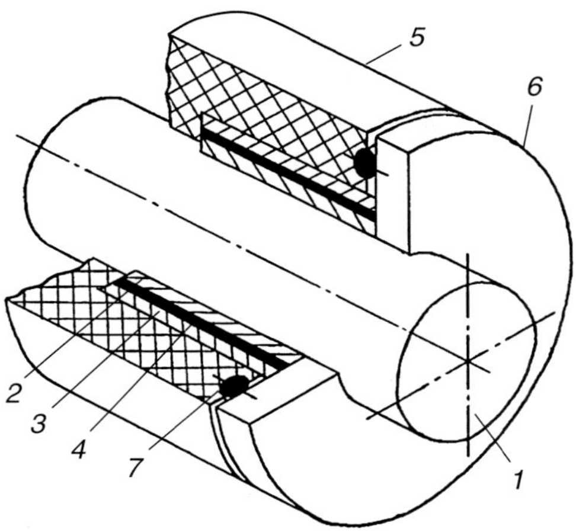

Fig. 4.20. A sliding friction joint

The friction joint has been tested using a shaft on a bush friction machine with 2 cm^2 friction area, 0.35 MPa load and 2.4 m/s sliding velocity. A 40-mm-diameter shaft has been made of a carbon steel of 40–45 HRC hardness and 0.8-1.0 µm surface roughness. The outer bush material was aluminum, the inner was copper, the polymer layer was 200-µm-thick PVB. A 0.1 N solution of NaCl was fed into the friction zone, the wear rate was determined by weighing. The test results are presented in Table 4.7.

It is evident that wearing of the shaft is reduced by as much as six times if the electrical circuit of the bushing is closed and the potential shear is realized against the bushing whose metal bushes are open, rather than through wear, and no current passes through the polymer layer. This wear reduction of the sleeve bearing, for which the shaft is the most critical and costly

Table 4.7. Mechanochemical wear values of the shaft

Electrical Circuit of the Bushing	Wear Value (10^{-6} g/m^2) at Test Duration, Hours				
	3	6	9	12	15
Open	1.7	2.8	3.2	4.0	4.2
Closed	0.3	0.5	0.6	0.7	0.7

part, elevates its life several times. Along with electrochemical protection of the shaft without application of external polarization sources, the described structure of the friction joint can be efficiently used for generation of polarization current. One more advantage is the absence of any seals, which prevents the conducting fluid from getting into the friction zone.

To lower the friction and mechanochemical wear in friction joints it is possible to use *solid lubricants* [71] containing fuming-oxides (FO) as an active modifier of nonferrous metals. The solid lubricants represent fine-dispersed sublimation products of lead smelting slag.

The lubricant is prepared by mechanical mixing of components followed by fusion at 220°C. The friction coefficient of the lubricant, *f*, is measured in the dry friction regime using a shaft-on-partial-insert geometry with sliding velocity $v = 0.15\,m/s$ and loading of $p = 5\,\text{MPa}$. Adhesion is characterized by the breaking stress σ by the method of normal separation of cylindrical samples of the lubricant from a steel plate preheated to 180°C and cooled to 20°C. The corrosion rate i_c of steel samples protected by a layer of the lubricant is recorded in a 0.1 N solution of Na_2SO_4 by the polarization resistance method. The optimum FO content made up 30–50% (Table 4.8). With increasing filling degree, service characteristics of the lubricant are impaired as a result of lowered plasticity of the binder. Paraffin content above 12% increases the friction coefficient and deteriorates lubricant adhesion. For comparison, properties of a lubricant containing molybdenum disulfide instead of FO are presented in Table 4.8.

As can be seen, introduction of FO into the lubricant composition improves adhesive and the antifrictional properties of the lubricating layer.

Table 4.8. Service characteristics of lubricants

Concentration of Components, wt. %			μ	σ, MPa	Corrosion Rate, g/(m^2·h)
Paraffin	FO	Polyamide			
12	30	58	0.03	0.06	1.35
10	40	50	0.03	0.07	1.25
8	50	42	0.03	0.06	1.30
12	30(MoS_2)	58	0.05	0.03	2.90

Their mechanism can be presented as follows. During thermal treatment, FO contact the polymer melt and the oxides partially reduce to metals [72]. This leads to the generation of microchains M1-P-M2 in the lubricant bulk and formation of an MPE as a result of electrochemical interactions of unlike metal components. As a consequence, adhesive interactions between the filler and binder intensify. In our view, the high level of antiscoring properties and low friction coefficient can be attributed to the catalyzing effect of the reduced metals on the formation of the friction polymers from the products of mechanical destruction of organic components of the lubricant.

One of the electrochemical methods of regulating friction processes in hostile media is protection as a particular case of cathodic polarization [73]. Cathodic polarization can be exercised by connecting the protected structure to either the negative pole of the external source or to a metal with a more electrically negative electrode potential. In the latter case, there is no need for an external current source as a galvanic element is formed in which the protected element becomes the anode and the more electronegative metal, called a protector, becomes the cathode. Alloys based on magnesium, aluminum or zinc are often employed as protectors of steels. The protectors are in their essence portable sources of electric energy. They are specifically useful when the electric energy should be transferred directly to the metal protected, particularly, in friction joints.

Protective metal coats and films on the working surfaces of rubbing bodies are frequently used as protectors of friction joints. These films of more electronegative or more electropositive metals than the substrate can be applied on the contact surface. The former function as protectors, the latter passivate the substrate material operating as sacrificial anodes.

To protect copper alloys (aluminum and tin bronze, brass) films of $Cu_5Zn_8, Cu_{31}Sn_6$, CuZn and other compounds [74] are recommended. These films can be applied onto the contact surface of protected parts by either chemical or electrochemical deposition, or directly in the friction process using metal cladding lubricants.

Notice that application of protecting films by metal cladding has a number of advantages. In particular, porous oil-absorbing films formed during this process are able to stabilize friction and possess good antiwear properties. Moreover, the antiwear properties of the lubricating material itself also improve significantly. High efficiency of the protective metal films in conditions of mechanochemical wear agree with the results of numerous investigations [74, 75].

Graphite impregnated with babbit [76] has been tested as a protecting material. In such systems babbit behaves like the anode towards steel. During friction of this pair in a 5% solution of H_2SO_4 the rate of mechanochemical wear of steel reduced by more than 10 times in contrast to friction of steel against the babbit-free graphite. It should however be born in mind that babbit dissolves strongly in acidic media.

4.3 Tribochemical Processes in Metal-Polymer Contacts

Many researchers believe that the products of tribochemical transformations of polymers and metals used as corrosion inhibitors may enhance the wear resistance of metal-polymer friction joints operating in hush media. A friction joint can be theoretically presented as a reactor continuously generating products able to lower mechanochemical wear. For practical purposes we should know the regularities of these tribochemical transformations in macromolecules and their relation with the nature and surface parameters of the frictional metal counterbody.

The present chapter presents investigation results on the mechanisms of tribochemical transformations in macromolecules, gives an analysis of the effect of metal on these processes and outlines the ways they are used to inhibit wear.

4.3.1 Tribochemical Transformations in Polymers

Tribochemical transformations of polymers are chemical reactions in the solid phase with participation of macromolecules taking place during frictional interactions between polymers or a polymer and a counterbody. We discriminate between two types of competing chemical reactions responsible for tribochemical processes in polymers, namely tribodestruction (decomposition) and tribological structurization (cross-linking). The systematic investigations in the field of tribochemistry of high-molecular compounds were first undertaken in the 1920s. It was established that machining of polymers, e.g. grinding of rubber, cellulose or polystyrene, results in decreasing molecular mass and destruction of macromolecules [3].

Destruction of macromolecules and distortion of the permolecular structure lead to changes in their properties, especially solubility, resistance to chemical agents, strength, fatigue and impact viscosity, as well as elasticity and plasticity. From the practical viewpoint, variations in polymer properties induced by the tribochemical destruction raise interest in two respects: first, as unavoidable phenomena that accompany any mechanical effect on the polymers during their treatment or operation and, second, as a deliberate change of properties of solid polymers to obtain materials with specific characteristics.

Elementary stages of tribochemical transformations. Tribochemical reactions in polymers usually follow a free-radical mechanism [78, 80]. As a result of mechanical interactions polymer or low-molecular radicals appear, taking part in substitution, addition, decomposition or other reactions. On breakage of a macromolecule, two primary radicals with free valence at the chain ends are formed. In the vinyl series these radicals are $\sim^{\bullet}CH_2$ and $\sim^{\bullet}CXY$ (where X and Y are hydrogen atoms and other side substitutes); in

Table 4.9. Structure of primary free radicals formed during mechanical destruction of polymers [3]

Polymer	Radical Structure
Polyethylene	$\sim CH_2 - H_2C^\bullet + {}^\bullet CH_2 - CH_2\sim$
Polytetrafluoroethylene	$\sim CF_2 - F_2C^\bullet + {}^\bullet CF_2 - CF_2\sim$
Polystyrene	$\sim CH(C_6H_5) - H_2C^\bullet + {}^\bullet CH(C_6H_5)CH_2\sim$
Polyamide	$\sim CH_2 - CH_2C^\bullet + {}^\bullet CH_2 - (NHCO)\sim$
Polyethylene oxide	$\sim O - H_2C^\bullet + {}^\bullet CH_2 - O\sim$

rubbers they are allyl radicals $\sim CX = CY -^\bullet CH_2$ and so on. The formulas of primary radicals appearing at destruction of certain polymers are given in Table 4.9.

During interaction with neighboring macromolecules end radicals may convert into central ones [79]:

$$\sim CH_2 - H_2C^\bullet + \sim CH_2 - CH_2 - CH_2\sim \ \rightarrow \ \sim CH_2 - CH_3 + \sim CH_2 - {}^\bullet CH - CH_2\sim . \tag{4.3}$$

Secondary radicals with free valence in the chain center ($\sim^\bullet CX\sim$) participate as a rule in the reactions of macromolecular decomposition and the transfer of H atom. Decomposition is the inverse to the reaction of addition. That is why, the lower the polymerization heat, the lower the activation energy of decomposition and the stronger the probability of macromolecular destruction on free valence localization. Polymers characterized by low polymerization heat of their monomers (e.g. the polymerization heat of PMMA is about 55 kJ/mole) are strongly decomposing, not only under mechanical effects, but also in other processes of radical formation.

The probability of decay of $\sim^\bullet CX\sim$ radicals in polyethylene and other polymers whose polymerization heat of monomers reaches 88–92 kJ/mole is less than that of other possible transformations of radicals.

Primary and secondary radicals with free valence localized on carbon atoms can easily have a molecule of oxygen added and be converted into a peroxide form

$$R^\bullet + O_2 \rightarrow ROO^\bullet . \tag{4.4}$$

This reaction is often restricted by the velocity of oxygen diffusion in the polymer. The peroxide radicals are commonly more active than the allyl radicals when transferring a hydrogen atom

$$ROO^\bullet + HR \rightarrow ROOH + R^\bullet . \tag{4.5}$$

Alterations due to oxygen addition reactions and transfer of a hydrogen atom leads to chain oxidation, where the chain usually consist of a few links.

Some peculiarities of the free radical processes that are of importance for tribochemical reactions should be mentioned.

The reactivity of macro-radicals formed under mechanical effects depends on their chemical structure and conformation. It is accepted that the reactivity of macro-radicals is in fact independent of their molecular mass (the Flory principle) [80]. The efficient activation energy of recombination and transformations of macro-radicals is within the 20–25 kJ/mole range. The recombination and transformations of macro-radicals are related to their displacements by the mechanisms of free valence migration or diffusion.

Rupture of chemical bonds and formation of macro-radicals is observed when the energy of the mechanical action located on some link exceeds the energy of this link. The resistance of polymers to tribochemical transformations is evident from the value of the dissociation energy of the links (Table 4.10).

Table 4.10. Dissociation energy of links in macromolecules at 25°C [80] (characterizes the link marked by pointer ▼)

Link	Dissociation Energy, kJ/mole
C – N	224
C ▼ C = C	231
C ▼ C	
C |	263
C C_6H_5	
C – Cl	293
C – O	314
C – H	370
C – C	372
► |	
C_6H_5	
C – C ▼ C	374
H – F	419
C = C	425
H – Cl	428
H – H	432
C – F	436
H – F	562
C = O	727

From the data cited in [80, 81] it follows that:

- on mechanical loading of carbochain polymers the chains break along C–C links
- polymers with side branching in macromolecules undergo destruction more readily

- a C–C link in the α-position towards a double link C = C is always weak and dissociates easily
- a C–C link with a quaternary atom of carbon breaks more easily than with a tertiary one
- polymers in which every second atom of carbon in the backbone chain is quaternary breaks through the depolymerization mechanism.

Physico-structural inhomogeneity. In contrast to low-molecular substances, the reactivity of solid polymers depends greatlyon the structure of the electron shells and to a greater extent on molecular dynamics and imperfections of the permolecular structure. Physical-structural inhomogeneity of polymers may bring about the following phenomena that are critical for tribochemical reactions [80].

A. Inhomogeneous distribution of additives and reagents in the polymers. Low-molecular substances (oxygen, oxidation products, inhibitors, Pl, fillers, etc.) accumulate in the amorphous regions of the polymers. The most reactive sections of macromolecules (oxidized groups, unsaturated links, branchings and so on) also localize here. Since local concentrations of reagents may essentially differ from the central values, the rates of tribochemical reactions in these regions of the polymer also differ.
B. The presence of spatially distributed areas in the polymer with different frequencies of molecular movements. This is the reason why the elemental reactions in solid polymers are described by a wide set of kinetic constants and activation energies. Molecular movements impose strong limitations on chemical reactions. Chemical reactions are intensified as a rule when molecular movements become defrosted. At low temperatures (below the glass transition point T_{g}) the reactions run slowly in polymers.

Consequently, physical-structural inhomogeneity of polymers leads to spatial localization of physical processes (sorption, diffusion, etc.) and chemical reactions in the areas of the polymers function like peculiar "microreactors".

The main role in the intensification of tribochemical reactions belongs to mechanical effects like pressure and shear deformation. Mechanical stresses affect the reactivity of substances as follows:

(1) elastic strains change the thermodynamic potentials of the reagents, the equilibrium constant of the chemical reaction and its activation energy
(2) during dissipation of elastic energy non-equilibrium intermediate states with high reactive capacity appear
(3) the particles are displaced in the frictional polymer by the deformations, intensifying transportation of the reagents

In real conditions different types of mechanical effects are commonly acting simultaneously. Because of the diversity of active particles participating in tribochemical reactions (radicals, ion-radicals, unstable molecules, etc.)

and the variety of mechanisms of the indicated processes, tribochemical conversions in polymers are highly intricate.

The kinetics of tribochemical transformations. The role of mechanical stresses consists of not only the initiation of radical processes but in kinetic changes of the elemental stages of the chemical reactions with the participation of macromolecules as well. Plastic deformations increase reaction velocities that have been limited by travel of the reacting particles in the material bulk. Shear and tension accelerate the decay of the radicals with a free valence in the chain center. Chemical reactions are also affected by variations in the intra-molecular mobility and permolecular structure during polymer deformation.

The kinetics of tribochemical transformations, i.e. the ratio of reaction velocity to temperature, the intensity of mechanical actions and other factors, is defined largely by two sequential stages of the mechanochemical process, namely: i) load application, deformations and distribution of elastic stresses in the system, and ii) the chemical reaction itself initiated by stresses.

In variable mechanical fields (the type of loading that occurs during friction) the slowest stage that defines the velocity of tribochemical transformations may turn out to be redistribution of elastic stresses in the polymer. In this case, the velocity constant K_i of the mechanical process will be proportional to the redistribution velocity $A(\mathrm{s}^{-1})$ and the probability of accumulating some critical deformation energy U^* on a given link sufficient for the reaction initiation [82] will be

$$K_i \approx A \cdot f\left(\frac{U^*}{\Delta U_{el}}\right) , \tag{4.6}$$

where the constant value A is determined by the relaxation properties of the material and frequency of the mechanical effect; $\Delta U_{el} = \alpha \tau_r I$ is the mean level of accumulated elastic energy; α is the constant of proportionality close to the process efficiency; τ_r is the relaxation time; and $I$is the- intensity of mechanical energy supply .

The form of the function $f(\frac{U^*}{\Delta U_{el}})$ is in a general case unknown. Since in a number of tribosystems the distribution of probabilities is described by an exponential law, (4.6) can be presented as

$$K_i \approx A \cdot \exp\left\{-\frac{U^*}{\alpha \tau_r I}\right\} . \tag{4.7}$$

This equation looks like the Arrhenius equation although the value and physical nature of the constants are different. Its exponent implies elastic rather than heat energy, and the pre-exponential factor A is close in value to the stress relaxation velocity (for the glassy polymers $A \sim 10^{-4}\,\mathrm{s}^{-1}$), which is many times less than the constant K_0 in the Arrhenius equation ($\sim 10^{13}\,\mathrm{s}^{-1}$).

Equation (4.7) does not contain the dependence of the process velocity on temperature in the explicit form. However, A and τ_r are dependent on temperature and change abruptly within the region of structural transitions, near the glass-transition temperature. As the relaxation time decreases with temperature growth, the velocity of mechanochemical transformations in variable mechanical fields is characterized by a negative temperature factor.

The cases considered above, when the reaction rate is restricted by the distribution of elastic stresses in the polymer, do not describe in full the kinetics of complex multistage tribochemical reactions in polymers. The limiting stage may happen to be the displacement of reacting particles, formation of a new friction surface and so on. The reaction velocity constant may vary during the friction process as it depends upon the polymer permolecular structure, molecular-mass distribution and other factors.

Structuring (cross-linking) or *tribochemical synthesis* is a process of macromolecular modification resulted in irreversible tribochemical reactions with macroradicals participation [3]. Macroradicals of each polymer formed during friction of polymer blends give rise to grafted and block-copolymers during recombination. Cross-linking can be induced by, e.g. interaction of macromolecules and active centers (functional groups of neighboring chains). The resultant products might undergo further tribochemical conversions. For instance, 3D structures are able to decompose under the action of mechanical stresses. As a result, a complex mixture of grafted polymers, block-copolymers, cross-linked and branched products can be found on the friction surface.

The brief information on polymer transformations under tribochemical effects cited above can be used for estimation of mechanochemical processes in the metal-inhibited plastic tribojoints only in the first approximation. A unified theory of tribochemical transformations of polymers has not yet been developed and the mechanisms of macromolecular conversions during friction are still to be studied comprehensively. Some investigation results of tribochemical phenomena in metal-polymer friction joints are discussed below.

4.3.2 Tribochemical Transformations in Polymer-Metal Systems

Complex studies of tribochemical transformations in metal-polymer friction joints have been a subject of a number of investigations. The team of V.V. Korshak (Institute of Element-Organic Compounds of Russian AS) has established that tribochemical processes might lead to the formation of secondary structures on the metal counterbody, a so-called "third body" or "transfer film" [83–88]. These structures exert a perceptible effect on further friction and wear of the joint. During friction of certain polymers and composites on their base a phenomenon of tribochemical lubrication has been established, which consists of the effect of gaseous and low-molecular products of tribodestruction on friction. Notice that the products of tribochemical reactions of the metal-polymer system of a certain structure and given conditions are able to function as wear inhibitors [89].

Let us examine the chief types of tribochemical transformations occurring during friction of different classes of polymers paired with a steel counterbody [88].

Olefins and their derivatives. Various destructive and oxidative processes occur during friction of UHMWPE against steel [90, 91]. A secondary film formed on the surface of the polymer part is inhomogeneous in structure due to different tribochemical processes varying across the depth of the sample. The thinnest, outside layer is depleted of oxidized groups (hydroxy1, carbony1, carboxy1, and etc) compared to the initial sample. At the same time, an insignificant amount of oxidized iron appears on the UHMWPE surface in the form of fine spherical cluster-type particles. Active oxidation taking place in the subsurface layers of the UHMWPE sample is encouraged by diffusion of oxidized particles into the sample. Tribochemical processes in the sample lead to abrupt impairment of its friction parameters up to seizure and augmented wear of the friction pair.

The high thermal resistance and antifrictional properties of PTFE are attributed to the presence of a large amount of negatively charged substitutes (fluorine atoms) in the carbon atoms that screen positively charged carbon and attenuate molecular interactions. This is why PTFE has low shear resistance and reduced friction coefficient when paired with metals. As PTFE and its copolymers wear, their macromolecules undergo thermal and oxidative destruction with rupture of C – C and C – F links, formation of olefin structures, metal fluorides and gradual amorphization of the surface layer. This brings about a new structure on the friction surface resembling that of thermotropic liquid crystals, which ensures easy sliding and improves wear resistance. Simultaneously, a transfer film is formed on the metal counterbody surface [92, 93], being a peculiar wear inhibitor of the metal-polymer friction pair [34].

Polyphenylens. In the course of friction of these carbocyclochain polymers at 200°C, their friction surfaces accumulate dypnone fragments formed as a result of condensation reaction of macromolecular end groups. In addition, more thermally stable triply-substituted benzol nuclei, formed during polycondensation trimerization with the participation of remaining end groups and dypnone fragments of the main chain, grow in concentration. The processes of tribochemical synthesis in the surface layer of polyphenylen result in the formation of thermally stable secondary polymer structure. Friction of completely solidified polyphenylen is accompanied by the destruction of the main chain along the phenyl oxide fragments. Materials based on polyphenylen containing 15% sol fraction are distinguished by a higher wear resistance [83, 94, 95].

Phenol formaldehyde resins (polymethylene phenylene) are polymers with irregular cross-linked structure. During friction of resol- and novolac-type phenol formaldehyde polymers two processes take place simultaneously, namely structuring (aftercure) via recombination of weak ester groups into methylene groups, and tribodestruction accompanied by phenol isolation [84, 96]. End groups, defective fragments and low-molecular products play an important role in cross-linked polymethylene phenylenes. They participate in triboengineering reactions leading to the formation of more stable secondary structures and low-molecular products. The latter carry out so-called tribochemical lubrication that lowers and stabilizes the friction coefficient.

Polyethers used as frictional materials are aliphatic polyethers (polyacetals) and aromatic polyethers (polyacrylates, polyphenylenoxides, polyarylene etheroketones, etc.). A major drawback of polyacetals and materials based on them is their low wear resistance. Therefore, copolymers of formaldehyde with 1,3-dioxolan (CFD) dominate as materials for friction joints. Friction of CFD goes hand in hand with tribochemical reactions leading to intensive isolation of destruction products. The chief one is formaldehyde, which is a proof of depolymerization of the backbone chain along the oxymethylene fragments ($-CH_2O-$) accompanied by the rupture of the end methylol groups.

In the case of friction of thermostable polyphenylen oxides (PPO) containing substituted aromatic nuclei in macromolecules, mainly cresol and small amounts of xylenol are derived. As a result, a secondary heat resistant structure is formed on the friction surface of PPO parts. The material with a hybrid structure of "caged snake" type based on PPO in combination with polymaleimide displays better wear resistance owing to its cross-linked structure. The tribochemical lubrication in such pairs is produced via tribological decomposition of less heat-resistant polymers [98].

Rubbing of polyarylene etheroketone (PAEK) containing ether links in its backbone chain, a carbonyl group and aromatic nuclei is accompanied by destruction along mainly ether links and structuring processes. Formation of stable transfer films on the steel counterbody leads to friction between polymer layers with the effect of self-lubrication that lowers the friction coefficient and heightens wear resistance of PAEK as compared to PPO [99].

Polyamides. Wear resistance of polyamides (PA) of different chemical structures is to a great extent determined by the nature of the tribochemical processes, which in turn depends largely on the chemical constitution of the PA links, location of the amide bonds, the structure of the end groups, chains and frictional conditions [86, 100, 101].

All aliphatic PA isolate nitrogen and oxygen-containing products during friction, formed on decomposition of the α-methylene link $NH-CH_2-$ and rupture of aliphatic chains. Depolymerization processes at chain ends generating monomer lactams (ε-caprolactam in PA and ω-dodelactam in PA-12) contribute substantially to the tribological decomposition of polymerized PA whose links alternate in the "head to tail" manner. A large amount of aliphatic amines is liberated during decomposition of polycondensed PA.

Unstable friction of PA is attributed to tribo-oxidation processes, namely processes with formation of peroxide groups at α-carbon atoms. As a result, a branched and partially cross-linked secondary structure is formed on the friction surface. A 3D rigid skeleton formed in materials on the base of polyamide blends with thermosetting polymers (phenol, melaminoformaldehyde, epoxy-novolac block copolymers and other types) improves the wear resistance of their compositions compared to PA [102].

Heterocyclochain polymers (polyimides, polyoxadiazoles, polyphenyl quinoxalines, etc.) display higher wear resistance in contrast to heterochain polymers not containing heterocycles in the main chain (polycarbonate, polyarylate, aromatic polyamide, polyphenylene oxide). This is because of the thermal stability of the chemical bonds of the main chain, the high density of cohesive energy and the ability to form thermostable cross-linked secondary structures during friction [103].

We know two major competing directions of tribodestruction of heterocyclic polymers: i) decomposition of a heterocycle (quinoxaline, imide, oxadiazole or others) as a result of the destruction of weak links of the backbone chain leading to a drop in the polymer molecular mass until the formation of low-molecular products and ii) the reactions resulting in formation of a branched and cross-linked polymer structure [97, 104–106]. Low-molecular products of macromolecular destruction (both gaseous and oligomer) may serve as tribochemical lubricants that reduce the friction coefficient and function as wear inhibitors.

Some ordered structure is formed in the surface layer adjoining the friction surface during friction of polyheteroarylenes of the chain and cyclochain structure (polycarbonate, polyarylates, polyimides, polyphenylquinoxalines, polybenzimidazoles and others). This results in stabilization of friction and reduction of the friction coefficient [107].

The above analysis shows that wear resistance of polymers and copolymers of different molecular structure depends largely on the nature of the triboengineering processes leading to destruction and cross-linking of macromolecules. These processes display, as a rule, a competitive behavior and the structure of the dominating product depends on the polymer material and its friction characteristics.

4.3.3 The Mechanochemical Concept of Wear of Metal-Polymer Friction Pairs

Tribochemical conversions of macromolecules take place in the polymer layers immediately in contact with the metal counterbody. Since the counterbody surface commonly shows certain chemical and catalytic activity, this contributes to the friction process of the contact chemical reactions.

The composition of the products of tribochemical reactions depends on the service load and velocity regimes of the metal-polymer friction joint and the origin of the contacting materials. As has been indicated earlier, the processes of cracking, destruction, splitting of side groups, dehydration and other processes occur in the polymer material during friction. The surface layers of the metal counterbody experience intensive changes, including oxidation, reduction, adsorption of low-molecular products, dispersion of the surface layers and transfer of metal particles to the polymer friction surface, and so on.

The analysis of wear debris and surface layers of contacted polymers and metal bodies has visualized the intensive tribochemical processes taking place in the friction zone. IR spectra of PA6, HDPE and CFD debris have displayed adsorption bands within the 1710–1740 cm^{-1} region. This suggests that acidic, aldehyde and ketone groups have been formed in the polymer chain. In addition, the adsorption bands in the 1000–1200 cm^{-1} region are intensified due to the formation of metal-containing compounds and alcohol-ester groups. The differential-thermal analysis has revealed that the wear debris has a lower T_m than the initial polymer and a larger amount of oxidized compounds [108].

The nature of the metal counterbody significantly influences the wear of contacting polymers (Table 4.11). Under low loads and friction velocities tribological characteristics of metal-polymer friction pairs depend only slightly on the metal nature. As soon as the load-velocity regime becomes more severe, the role of this factor increses drastically [108].

Table 4.11. The effect of the nature of the metal counterbody on the wear rate of the polymer samples during dry friction (friction regime: $p = 0.5$ MPa, $v = 0.625$ m/s)

Counterbody Metal	Wear Rate ($I \cdot 10^8$) of Samples			
	HDPE	PA6	CFD	PTFE
Cu	0.02	0.9	0.02	1.5
Al	0.07	0.12	0.23	–
Zn	1.24	0.31	1.52	–
Steel 45	0.05	0.22	0.12	0.42
Ti	0.36	0.8	0.43	–

The nature of the metal predetermines properties as well as the composition of the wear debris. As PA6 rubs against copper, the amount of oxidized compounds in the debris is considerably higher than during friction against steel 45. Debris of HDPE and CFD formed when rubbing against copper contains, on the contrary, low amounts of oxidized matter, besides they are more heat resistant [109].

The metal counterbody surface undergoes severe changes during the course of friction. X-ray structural analysis has proved a high content of CuO, Cu_2O, FeO, and Fe_2O_3 in the surface layer after friction. Along with this, the counterbody surface layers become strongly dispersed, especially for Al, Cu, Zn, Ti. Metal particles and their oxides are transferred onto the mated polymer friction part.

The XPE spectroscopy of transfer films (third body) formed when rubbing PTFE against steel 65 has shown the presence of new compounds [112]. The chemical composition of the transfer films and its dynamics are dependent on the friction regime, and is different on the metal counterbody and the polymer part. A PTFE film can be detected on the counterbody surface already after ten seconds of friction. As friction time reaches two hours, such compounds as $-CF_2-CF_2-$ and $-CF_2-CFH-$ appear on the counterbody surface. This is proof of the presence of tribodestruction products of PTFE in the transfer film as a result of separation of fluorine atoms in the PTFE macromolecules and their substitution by hydrogen. The XPE spectra of transfer films on the metal counterbody display a peak with an energy link 689 eV in Fls spectrum of the sample after six hours of friction. After ten hours of friction this peak intensities and becomes several times stronger than that with 689 eV energy characterizing C–F links in the PTFE molecules. According to [113], a component of the spectrum with an energy of 685 eV characterizes the bond of fluorine atoms with the metal in compounds of the FeF_2 kind. Hence, we can anticipate that the bond of PTFE transfer film with the counterbody metal is realized via fluorine atoms chemically linked with the metal [112].

The *failure mechanism* of metal counterbody surface layers by tribocracking and tribooxidation products of the polymer is analogous to that of mechanochemical corrosion. In other words, one of the main processes occurring during the frictional interaction of the polymer and metal (at sliding velocities up to 1 m/s and loads of the order of 1 MPa) is thermooxidation induced by mechanical effects [110]. The resultant polar oxygen-containing groups intensify adhesive interactions of the metal and polymer, which commonly increases the friction coefficient. Note that the products of thermally oxidative destruction of macromolecules in depth gain a lubricating property that improves wear resistance of metal-polymer friction pairs. It is quite evident that monitoring of tribooxidation processes in the metal-polymer system is one of the means of raising its wear resistance [111].

The friction zone medium significantly affects wear of the metal counterbody. Wear in the oxygen medium is, as a rule, lower than in an inert medium [81], which is explained by the fact that metals fail under friction largely as a result of their interaction with the products of macromolecular decomposition. By insulating the metal surface, oxygen passivates it and lowers the probability of chemical reactions during adsorptive interactions with destruction products of macromolecules. Both electrochemical corrosion and mechanochemical wear of the counterbody may take place in the electrolyte medium. However, the incentive for electrochemical processes is in this case the tribodestruction of macromolecules [81].

Since friction is accompanied by heat generation, so micro-areas can be formed in which the polymer in the metal-polymer junction is in a viscous-flow state. Therefore, surface layers of the metal counterbody may dissolve and the metal can be transferred onto the mating polymer part. This *dissolution of the metal* is characteristic of both static (adhesive) and dynamic (frictional) contacts [114]. As the polymer melt contacts metal alloys, they selectively dissolve since their rates of dissolution are different. For instance, predominantly zinc is dissolved when brass contacts pentaplast melt and the surface layer of the brass sample becomes enriched with copper. As PE contacts bronze, mainly lead is dissolved.

When a polymer melt is brought into contact with an oxidized metal surface, the metal and its oxides undergo instantaneous reduction [72, 114]. Note that even very thin polymer films (of micron and submicron thickness) possess this reducing property [115].

Thus, it is evident that tribochemical reactions are critical for the formation of antifrictional films on the friction surface of the metal counterbody. A new phase (transfer film or third body) functioning as a wear inhibitor protects metal parts first of all from fatigue failure as well as against adhesive and mechanochemical wear.

The third-body effect on wear is analogous to that of corrosive media on decomposition of heterogeneous systems where the products of tribochemical reactions play the role of a corrosion-active medium. By accepting that wear is the loss of functional properties of a system, we imply the existence of a certain relationship between kinetic regularities of the metal-polymer friction pair and corrosion processes in the contact zone. The intensity of corrosion processes is a function of contacting bodies properties and mechanical factors of the frictional interaction, including relative motion velocity, load, roughness, etc. Hence, wear of the metal-polymer friction pair can be viewed as a corrosion and mechanical process whose kinetics depends on a combination of interrelated static (initial) and dynamic (tribochemical) requisites (Fig. 4.21) [110, 116].

This interrelation of corrosion-induced failure processes in metals and thermomechanical destructive processes in polymers can be observed in real metal-polymer systems. For example, during friction of PA6 against steel 45, tribocracking products of the polymer material and metal surface layers form oxide and metal-containing compounds. The products of contact reactions, which are in fact corrosion products, catalyze thermooxidative and destruction processes in the surface layers of PA6. This results in a closed cycle of reactions that intensifies corrosion processes and finally increases the total wear. Similar cycles stimulate the formation of active gaseous products of tribocracking and thermal destruction of polyamide ($H_2, NH_3, CO, CO_2, O_2, H_2O$) leading to the dispersion of the steel 45 surface and tribocracking catalysis. The resultant oxide compounds of PA6 intensify thermally oxidative destruction of the polymer binder and dispersion of the metal surface layers.

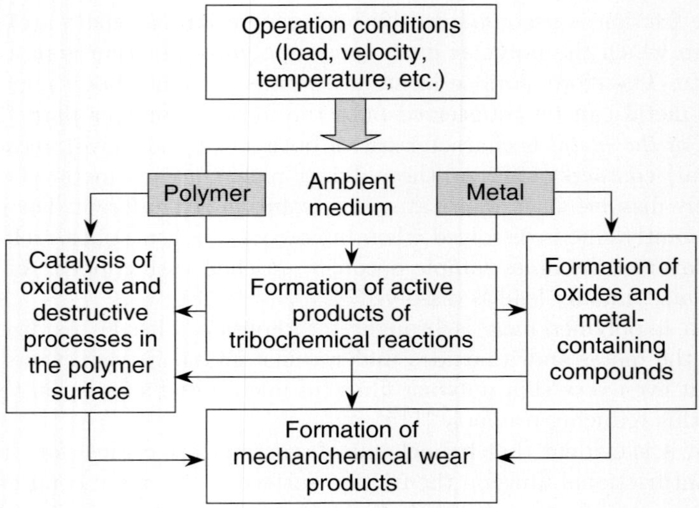

Fig. 4.21. Functional interrelation of the factors promoting mechanochemical wear of the metal-polymer friction pair

This closes the cycle of unfavorable processes in the system and intensifies wear of the friction pair as a result of thermal and mechanical destruction of polyamide.

Analogous cycles are observed in the polyacetal-steel and polyolefin-steel friction pairs. Physical-chemical processes can exert a favorable effect on the metal-polymer friction pair as well. For instance, metal-containing compounds of copper stearate and oleate types are formed in the contact zone as a result of tribochemical reactions when CFD copolymer rubs against copper. These compounds retard thermally destruction processes in CFD, serving as wear inhibitors (WI) for this system.

As can be seen, the problem of improving resistance of the metal-polymer friction pair to mechanochemical wearing is determined by the governing cycle of tribochemical reactions that intensifies wear and elimination of conditions stimulating their generation.

In practice, the choice of WI is based on the analysis of service, design, economic and other factors. An efficient means of improving wear resistance of the polyamide-steel 45 friction pair turns out to be inhibition of the thermally oxidative and destructive processes in the polymer surface layers to avoid the formation of corrosion-active oxide compounds. It is possible to break this unfavorable cycle by the introduction of antioxidants into the polymer composition, thus disabling macroradicals through the reactions of mechanochemical synthesis, polyamide alloying by functional additives forming separating layers or more thermally stable products, and so on [108]. Application of WI with this aim abates undesirable thermally oxidative processes in polyamide and retards failure of the mated counterbody. These combined influences of CI results in essential improvement in wear resistance of the metal-polymer friction joint during operation in hush media (see Sect. 4.4).

Wear resistance of the polyacetal-metal friction pair can be improved considerably by the introduction of higher fat acids or realizing their synthesis conditions in the friction zone. Passivation of metal surface layers by phosphating formulations and epilamens may elevate wear resistance of friction bodies in which polyacetal, polyamide, fluoroplastics, and other substances rub against copper alloys, aluminum, chrome or titanium [108, 117, 118].

Proceeding from the above, it can be stated that the concept of mechanochemical wear of metal-polymer friction pairs put forward in [110, 116] can serve as the basis for expanding the range of methods for fighting wear.

4.4 Suppression of Mechanochemical Wear of Metals by Inhibited Plastics

Inhibited plastics represent a novel type of polymer composite materials containing metal CI. The CI are commonly impregnated into compositions as a solution in Pl, i.e. in the liquid phase. Plastics with a liquid inhibitor acquire a specific structure with a polymer matrix containing an inhibiting liquid in its pores. This provides grounds for both volume and local modification of the plastics with the help of various engineering methods. The presence of a mobile liquid phase assists in the regulation of the structure of composites in the course of their molding into products and during operation by such effects as mechanical, electromagnetic, thermal, etc.

4.4.1 Structure and Properties of Inhibited Plastics

The following fundamental concepts lie at the base of designing inhibited plastics:

- thermodynamic compatibility of CI (or its solution in Pl) with the polymer under processing temperature or restricted compatibility at operation temperatures of composite articles
- joining of the functions of a skeleton bearing the load and a vessel containing a CI
- use of synergetic processes occurring in materials with restricted compatibility of components for transporting CI to the contact with the metal

These concepts are realized to a greater extent in composites based on thermoplastic polymers incorporating low-molecular plasticizing liquids and CI [119–121]. Compositions based on HDPE, mineral oil MS-20 (MO) and oil-soluble CI of metals, e.g. GRM (tar of vegetable oils and inedible fats), represent a handy model for studying the structure and properties of the materials concerned.

If the concentration of low-molecular liquids (solvents) in the polymers surpasses their compatibility limit, they isolate and form spherical arrangements with a size of 10–20 μm in the polymer structure. When the solid phase volume exceeds that of the liquid, the formed structures are of the closed-pore kind and the liquid phase is distributed within the solid phase as local spherical inclusions [122]. As soon as the liquid phase content surpasses that of the solid, a new honeycomb structure with communicating cavities is formed whose solid phase builds up thin walls that separate the cells. This feature is to a greater extent typical of tough and crystallizable polymers. This is also relevant for systems like PE-MO where honeycomb structures with a pore size of up to several micrometers can be formed under certain conditions (Fig. 4.22) [123]. Such porous structures are perfect for the impregnation of modified additives, e.g. CI.

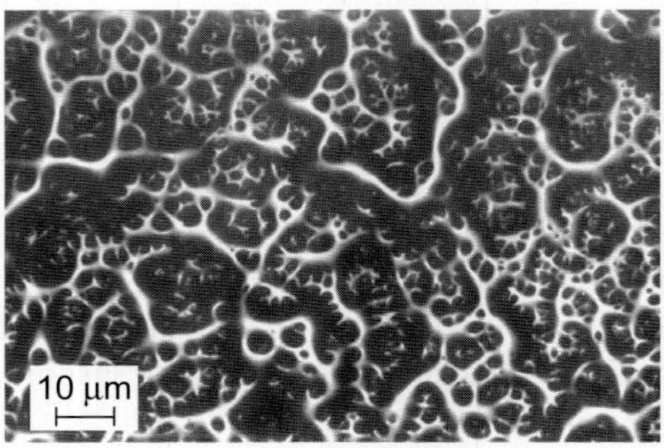

Fig. 4.22. Plasticized PE structure with 50 wt% of mineral oil

A characteristic feature of plasticized polymers is *syneresis*, the spontaneous liberation of the liquid phase. Syneresis may take place during both molding of the samples and in complete products.

Of specific interest from the view point of designing inhibited plastics is the examination of syneresis in systems with a polymer and mixture of plasticizers. It has been stated in [119] that syneresis may increase or diminish during combining of Pl or may stay at some average level, in contrast to syneresis of polymer blends with each plasticizer individually. The latter case is illustrated in Fig. 4.23, in which the dependencies of the liquid phase liberation from different PE-based blends over 18 days are presented.

Syneresis in the samples based on PE and pil (curve 1) is already observable at 20% concentration of the liquid phase in the composition. In samples containing up to ∼70% of CI (curve 3) syneresis is three to four times lower. This confirms the better compatibility of CI with PE than with oil, which is attributed to its higher viscosity. An abrupt increase of CI liberation at

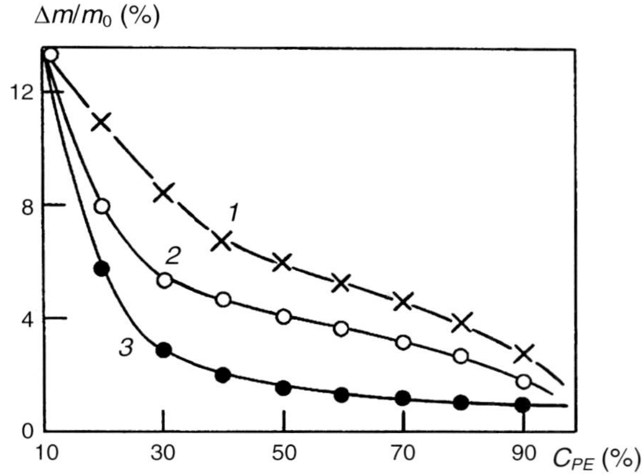

Fig. 4.23. Liberation of the liquid phase from PE-based materials containing: (1) MO; (2) MO + GRM (1:1); (3) GRM

high content in the material is evidence that the formation of a developed system of pores as a result of phase separation has terminated. Syneresis in the samples containing a mixture of named liquids (curve 2) occupies an intermediate position, approaching in intensity that of the oil at relatively low liquid phase concentration (below 50%) and is closer to the syneresis of CI at higher concentrations.

Pl incorporated in inhibited plastics and CI significantly influence the *physical-mechanical and physical-chemical characteristics* of materials. The three-component composition-property charts presented in Fig. 4.24 help to estimate the effect of CI and Pl on the mechanical and antirust properties of composites.

Notice that the charts describing the liquid phase separation and those of the material strength at tension (Fig. 4.24a and b) have much in common. Apparently, syneresis and strength of the composites are dependent on the same group of factors. Among these factors we can with certainty name the stress relaxation time in the composite, which changes on filling with low-molecular matter and exerts a considerable effect on both syneresis [122] and the strength of the composites. It should be mentioned that, when the oil content is below 10%, increasing the GRM concentration from 20% to 70% does not in fact influence the strength characteristics of the material (Fig. 4.24b). In contrast, even insignificant increases in oil content drastically changes the strength of the material.

The mechanical parameters of the plasticized polymers can be improved by making use of the so-called *antiplastification* effect, which occurs at relatively low Pl content. This term implies extreme variations of a series of polymer properties on the introduction of small amount of Pl. Under certain

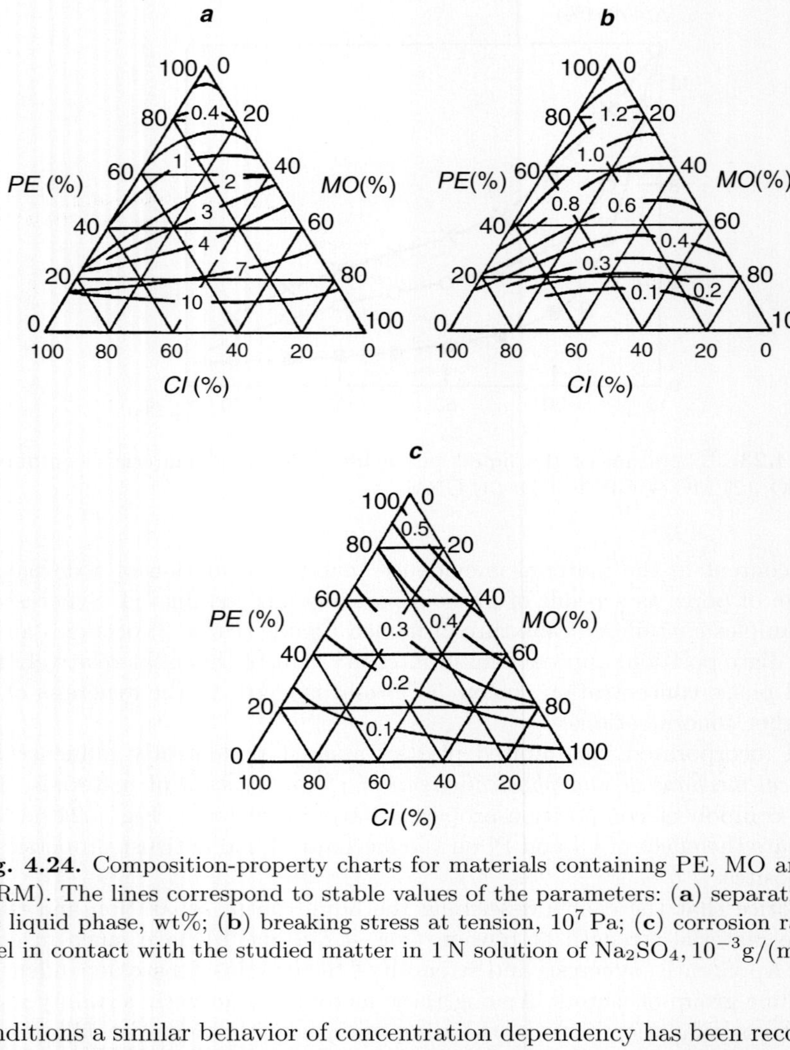

Fig. 4.24. Composition-property charts for materials containing PE, MO and CI (GRM). The lines correspond to stable values of the parameters: (**a**) separation of the liquid phase, wt%; (**b**) breaking stress at tension, 10^7 Pa; (**c**) corrosion rate of steel in contact with the studied matter in 1 N solution of $Na_2SO_4, 10^{-3} g/(m^2 \cdot h)$

conditions a similar behavior of concentration dependency has been recorded for the polymer strength, relative elongation, toughness, elasticity modulus, crystallinity, ordering degree of the permolecular structure, and molecular mobility [124–126]. It has been shown in [127] that the elasticity modulus of compositions based on PE and oil changes extremely (Fig. 4.25). The maximum corresponds roughly to a 20% content of oil in the composition. Analogous variations are observed in the degree of crystallinity of the material determined by two independent methods, namely IRS and DTA.

Absence of any changes in IR spectra visualizing interactions between components gives grounds for the assertion that these results are a consequence of variations solely in the ordering of the structural elements of the

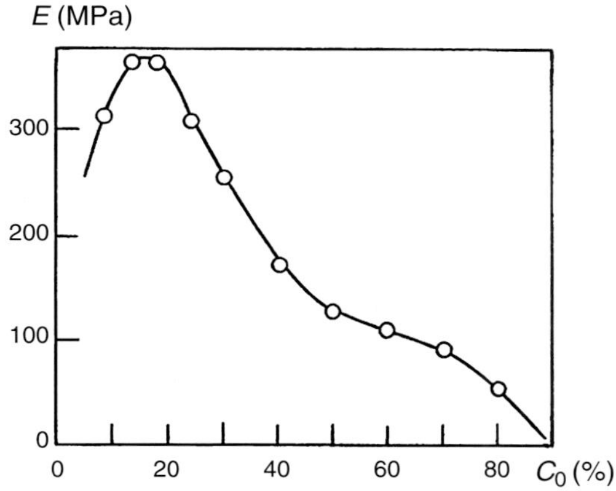

Fig. 4.25. Variations of elasticity modulus (E) versus mineral oil content (C) in PE-based material

polymer on impregnation of a small amount of Pl. This is why a material composition can be made so as to combine the syneresis phenomenon with the antiplastification-induced strengthening effect.

4.4.2 The Effect of Corrosion Inhibitors on Thermal Oxidation of Polymers

Due to the presence of amine compounds in formulations of a number of corrosion inhibitors of metals, they acquire an antioxidant property for thermoplastics entering into chemisorption interactions with metals. The effect of CI (trade marks GRM, VITAL, IFHANGAZ, MSDA, NDA, VNHL-20) on thermal oxidation and thermal resistance of polymers has been described in [34]. For investigations 5–20 wt% of CI were introduced into the composite via mechanical mixing with dispersed polymers followed by injection molding or hot pressing of the composition.

Introduction of CI into a thermoplastic composition changes the kinetics of its thermal destruction processes. This is expressed in variations of the thermal resistance temperatures (T_t) and destruction initiation (T_d) recorded by DTA. T_t and T_d for the PE-based compositions shift to the region of lower temperatures by, correspondingly 1–50°C and 5–20°C. The activation energy (E_a) and macromolecular destruction found by DTA also diminish. Compositions based on HDPE show a reduction of E_a from 266 to 100–150 kJ/mole, and the more efficient the antioxidizing effect of CI towards HDPE the lower the E_a. This is probably because the lower amount of oxide compounds in the polymer volume retards cross-linking of macromolecules and results in E_a reduction.

The T_t, T_d and E_a of the materials based on CFD increase with the introduction of CI except for VNHL-20, which shows a reduced CFD T_t, and T_d, evidently due to its own low thermal resistance. The inhibitors are likely to assist in binding of formaldehyde formed during CFD destruction and lead to decomposition of the copolymer macromolecular chain. In compositions on a polyamide base, CI inhibit thermally oxidative processes that accelerate thermal decomposition of the polymer, elevating T_t and T_d by 5–10°C.

It turns out that thermal effects on the mixture of thermoplastics and CI promote suppression of thermal oxidation in the polymer. CI behaves analogously in extruded inhibited thermoplastics. T_d of HDPE-based extrudates is within the range 350–370°C, while that of the molded compositions is in the 340–395°C range. In PA6-based compositions, these parameter are within the ranges 343–395°C and 351–375°C, respectively, and in polyacetals they are, correspondingly, in the ranges 226–250°C and 240–258°C. It appeared that the studied CI preserve their stabilizing properties in PA6-and HDPE-based compositions extruded under optimum temperature regimes.

The introduction of CI into structural thermoplastics ameliorates their resistance to thermally oxidative aging in air. As a result, inhibited plastics tend to keep their physical-mechanical properties during operation for a sufficiently long period. Particularly, impregnation of 1–5 wt% of MSDA into PA6 keeps breaking stress of the inhibited plastics $\sigma \geq 60$ MPa after thermal treatment in air at $T = 100$°C for 500 hours [121].

As can be seen, CI for metals can be efficiently used for modifying structural thermoplastics to improve their service characteristics, resistance to elevated temperatures in the first place and also their wear resistance.

4.4.3 The Kinetics of Electrochemical Processes During Friction of Inhibited Plastics

As has been indicated previously (see Sect. 4.2), polymer parts in friction joints add specific features to electrochemical processes in the friction zone. The polymer components acquire the properties of surfactants during rubbing against metals, which change the electrochemical activity of the metals. In this connection, the effect of the liberation of CI from inhibited plastics on electrochemical processes in the friction zone has been studied [128].

A pendulum tribometer (see Fig. 4.14) whose prism was made of a PE-based composite, mineral oil and CI were used in the investigations. CI of the class of surfactants with a carboxyl or hydroxyl group (inhibitors of series A), as well as organic amines and sulfur acids (B set) readily dissolving in oils thanks to the hydrocarbon part of macromolecules were used in the experiments. Compositions with different PE content were prepared for each inhibitor type. Formulation 1 had 36 wt %, formulation 2 had 45 wt % and formulation 3 had 65 wt %. The trays (prism support) were made of copper, aluminum, and carbon steel. 0.1 N solution of Na_2SO_4 was used as the electrolyte. The support was polarized in galvanostatic regimes following a three-electrode scheme with the application of a potentiostat.

At the time $\tau_1$of setting the pendulum into motion the electrode potential of the tray ΔU shifts to the positive side (Fig. 4.26). This is because of the mechanochemical activation of metals and damage to the film of the products of electrochemical reactions on the tray surface. At time τ_2 the pendulum stops. Time $\Delta\tau$, during which the potential returns to its initial value (before actuation), is seen to reduce with increasing liquid phase in the composite. This regularity is attributed to structural features of the studied composites, which is characterized by a porous polymer matrix filled by the liquid phase liberating from the material in the process of syneresis.

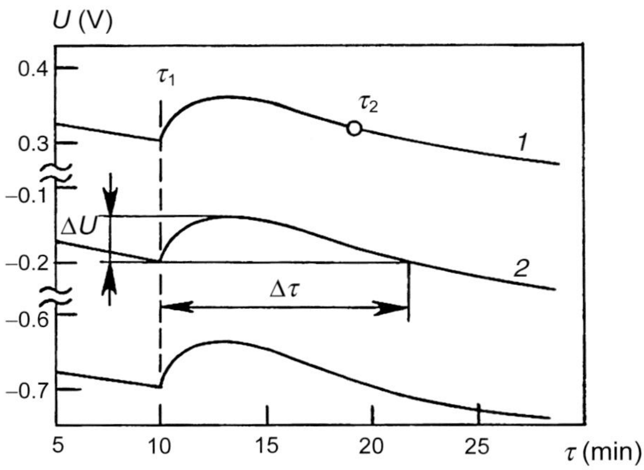

Fig. 4.26. Variation of electrode potential of the pendulum support (copper) depending on polarization time and polarization current density (μA/cm^2); (1) (+2.6); (2) 0; (3) (−2.6)

The electrode potential of the metals found in a static contact with materials under study can be changed perceptibly by regulating the CI content in the composite. Evidently, areas of the tray cleaned during friction passivate more quickly during dynamic contact due to the adsorption of contact inhibitors separated at syneresis than during polarization (see Sect. 4.2.3).

Date presented in Table 4.12 show that the liquid phase content affects the electrode potential shift value of the support, all other conditions being equal. This is because of the concentration dependence of the deformation and strength properties of the composites considered. Higher liquid phase content brings the composite closer to a liquid. Therefore, the contour area of the prism contact with the support is much larger for the composites of the first formulation compared to the third one. Correspondingly, the film of the products of electrochemical reactions on the tray surface is damaged over a

Table 4.12. Effect of polarization current and material types of the prism and support on oscillation time τ of the pendulum (min) and electrode potential shift of the support ΔU (mV)

Support Material	τ, min; ΔU, mV	Prism Material: PE (wt%) + MO (10) + MSDA (the rest)								
		1(36)			2(45)			3(65)		
		Polarization Current, μA/cm^2								
		0	−2.6	+2.6	0	−2.6	+2.6	0	−2.6	+2.6
Cu	τ	5.66	5.17	5.20	4.83	3.83	4.50	6.1	6.0	5.5
	ΔU	35	34	31	35	32	31	28	2.4	22
Al	τ	8.0	8.48	8.03	7.94	8.44	7.96	8.00	8.43	8.02
	ΔU	5	4	4	4	4	4	4	4	4
Steel	τ	8.15	7.2	6.01	7.1	7.33	7.33	8.33	7.33	7.05
	ΔU	21	15	15	19	14	12	13	8	10

greater area by the prism of formulation 1. The degree of approach of the electrode potential of the support towards a stationary potential characterized by ΔU also increases in this case.

Change in formulation of the composite operating in pair with aluminum does not in fact affect friction, characterized by the time of pendulum oscillations. Friction against steel becomes milder when moving from the less hard composites of formulation 1 to the harder ones of formulation 3. Copper shows the most intensive friction when paired with the composite of formulation 2. Evidently, the hard composite of formulation 3 creates less resistance to friction, while increased liquid content in the composites of formulation 1 intensifies syneresis of the inhibited liquid, which results in lowered friction in the studied pair.

The electrode potential shift of the support at pendulum start is affected by the inhibitor's nature. It is justified to consider the ΔU value as a measure of the speed of response of the inhibited composite. The speed of response depends first of all on the amount of inhibitor isolated from the composite during syneresis and transferred to the mated surface during friction. The next factors in importance are viscosity and the adsorption characteristics of the inhibitor towards the metal of the support. Composites containing inhibitors that combine the cathodic effect (amine groups) and a general screening effect (synthetic fatty acids) show the minimum ΔU [128]. The composites with inhibitors intended for ferrous metals display lower friction when paired with steel than with copper. And finally, frictional characteristics vary only insignificantly when moving from one metal to another for composites containing inhibitors with a complex effect and that protect both ferrous and non-ferrous metals.

Proceeding from the above data we can state that electrochemical processes in the metal-polymer friction zone can be regulated with the help of CI, and hence friction and wear of metal-polymer joints can be monitored as well.

4.4.4 Wear Resistance of Inhibited Plastics

Wear of inhibited plastics has been studied on model systems (PE + MO + CI) [128] in pair with a steel counterbody. The wear rate has been estimated by weighing using a friction machine fitted with a vessel for friction in an electrolyte following the shaft in a bush friction geometry ($p = 0.35$, MPa, $v = 2.4\,\text{m/s}$, with a friction area of $2\,\text{cm}^2$).

The effect of operation factors on the friction joint intensifies syneresis in the form of the liquid phase liberation from the plasticized PE structure. The liberated liquid phase migrates under the action of the load and temperature gradient to the friction surface and creates a lubricating layer containing metal CI. This retards the corrosion processes in the metal-polymer friction joint, and hampers mechanochemical wear. As a result of Red-Ox processes, CI form chemisorbed compounds on the metal surface thus reducing its catalytic activity in relation to thermal oxidation of the polymer. On the other hand, CI inhibit thermal oxidation of the polymer matrix via interaction with macromolecular active centers, and improve polymer resistance to tribocracking. Supply of the liquid-phase components begins with their consumption in the friction zone. This is how the mechanism of self-regulation of both triboengineering and anticorrosion properties of composite materials operates [34].

Information on wear of a steel-polymer composite friction pair in a solution of sulfuric acid and in air is illustrated in Fig. 4.27. The intensity of mechanochemical wear of steel against the inhibited composite (a, curve 3) is perceptibly lower than when paired with PE (curve 1) and the inhibited composite with the liquid phase amount corresponding to formulation 3 (curve 2). Apparently, the inhibitor transferred to the steel sample surface affects essentially growth and decomposition kinetics of the products of reactions between the acid and steel. In conditions of dry friction, frictional transfer of the studied polymer materials on the steel counterbody is observed under the preset operation regimes (Fig. 4.27b). Therefore, the steel wear acquires a negative sign. The least transfer is observed in the pair with the plasticized PE (curve $2'$) from which the lubricant (MO) is isolated during friction, the highest transfer is displayed by the non-modified PE (curve $1'$). The inhibited composite (curve $3'$) contains a surfactant interacting with the metal and occupying an intermediate position between above two cases.

The value of polymer part wear in acid is roughly of the same order as in air. The behavior of wear curves of the polymer samples during dry friction is attributed to frictional transfer (curves $4'$–$6'$) in which case the curves are arranged in order of material strength: the strongest non-modified PE wears less. Impregnation of the mineral oil and CI improves chemical resistance of PE and it therefore wears less (curves 5 and 6) than the pure PE (curve 4).

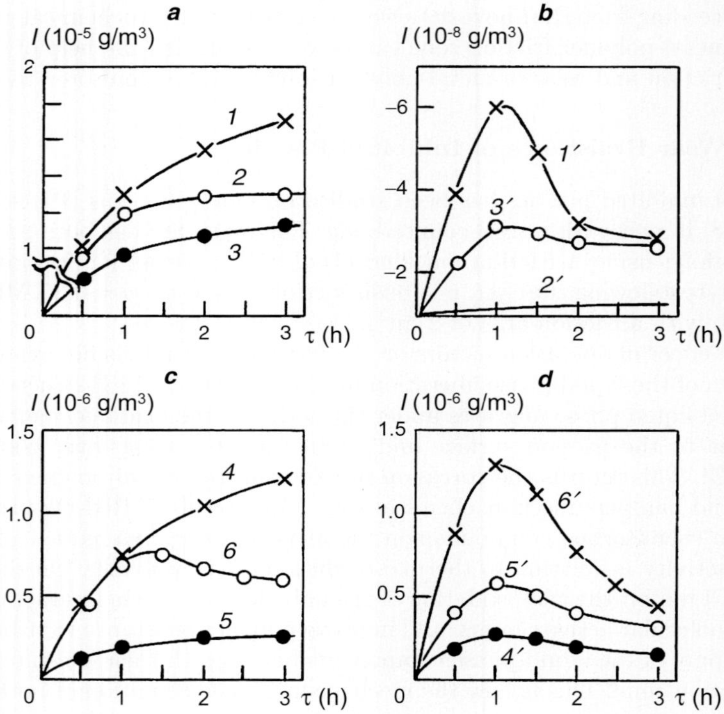

Fig. 4.27. Wear rate of steel (**a**, **b**) and polymer materials (**c**, **d**) in 0.1 N solution of H_2SO_4 (**a**, **c**) and in air (**b**, **d**). Polymer materials: $1, 1', 4, 4'$ – PE; $2, 2', 5, 5'$ – PE (65 wt%) + MO; $3, 3', 6, 6'$ – PE (65 wt%) + MO + CI

An analogous effect is observed for the PTFE-chromium pair under lubricated friction with CI [121]. Wear of both components diminishes noticeably in this case. This is due to passivation of the metal friction surface induced by desorption of CI components containing active functional groups. As a result, the catalytic effect of chromium oxide compounds on tribocracking of PTFE macromolecules diminishes, leading to a decrease in the friction pair wear.

It is quite apparent that CI are able to function as efficient wear inhibitors and contribute to extending the lifespan of metal-polymer friction joints.

4.5 Inhibited Antifrictional Materials

Use of polymers in their original composition cannot fit specific requirements imposed on the reliability of bearings, movable seals and other friction members of machines. Friction joints operating in hostile media undergo elevated wear with time whose signs can be observed on both metal counterbodies and, as a consequence, on polymer parts. Machines and mechanisms incorporating metal-polymer friction joints are similarly damaged during prolonged storage or preservation. Corrosive damage of a metal part, e.g. shaft, may lead during operation to seizure followed by failure of the whole joint.

The development of inhibited polymer materials for antifrictional purposes has been a new approach to the problem of raising performance and durability of friction joints operating in hostile environments. As experience has shown, protective inhibited lubricants are not always efficient for movable joints where the lubricant is squeezed with time from the friction zone. During long-term operation of a friction joint it is necessary that CI be continuously supplied to the friction zone and the inhibited antifrictional plastics have proved to be most suitable for this purpose [37].

Various polymer binders and their blends are employed (PE, PA, CFD, polyacetal, thermoelastoplasts, and so on) for the inhibited antifrictional materials of INHAM trade mark. In addition, oil-soluble inhibitors of donator types (KP, KSK, MNI, etc.), acceptor types (MSDA, BMP, SIM, etc.) and their synergetic blends with complex effects (AKOR, NG and others) are used as their CI. The inhibitors are impregnated into the INHAM materials in the form of a solution in Pl (mineral or silicone oils) and may contain various fillers and modifiers depending on the purpose.

4.5.1 Compositions and Properties

INHAM materials are related to self-lubricating antifrictional composites intended for friction parts experiencing the effect of harsh liquid media.

For plain bearings operating in dry friction regimes or scarce lubricant supply, a PA-based material *INHAM-1* [129] also containing polyolefins, lubricant oil and either donator- or acceptor-type CI can be used. The composition is injection molded into parts having a surface layer enriched with a lubricant-plasticized polyolefin. The principal characteristics of the material are given in Table 4.13. Among its advantages we can name the following:

- capability of hindering crevice corrosion in friction joints operating periodically (with long preservation time), reducing the intensity of mechanochemical wear of conjugated metal parts
- improved stability of physical-mechanical characteristics at elevated temperatures against aging of the polymer binder
- ability to operate without lubrication at liquid and boundary lubrication regimes
- satisfactory physical-mechanical and frictional characteristics sufficient for its use as a substitute of polyamides in friction joints

PA-based inhibited antifrictional materials display similar properties and can serve as: (i) a sealing material [133] containing additional dry lubricants (graphite, molybdenum disulfide and so on), and (ii) an antifrictional sealing compound [134] incorporating pine resin as a CI. The resin is a product of resinous piney wood distillation consisting of a mixture of tar and fatty acids and their decomposition products.

INHAM-2 [130] has been developed on the base of PA and finely dispersed vulcanized rubber for operation in friction joints under the simultaneous effect of hostile and abrasive environments. The material contains polyacetal of the CFD or CTD series as a functional additive and an oil mixture with CI as a lubricant.

The physical-mechanical characteristics of the material are improved due to the thermodynamic compatibility between PA and polyacetal. Finely dispersed rubber based of chloroprene or butadiene-nitrile rubber with 5–300 μm particle size possesses a high adsorptive capacity and is able to absorb the inhibited oil. A mixture of inhibited oil decreases the melt viscosity during injection molding of the materials. In a complete article excess oil isolates as a result of syneresis and elevated temperatures in the friction zone thus improving its wear resistance. Besides, rubber contains numerous active links through which PA macromolecules interact with the filler. The CI chosen for a given composite induce a stabilizing effect and raise resistance of the material to high temperatures. These peculiarities form in combination a complex of positive properties of the material (see Table 4.13).

Metal-polymer antifrictional materials have been specifically developed for friction parts working in severe conditions (joined effect of corrosive and thermooxidative media, low temperatures, vibration, etc.).

INHAM-3 [131] contains up to 97 wt% PE and copper in the metal-polymer with the formula $[-CH_2-CH_2-]_nCu_m$, dispersed rubber, CI, unsaturated hydrocarbons of oil and PE copolymer with polyformaldehyde. This material shows high impact viscosity (see Table 4.13), resistance to thermooxidative aging (the strength does not impair upon thermal aging) and frost resistance (down to −70°C).

INHAM-4 [132] is also a metal-polymer material containing PA and copper of the formula $[-NH(CH_2)_5 - CO-]nCu_m$, as well as polyolefin, polyacetal, unsaturated oil hydrocarbons and a CI and is highly resistant to hostile media. Upon prolonged thermal treatment in air the material does not in fact lose its strength (see Table 4.13), preserves a rather low friction coefficient and elevated thermal stability (430−450°C).

These antifrictional materials of INHAM grade display higher wear resistance under friction in the liquid hostile media compared to non-inhibited plastics. For comparison, the properties of an antifrictional material based on PA6 and molybdenum disulfide without a CI in its composition are presented in Table 4.13. Its wear resistance is seen to fall short of the INHAM materials since corrosion processes are not suppressed in the friction zone of this material with the metal.

Table 4.13. Service characteristics of inhibited antifrictional materials

Property	INHAM-1	INHAM-2	INHAM-3	INHAM-4	PA6 + MoS_2
Breaking stress, MPa:					
– under compression	55–90	80–140	25–30	–	60–70
– under tension	45–70	60–120	23–27	53–60	40–45
– under tension after thermooxidation at 150° for 200 hours	40–70	50–90	23–28	50–53	–
Friction coefficient against steel ($p = 2.5$ MPa, $v = 0.1\,\mathrm{m/s}$)					
– initial	0.30–0.35	0.20–0.24	0.5–0.6	0.21–0.25	0.18–0.22
– upon thermooxidation	0.30–0.35	0.22–0.26	0.50–0.56	0.21–0.27	–
Wear rate. 10^{-8}:					
– at $p = 2.5\,\mathrm{MPa}$ and $v = 0.1\,\mathrm{m/s}$	1.0–2.3	1.8–2.0	1.4–1.6	1.2–2.2	2.8
– at $p = 0.5\,\mathrm{MPa}$ and $v = 1\,\mathrm{m/s}$	2.0–3.1	1.6–2.0	1.4–1.8	1.6–2.4	4.6
Corrosion rate of a steel plate pressed against a studied material sample, $\mathrm{mg/(m^2 \cdot h)}$:					
– in 0.1 N solution of HCl	10–40	20–35	4.2–5.8	–	120–150
– in 1.0 N solution of H_2SO_4	1.2–4.8	1.2–2.6	–	0.4–0.9	60–90
Impact viscosity, $\mathrm{kJ/m^2}$:					
– initial:					
• without incision	30–35	37–40	Does not break	–	60–80
• with incision	–	4–9	14–16	–	3–7
– upon thermooxidation:					
• without incision	30–35	32–36	–	–	–
• with incision	–	2–8	13–20	–	–

OILES Corp. (Japan) has elaborated inhibited materials for journal bearings operating in aggressive environment that are based on polyacetals (grade *OILES 80*) and containing lubricants (oil, antirust grease) evenly distributed within the material [138]. The distinguishing features of these materials are:

- usable without lubrication
- provide high load capacity and excellent wear resistance
- superior in high velocity properties with coefficient of friction being low, $\mu = 0.05 - 0.1$
- free from stick-slip and creaking noise
- more economical than oil-containing sintered bronze bearings

The results obtained using a journal bearing tester indicated that OILES 80 had a higher wear resistance than other types of bearings including PTFE with fillers, oil-containing sintered bronze bearings, and nylon with MoS_2 (Fig. 4.28). OILES 80 showed a constantly low coefficient of friction (Fig. 4.29).

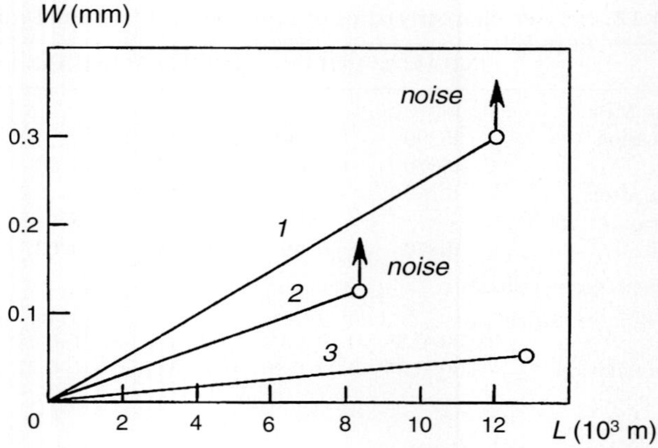

Fig. 4.28. Wear dependence (W) on sliding distance for self-lubricating materials: (1) Nylon with MoS_2; (2) acetal resin + PTFE; (3) OILES 80. Test conditions: mating material S45C (surface roughness $R_a = 3\,\mu m$); load 3 MPa; velocity 0.167 m/s (10 m/min)

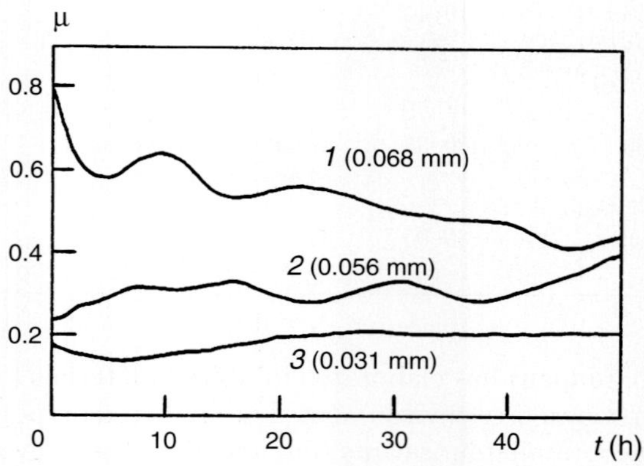

Fig. 4.29. Variation kinetics of the friction coefficient for materials: (1) Nylon 66; (2) Polyacetal; (3) OILES 80. Wear values are given in brackets. Friction conditions are similar to Fig. 4.28

Sealing of friction joints is one of the most efficient filed of the inhibited antifrictional materials use. Seals are precise and costly machine members, which are critical for safe operation of equipment.

The specifics of the operation and structure of sealing units may bring about intensive crevice corrosion in the seal joint arising from looseness, stagnant zones and other sources of differential aeration. Hence, a seal joint presents a hotbed of corrosion stresses leading to untimely failure of the whole joint. Therefore, sealing of joints operating in contact with hostile media necessitates a complex approach to raising the tightness of machine junctions, which includes packing of clearances and depressions in combination with inhibiting corrosion processes on conjugated surfaces. Inhibiting composite materials based on plasticized thermoplastics are intended just to solve this problem.

Low-modular composite materials based on plasticized thermoplastics belong to a group of sealing materials that stands in between structural and sealing materials. The fact that these materials display elasticity modulus, hardness and toughness close to the initial polymer allows their use in parts that are to bear considerable mechanical loads and other effects, and in particular in sealing joints of machines and equipment [123].

Stiff enough joints do not restrict serviceability of the whole design where seals of low and average strength are placed between conjugated parts; stuffing box seals with a wadding belong to this class [137].

A thermoplastic molding composition [135] on a base of polyurethane contains CFD, fine-dispersed rubber and CI solution in mineral oil or organosilicone liquid. It is highly stable to abrasive wear in hostile media, is perfectly pliable and forms tight contact with the conjugated part.

A *sealing material* [136] based on a polymer binder (PE) and a fibrous filler contains a polyamide network, CI, crezol or resorcin. The netting present in PA serves for the material as an antifrictional wear resistant component. It can be made as a shell containing a plastic mass from other ingredients. The PE plasticized by a liquid-phase inhibitor becomes a hydrophobic binder and adds plasticity to the material as well as resistance to the chemical matter. Its fibrous filler (asbestos, textile wastes, polymer chips, etc.) imparts shape stability and structural definiteness to the sealant, cuts polymer binder consumption and helps to regulate the strain and strength characteristics of the sealing material. Elevated wear resistance of the material makes its sealing capacity rather high and improves durability of the seals.

4.5.2 Processing Into Goods

Processing of compositions containing liquid-phase ingredients presents certain difficulties. In the presence of a liquid the polymer powder loses its friability, coagulates and sticks to the processing equipment members. The granulated polymer adsorbs the liquid poorly and it trickles down the granules presenting difficulties in reaching the necessary composition. The initial CI and Pl are largely low boiling and flashing, which bounds their application in composite processing with the polymer binder.

Serious difficulties are encountered too when processing compositions with readily evaporating or sublimating components. Such materials cannot in fact be extruded or injection molded because of the large amount of resultant defects [123].

Let us consider some processing peculiarities of inhibited polymer composites by different techniques.

Injection molding presents certain interest as a most productive technology of inhibited plastics.

Modern modifications of injection molding (vibromolding, intrusion, pultrusion, foam molding, etc.) differ by the method and degree of heating of the material, its injection and peculiarities of forming [141, 142]. In spite of the diversity of technological schemes of injection molding all these methods include the following chief operations: material heating (plasticization), injection (mold filling), pressure endurance and cooling (curing).

These operations are carried out in a molding machine that also prepares the melt for injection into the mold, the injection into the forming cavity, forming and removal of the ready cast from the mold. Contemporary molding equipment is distinguished by sufficiently high universality, productivity, capability of regulating preset parameters and individual machine units.

The injection molding specifics of inhibited composites consist first of all in the necessity of accurate metering of components and preparation of the extruded composition containing a liquid-phase ingredient. The simplest and nonetheless most efficient method of preparing the composition is blending of the polymer powder with the liquid PI followed by addition of granules of the same polymer [139].

The polymer powder absorbing the liquid-phase modifier represents a kind of a vessel while its granules add friability to the composition, facilitating charging and transit. The term granules characterizes in this case a friable granular product consisting of particles homogeneous in shape and size (1-5 mm) intended for injection molding. If the granules are excluded from the compositions, it loses friability, becomes lumpy and sticks to the working parts of the equipment as the PI content reaches 10 wt%. It also makes no sense to blend the granules with the liquid PI in the absence of the powder since the granules absorb only a negligible amount of the liquid onto their surface while the rest drains. If the polymer is perfectly compatible with the liquid, the surface layer of the granules becomes sticky and they lose their friability.

The combination of granules and powder mixture with the liquid that is best from the processing viewpoint is that in which the mass ratio is 2:1, although the range 2:0.5 to 2:1.5 is satisfactory. The most homogeneous blend is obtained when the polymer powder is sequentially mixed with the liquid-phase components and this composition is then mixed with the granules. Granules above 5 mm in size are undesirable since they impede the work of pneumochargers and batchers. If the granules have a size below 1mm, the composition looses its friability. The optimum granule size is within 2–4 mm, which is characteristic of most extruded polymer materials.

The method considered above employs equipment intended for mainly granulated polymers. It is mostly adopted in enterprises for processing thermoplastic polymers. This process is not modified substantially for processing polymer compositions with up to 10 wt% and higher liquid-phase modifier content.

The described procedure and its modifications are extensively used for processing materials of the INHAM series. To prepare INHAM-1 [129], the CI solution in oil is mixed with polyolefin powder into which PA granules and dry lubricant powders are added. Upon mixing, the blend is injection molded into various goods. INHAM-2 [130] is produced by mixing CI solution in oil with rubber powder into which polyamide and polyacetal granules are added. Thermoplastic sealing compositions [135] can be formed analogously. The preparation process of INHAM-3 and INHAM-4 materials is somewhat more complex. First, a metal-polymer is produced in a worm granulator by thermolysis of copper salts in PE [131] or PA [132] melt. Then the metal-polymer granules are mixed with pre-formed powdery blends of saturated hydrocarbons with CI.

The injection molding regimes are defined proceeding from the material composition and correlation of ingredients. A peculiarity of liquid-filled compositions is slow curing in the mold and gate channel due to their high heat conductivity. To prevent the formation of blisters or pits on the surface of products the time of their endurance in the mold under pressure should be extended.

Injection molded polymer composite products containing liquids are characterized by minimal residual stresses. Cracking under internal stresses is not evidently characteristic for this type of materials at all. Shrinkage of the articles of inhibited composites depends on the liquid phase content and does not surpass 2–3% for PE- and PA-based materials.

If the amount of modifying additives in the material does not have to be considerable for the service conditions, they can be impregnated into the surface layer during thermal treatment. For this purpose the part molded of solid-phase components is held in a modifying liquid consisting of a lubricant oil and oil-soluble CI. The composition temperature surpasses the polyolefin melting point but is below that of PA, which is taken as the composition base [129]. The process time depends on the required depth of CI impregnation. For a 500 μm depth the saturation rate is 8-10 μm/min. The procedure described assists in preservation of the strength of the material volume and imparting antifrictional and antirust properties to the surface layer only.

Another method for manufacturing inhibited antifrictional materials on a PA base uses phenol formaldehyde-silicate of copper as a powder filler and benzoldehyde as the liquid medium. The formed articles are cooled in morpholine [140]. The strength and heat resistance of the articles produced are noticeably improved.

Compression-molding is one of most applicable methods of processing plastic masses [143]. The press material, in the form of a powder, granules or pellets, is charged into the mold and subjected to heat and pressure effects. The article is molded and cured in the heated state. The molding pressure should reach a level at which the material is compacted, shaped and any excess of gaseous products and vapors is removed.

By compression-molding in heated molds it is possible to process composites containing large amounts of the liquid-phase components (up to 50 wt %). Liquid plasticizers are able to lower the polymer melting point and, consequently, that of the composite processing, together with the compaction pressure due to high fluidity of the composite melt.

Composites containing liquid ingredients are distinguished by low friability and sticking to the walls of the processing equipment, which impedes their transportation and charging into molds. To raise process the effectiveness of charging a procedure has been proposed [144], during which the composition is dispersed in a low-molecular service liquid that is thermodynamically incompatible with the polymer and liquid-phase components of the composition (Pl, CI and so on). Upon charging the mold, the service liquid is removed from the dispersion.

During this procedure the following phenomena take place. The polymer powder with a highly developed surface adsorbs the liquid-phase components during mixing of the composition. As a result, the composition is dispersed by the service liquids over a large area due to their interaction with Pl. This leads to the formation of Pl emulsions in the service liquid. In this manner, low-molecular liquids that are compatible separately neither with the polymer nor with Pl alleviate dispersion of the compositions where the polymer powder functions as an emulsifier. The service liquid is usually a low-molecular one and is chosen from economical reasons.

A distinctive feature of this method is exclusion of absorption of the service liquid onto the powder during preliminary mixing, which is why the liquid can be easily removed from the composition in the following stages.

The blend of the press composition with service liquid is poured into the molds, after which most of the liquid is withdrawn by free draining or prepressing. The remaining liquid fills the gaps between the powder particles thus increasing the heat conductivity of the blend. This cuts the time for heating the molds and raises productivity. Service liquids are also used for impregnation of CI, dyes, foaming agents, metal-containing fillers, monomers and other matter in the form of solutions, suspensions or emulsions. The described method of impregnating fillers ensures uniform distribution through the material bulk even if the filler is poorly compatible with the Pl.

Compression-molding is easily handled, reliable, and employs simple equipment. Its productivity is, nevertheless, rather low while it is quite labor intensive, so this method is used comparatively seldom for manufacture of inhibited antifrictional materials.

Other methods of processing inhibited polymer composites are intended for special purposes. For instance, antifrictional products with cylindrical configuration (especially large size products) can be molded by a centrifugal method. In this case, compositions on a based of thermosetting plastics are employed, e.g. unsaturated polyester resins [145]. This composition can be also used for molding reinforced products by impregnating polyamide, cotton or glass fabric to reach extremely strong designs for antifrictional purposes.

One of the ways of improving wear resistance and reliability of sealing elements is filling of porous semis by components able to form a secondary porous system [146]. For example, a blank of a polyurethane foam is impregnated with a mixture of dispersed PE with an inhibited lubricant. The formation of a gel under certain temperature regimes is accompanied by phase distribution within the material. The contact surface of the blank is cleaned by acetone to remove the lubricant. As the acetone evaporates, the prepared zones are zinc-plated and pores freed from the lubricant become filled with zinc. This makes the coating adhere strongly to the surface and allows zinc to penetrate to a depth acceptable for wear limits.

The material obtained combines elasticity of the polyurethane skeleton and wear resistance of the zinc component with the plasticity and lubricity of a gel-like filler. The presence of CI in the material bulk and their isolation in response to mechanical loading of the seal contributes to the improved antirust properties of conjugated parts.

In small-batch conditions of and in overhaul enterprises, the adoption of vertical plunger-type molding machines (grade PL presses) makes production more efficient. As investigations have shown, the physical-mechanical characteristics of products obtained by an automatic thermoplast machine do not differ in principle from those manufactured on a PL press. What is more, a large journal bearing injection-molded on a PL press shows higher wear resistance compared to that produced using the automated thermoplast method. This is because of the molding regimes of the thermoplastic blends with different melting temperatures. The injection pressure creates the conditions for shaping external parts from easily melted materials. As a result, the article acquires an outer layer consisting predominantly of the inhibiting antifrictional component, whereas the inner layer ensures the given design stiffness.

4.5.3 Application

A promising trend of extending life of friction joints operating in hostile environments is the application of INHAM kinds of materials based on different PA grades with the addition of lubricants, dispersed fillers and contact CI.

INHAM materials underwent testing as seals of *slush pump rods* used in pumping of aggressive media. Rod seals found in reciprocation motion are considered to be the most vulnerable joints of slush pumps, whose lifespan conditions the productivity of borehole drifting. One of the most efficient designs of this type is an equiloaded seal having a preloaded ring of structural plastics on each gland. The problem posed was to combat friction and mechano-chemical wear of the rod as a most critical and expensive member of

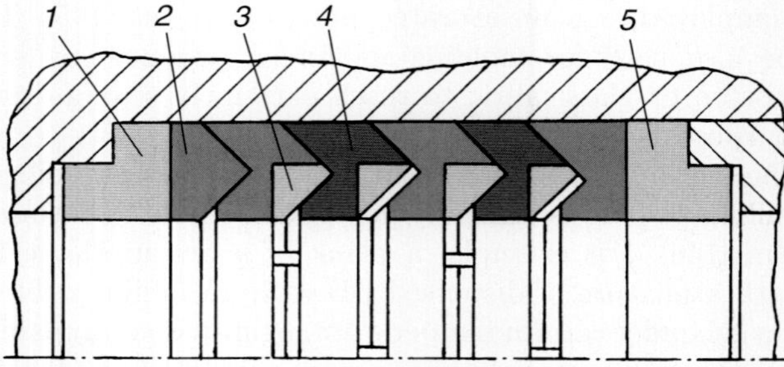

Fig. 4.30. Rod seal of slush pump U8-6M: (1) bushing; (2) mantle ring; (3) bearing collar; (4) rubber gland; (5) thrust collar (parts 1, 2, 3, 5 are made of INHAM)

the pump. This problem has been solved by using sealing plastic rings made of INHAM material (Fig. 4.30).

It has been established during testing that the rod life of a U8–6M slush pump with inhibited rings was extended from 76 to 300 hours. As a result of the lowered mechano-chemical wear of steel rods, the life of the seals was prolonged 3.5–4 times compared to rubber gland seals. The new seals have been adopted in oil-extracting enterprises in Belarus, Ukraine and Russia.

The INHAM material is also employed in *drilling slush pumps* [149]. In Fig. 4.31 a modification of a piston is illustrated, consisting of a metal core 1 which has a cylindrical collar 2 with ring glands 3 and seals 4 inserted from both sides.

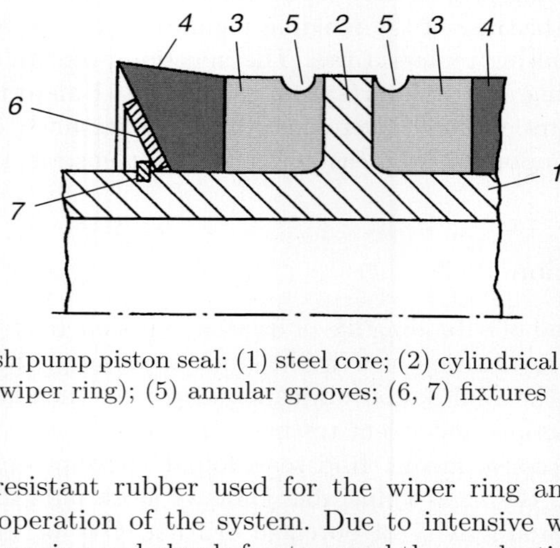

Fig. 4.31. A slush pump piston seal: (1) steel core; (2) cylindrical collar; (3) rubber glands; (4) seal (wiper ring); (5) annular grooves; (6, 7) fixtures

The petrol-resistant rubber used for the wiper ring and gland cannot ensure reliable operation of the system. Due to intensive wear by abrasive particles the wiper ring and glands fracture and the productivity of the pump drops abruptly. To enhance the durability of the seal the wiper ring is made of an INHAM material. The oil solution of CI isolating from the seal during friction provides a self-lubricating effect and reduces mechano-chemical wear. At the same time, annular grooves made in the rubber glands diminish erosive wearing arising from the piston reciprocation-induced cavitation [150].

Inhibited polymers have found application as modifiers of antifrictional materials based on wood. The simplest method of producing a self-lubricating compressed wood is known to be impregnation with oils [147]. Journal bearings made of such wood may operate for protracted periods without lubrication under up to 2 MPa specific loads and up to 0.5 m/s sliding velocities. Under tougher regimes, when the contact temperature exceeds 60–70°C, the viscosity of the oil drops abruptly and it starts to be released from the wood pores thus violating the friction joint operation regime.

To widen the application range of compacted wood more efficient self-lubricating materials of APD grade have been developed (antifrictional compressed wood). They are multicomponent systems of compacted wood impregnated with active plastic modifiers [147]. By modification of wood by plasticized polymers (based on PE and CI solutions in mineral oil) a new type of material that is highly wear resistant during operation in hostile and abrasive environments has been created. The lubricity of such materials does not in fact change under different regimes thanks to the high viscosity and low friction of plasticized PE at increasing temperature. Combined journal bearings of APD material are used instead of steel bushes on a spudding beam of timber truck drags operating in conditions of variable humidity, temperature and abrasive environment. Their service life surpasses that of steel bushes while their cost is two to three times lower.

Journal bearings made of a modified APD material have substituted rolling bearings in friction joints of drying drum wringer rolls and other equipment exposed to hostile media at 60–90°C temperatures. The service life of APD-based bearings exceeds that of roller bearings 1.5 times with a cost reduction factor of two.

The service characteristics of the products made of inhibited composites on a PE base were tested at Bealarusneft Co. (Belarus) and Ukrneft Co. (Ukraine). These compositions were used for the manufacture of *protecting and sealing rings* for clutch couplings of oil-well tubing (OWT) to abate turbulence of the liquid flow transported over the tubing (Fig. 4.32).

These rings should seal the space between the coupling and the ring where corrosion initiation is most probable, and be sufficiently deformable so as not to violate the specified process regulations of the screw joints between the tubes and the coupling, observing the approach of the tube butts, torque value, and so on.

Admissible deviations of the threaded joint of OWT may result in loosening of the clearance between butts. This loosening can be compensated for by optimizing the inhibited composite in terms of its deformation, strength properties and design. Towards this aim, a facet is made inside the ring faces

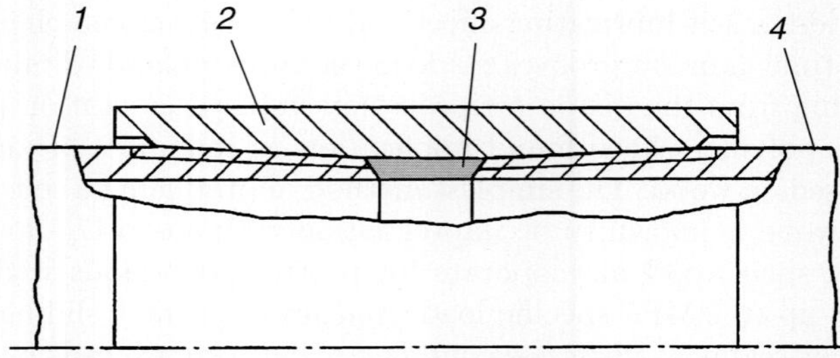

Fig. 4.32. Clutch coupling of oil-well tubing: (1) and (4) connected tubes; (2) coupling; (3) sealing ring

that alleviates deformation of the faces so that the ring reproduces the Laval nozzle shape when the tubes are screwed into the coupling.

The rings tested at Ukrneft Co. have shown tight sealing of the tubing butts, are not subjected to paraffin deposition, reduce resistance to liquid flows and are operable in temperatures up to 100°C [148]. The rings of inhibited composites suppress crevice corrosion on pipe faces and prolong their life by 1.5–2 times.

Developed composite materials have found application in wellhead seals of pumper oil wells. The main drawbacks of sealing materials for packing seals available today are their poor strength, susceptibility to delamination, wash-out of the lubricant by process liquids and intensive mechano-chemical wear of the counterbody. Adoption of inhibited materials in engineering is able to eliminate some of these disadvantages and prolong the service life of reciprocating seals.

The stuffing box [136] already mentioned produced using an inhibited material based on PE is encapsulated in a stiffening polyamide shell. Such stuffing boxes are installed in the form of rovings in wellhead seals of well pumps transmitting a mixture of oil, gas and highly mineralized water. According to the test results the new seal can reliably operate for 22 to 26 days, which exceeds the life of earlier used asbestos-based seals by 2–2.5 times.

References

1. P.I. Blau. *Friction and Wear Transitions of Materials: Break-in, Run-in, Wear-in* (Materials Sci. and Process Technol. Series). Park Ridge, New Jersey, Noyes Publications, 1989.
2. G.E. Lazarev and G.A. Preis. Friction and Wear in Aggressive Media. In: *Tribology – Lubrication, Friction, and Wear*, Eds. I.V. Kragelsky and V.V. Alisin, Professional Engineering Publishing. London and Bury St Edmunds, UK, 2001, pp. 371–389.
3. G. Heinicke. *Tribochemistry.* Berlin, Akademic-Verlag, 1984.

4. M.J. Furey. Tribology. In: *Encyclopedia of Materials Science and Engineering*, ed. M.B. Bever. Oxford, Pergamon Press, 1986, pp. 5145–5157.
5. D. Rigney and R.G. Bayer. *Wear of Materials*. Amsterdam, Elsevier, 1997.
6. R.Y. Lee and Z. Eliezer. On the critical thickness of protective films at sliding interfaces. *Wear,* 1984, Vol. 95, No. 2, pp. 165–175.
7. J. Molgaard. A discussion of oxidation, oxide thickness and oxide transfer in wear. *Wear*, 1976, Vol. 40, No. 3, pp. 277–291.
8. P.J. Blau and C.D. Olson. Application of thermal wave microscopy to research on the sliding wear break-in behavior of tarnished Cu – 15 wt % Zn alloy. *Wear of Materials*, Proc. ASME Conf., ASME, New York, 1985, pp. 425–431.
9. T.S. Eyre and D. Maynard. Surface aspects of unlubricated metal-to-metal wear. *Wear*, 1971, Vol. 18, No. 4, pp. 301–310.
10. N.C. Welsh. Frictional heating and its influence on the wear of steel. *J. Appl. Phys.*, 1957, Vol. 28, No. 9, pp. 960–968.
11. T.F.J. Quinn. The effect of "hot-spot" temperatures on the unlubricated wear of stells. *ASLE Trans*, 1967, Vol. 10, pp. 158–168.
12. T.F.J. Quinn, J.L. Sullivan and D.M. Rowson. New developments in the oxidational theory of the mild wear of metals. *Wear of Materials*, 2nd Intern. Conf., ASME, New York, 1979, pp. 1–11.
13. J. Glascott, F.H. Stott and G.C. Wood. The transition from severe to mild sliding wear for Fe – 12% Cr – base alloys at low temperatures. *Wear*, 1984, Vol. 97, No. 2, pp. 155–178.
14. P.T. Newman and J. Skinner. The high temperature sliding wear of stainless steel in CO_2 atmosphere – The effect of adding low concentrations of O_2. *Wear*, 1986, Vol. 112, No. 3–4, pp. 291–325.
15. D. Krajcinovic. *Damage Mechanics* (North-Holland Series in Applied Mathematics and Mechanics, Vol. 41). Amsterdam, North-Holland Elsevier Sci., 1996.
16. I.V. Kragelsky. *Friction and Wear*, Pergamon, Elmsford, 1982.
17. G.A. Preis. On the nature of corrosion-mechanical wear of metals. *Soviet J. Friction and Wear*, 1987, Vol. 8, No. 5, pp. 16–20.
18. G.A. Preis and A.G. Dzyub. Electrochemical phenomena in metal friction. *Soviet J. Friction and Wear*, 1980, Vol. 1, No. 2, pp. 18–31.
19. H.H. Uhlig and R.W. Revie. *Corrosion and corrosion control. An introduction to corrosion science and engineering*, 3rd edn., NY, John Wiley and Sons, 1985.
20. R.W. Staehle. *Fundamental aspects of Stress Corrosion Cracking*. Houston, Texas, National Association of Corrosion Engineers, 1969.
21. O. Devereux, ed. *Corrosion fatigue: Chemistry, mechanics and microstructure*. Houston, Texas, National Association of Corrosion Engineers, 1972.
22. S.N. Postnikov. *Electrofhysical and Electrochemical Phenomena in Friction, Cutting, and Lubrication*. NY, Van Nostrand Reinhold Co., 1978.
23. G.D. Parfitt and C.H. Rochester, eds. *Adsorption from Solution at the Solid/Liquid Interface*. London, NY, Academic Press, 1983.
24. G.P. Shpenkov. *Friction Surface Phenomena* (Tribology Series, Vol. 29). Amsterdam, Elsevier Science, 1995.
25. F.P. Bowden and D. Tabor. *The Friction and Lubrication of Solids*. Oxford, Clarendon Press, 1986.
26. D.N. Garkunov, I.V. Kragelsky and A.A. Polyakov. *Selective Transfer in Friction Joints*. Moscow, Transport, 1969.

27. A.A. Polyakov, D.N. Garkunov and I.V. Kragelsky. Physico-chemical mechanics of inhibiting wear in the phenomenon of selective transfer. *Proc. USSR AS*, 1970, Vol. 191, No. 4, pp. 821–823.
28. A.S. Kuzharov, R. Marchak, Ya. Guzik, et al. Self-organization in a brass-glycerin-steel tribosystem. *J. Friction and Wear*, 1996, Vol. 17, No. 1, pp. 107–115.
29. A.S. Kuzharov, V.E. Burlakova, R. Marchak, et al. Influence of friction on the electrochemical characteristics of the friction contact at the initial stage of selective transfer. *J. Friction and Wear*, 1998, Vol. 19, No. 1, pp. 73–76.
30. G.E. Lazarev, K.I. Afanas'ev and N.I. Gamazov. Study of the corrosion-mechanical wear process. *Soviet J. Friction and Wear*, 1984, Vol. 5, No. 5, pp. 99–104.
31. G.E. Lazarev, T.L. Kharlamova and V.I. Vereikin. Friction and wear of materials in aggressive media. *Soviet J. Friction and Wear*, 1981, Vol. 2, No. 1, pp. 32–39.
32. I.V. Hurey and M.I. Pashechko. Electrochemical characteristics of secondary structures formed in sliding friction of carbon steels. *J. Friction and Wear*, 2000, Vol. 21, No. 2, pp. 71–74.
33. G.E. Lazarev. Basic patterns of wear of corrosion-resistant steels and alloys during friction in electrolytes. *Soviet J. Friction and Wear*, 1987, Vol. 8, No. 2, pp. 22–28.
34. V.A. Goldade, V.A. Struk and S.S. Pesetski. *Wear Inhibitors of Metal-Polymer Systems*, Moscow, Khimia, 1993.
35. G.A. Gorokhovsky. *Surface Dispersion of Dynamically Contacting Polymers and Metals.* Kiev, Naukova Dumka, 1972.
36. E.A. Stambursky, A.I. Beil, V.P. Karlivan and Yu.A. Bespalov. *Wear of Equipment at Plastics Processing*, Moscow, Mashinostroenie, 1985.
37. L.S. Pinchuk and V.A. Goldade. Wear inhibitors for metals in polymer-metal friction pairs. *J. Friction and Wear*, 1994, Vol. 15, No. 6, pp. 75–81.
38. V.A. Belyi, A.I. Sviridenok, M.I. Petrokovets and V.G. Savkin. *Friction and Wear in Polymer-Based Materials.* Oxford, Pergamon Press, 1982.
39. J.S. Williams, R.G. Elliman and M.C. Ridgway. *Ion Beam Modification of Materials.* Amsterdam, North-Holland, 1996.
40. L.B. Loeb. *Ststic Electrification.* Berlin, Springer-Verlag, 1958.
41. B.V. Deryagin, N.A. Krotova and V.P. Smilga. *Adhesion of Solids.* Moscow, Nauka, 1973.
42. M.I. Kornfeld. Frictional electrification. *J. Phys. D: Appl. Phys.*, 1976, Vol. 9, No. 8, pp. 1183–1192.
43. V.E. Shashoua. Static electricity in polymers. I. Theory and measurement. *J. Polymer Sci.,* 1958, Vol. 33, No. 126, pp. 65–85; Static electricity in polymers. II. Chemical structure and antistatic behavior. *J. Polymer Sci., Part A: General Papers,* 1963, Vol. 1, No. 1, pp. 169–187.
44. G.S. Rose and S.G. Ward. Contact electrification across metal-dielectric and dielectric-dielectric interfaces. *Brit. J. Appl. Phys.*, 1957, Vol. 8, No. 3, pp. 121–126.
45. G.M. Sessler, ed. *Electrets.* 2nd edn., Berlin, Springer-Verlag, 1987.
46. J. Van Turnhout. *Thermally Stimulated Discharge of Polymer Electrets.* Amsterdam, Elsevier Publ. Co., 1975.

47. A.I. Sviridenok, A.E. Klimovich and V.N. Kestelman. *Electrophysical Phenomena in the Tribology of Polymers.* Amsterdam, Gordon and Breach Sci. Publ., 1999.
48. V.S. Mironov. Influence of sliding speed on parameters of triboelectret state in polymer coatings. *Soviet J. Friction and Wear*, 1989, Vol. 10, No. 1, pp. 69–72.
49. S.I. Guzenkov, Yu.V. Gromyko and A.F. Klimovich. On electret-triboelectrization superposition in friction of a composite with electret filler. *Soviet J. Friction and Wear*, 1987, Vol. 8, No. 1, pp. 107–110.
50. A.K. Vijh. Electrochemical effects as the source of electromotive force in M1-polymer-M2 systems. *J. Appl. Phys.*, 1978, Vol. 49, No. 6, pp. 3621–3624.
51. J.P. Crine and A.K. Vijh. The electrochemical origins of open-circuit voltage observed during heating of metal-polymer-metal systems. *Materials Chem. and Phys.*, 1984, Vol. 11, pp. 85–98.
52. R.P. Bhardwaj, J.K. Quamara, K.K. Nagpaul and B.L. Sharma. Thermally stimulated spontaneous current from Kapton polyamide-metal contacts. *Phys. Stat. Sol. (a)*, 1983, Vol. 80, pp. 325–332.
53. C.W. Cien and J.B. Schelenoff. Limiting battery performance parameters for polyacetylene. *Nature*, 1984, Vol. 311, No. 5984, pp. 360–363.
54. V.A. Goldade and L.S. Pinchuk. *Electret Plastics: Physics and Materials Science.* Minsk, Nauka i Tekhnika, 1987.
55. V.N. Kestelman, L.S. Pinchuk and V.A. Goldade. *Electrets in Engineering: Fundamentals and Applications.* Boston/Dordrecht/London, Kluwer Academic Publ., 2000.
56. M. Ieda, G. Sawa, Sh. Nakamura and Y. Nishio. A spontaneous current from the metal(1)-polymer-metal(2) system. *J. Appl. Phys.*, 1975, Vol. 46, No. 6, pp. 2796–2798.
57. L.S. Pinchuk, V.A. Goldade and A.S. Neverov. About electrical phenomena in metal-polymer moving joints. *Proc. of Soviet Conf. "Theory of friction, wear and lubrication"*, Tashkent, 1975, Vol. 2, pp. 145–146.
58. A.U. Shareef, K.K. Saraf and A.P. Srivastava. Spontaneous electric current from polyvinyl alcohol induced by the first heating run. *Phys. Stat. Sol. (a)*, 1983, Vol. 77, pp. 381–386.
59. V.A. Goldade. *Investigation of some physical-chemical phenomena during design and exploitation of metal-polymer systems.* Ph. D Thesis, Riga, 1980.
60. V.A. Goldade. About electrochemical mechanism of metal-polymer electrets formation. *Vysokomolecul. Soedinenia*, 1989, Vol. B31, No. 1, pp. 13–17.
61. I.M. Vertyachikh, V.A. Goldade, A.S. Neverov and L.S. Pinchuk. Influence of electric field of polymer electret on the sorption of organic solvent. *Vysokomolecul. Soedinenia*, 1982, Vol. B24, No. 9, pp. 683–687.
62. L.S. Pinchuk, V.A. Goldade and A.S. Neverov. On the kinetics of electrochemical processes during metal-polymer pair friction in electrolytes. *Soviet J. Friction and Wear*, 1980, Vol. 1, No. 4, pp. 88–91.
63. L.S. Pinchuk, A.S. Neverov and V.A. Goldade. On some possibilities of friction pair polarization. *Soviet J. Friction and Wear*, 1980, Vol. 1, No. 6, pp. 105–107.
64. Yu.I. Voronezhtsev, V.A. Goldade, L.S. Pinchuk and G.V. Rechits. Influence of bulk charge distribution on friction characteristics of polymer electrets. *Soviet J. Friction and Wear*, 1984, Vol. 5, No. 1, pp. 110–113.

65. I.L. Rozenfeld, F.I. Rubinstein and K.A. Zhigalova. *Protection of Metals from Corrosion by Paint Coats.* Moscow, Khimiya, 1987.
66. V.A. Belyi and Yu.M. Pleskachevsky. *Metal-Polymer Systems.* Moscow, Znanie, 1982.
67. V.A. Goldade, L.S. Pinchuk and V.V. Snezhkov. Structural orientation and electric polarization of polymers in stationary magnetic fields. *Proc. 7th Intern. Symp. on Electrets*, 1991, pp. 316–321.
68. U.S.S.R. Patent 1165588, B 29 C 33/00. *Device for production of goods from polymer materials.* I.M. Vertyachikh, Yu.I. Voronezhtsev, V.A. Goldade and L.S. Pinchuk, 1985.
69. U.S.S.R. Patent 1014844, C 08 J 5/16. *Method of manufacturing articles of polymer compositions.* I.M. Vertyachikh, V.A. Goldade, A.S. Neverov and L.S. Pinchuk, 1983.
70. U.S.S.R Patent 1055913, F 16 C 33/04. *Sleeve bearing.* L.S. Pinchuk, I.M. Vertyachikh, Yu.I. Voronezhtsev, et al., 1983.
71. U.S.S.R Patent, C 10 M 7/02. *A lubricant for friction joints.* V.A. Goldade, A.S. Neverov, L.S. Pinchuk, et al., 1980.
72. V.A. Goldade. On adhesive strength and protective properties of polymer composite coatings. *Mechanics of Composite Materials*, 1981, No. 5, pp. 841–845.
73. V. Ashworth and C.J.L. Booker, eds. *Cathodic Protection Theory and Practice.* Ellis Horwood, Chichester UK, 1986.
74. O.S. Zheglov, V.Yu. Sokolov and V.M. Kremeshnyi. Service life improving of partially mobile hinges under conditions of corrosion-mechanical wear. *Soviet J. Friction and Wear*, 1982, Vol. 3, No. 5, pp. 57–65.
75. K.N. Stafford, P.K. Datta and C.G. Googan, eds. *Coatings and Surface Treatments for Corrosion and Wear Resistance.* Ellis Horwood, Chichester UK, 1984.
76. G.E. Lazarev, T.L. Kharlamova and V.I. Vereikin. Electrochemical methods for improving the wear resistance of graphite-steel friction pairs lubricated by liquid aggressive media. *Soviet J. Friction and Wear*, 1985, Vol. 6, No. 1, pp. 89–92.
77. V.N. Kestelman. *Physical Methods of Polymer Materials Modification.* Moscow, Khimia, 1980.
78. V.A. Zakrevsky, E.U. Tomashevsky and V.V. Baptizmansky. EPR of places of macromolecular rupture in polyamides subjected to mechanical breakage. *Solid State Physics*, 1967, Vol. 9, No. 5, pp. 1434–1441.
79. P.Yu. Butyagin, A.M. Dubinskaya and V.A. Radtsyg. EPR spectra, conformation and chemical properties of free radicals in solid polymers. *Successes in Chemistry*, 1969, Vol. 38, No. 4, pp. 593–623.
80. N.M. Emanuel and A.L. Buchachenko. *Chemical Physics of Aging and Stabilization of Polymers.* Moscow, Nauka, 1982.
81. N.K. Baramboim. *Mechanochemistry of High-Molecular Compounds.* Moscow, Khimiya, 1978.
82. P. Yu. Butyagin. *Mechanochemistry of Polymers. In: Encyclopedia of Polymers.* Moscow, Soviet Encyclopedia, 1974, Vol. 2, pp. 241–247.
83. V. Korshak, I. Gribova and A. Krasnov. Tribochemical reactions and wear resistance of polymers of the polyphenylene type. *Intern. Conf. Wear of Materials*, Houston, Texas, 1987, Vol. 1, No. 1, pp. 435–438.

84. V. Korshak, I. Gribova, A. Krasnov and G. Mamatsashvili. Tribochemical transformations and wear resistance of materials based on the cross-linking polymers. *Intern. Conf. Evrotrib-81*, 1981, Vol. 1, p. 189.
85. V.V. Korshak, I.A. Gribova, A.P. Krasnov, et al. Study of tribochemical transformations in friction of copolymer of formaldehyde with 1,3-Dioxolan. *Soviet J. Friction and Wear*, 1988, Vol. 9, No. 5, pp. 1–5.
86. V.V. Korshak, I.A. Gribova, A.P. Krasnov, et al. Tribochemical transformations in polycaproamide. *Vysokomolecul. Soedinenia*, 1987, Vol. A29, No. 8, pp. 1699–1704.
87. V.V. Korshak, I.A. Gribova, A.P. Krasnov, et al. Tribochemical processes and wear resistance of polymeric materials. *ASLE Trans.*, 1983, Vol. 26, No. 3, pp. 401–404.
88. A.P. Krasnov, I.A. Gribova and A.N. Chumayevskaya. Chemical structure of polymers and tribochemical transformations in polymers and filled systems. *J. Friction and Wear*, 1997, Vol. 18, No. 2, pp. 110–126.
89. V.A. Struk. Role of the tribochemical factor in metal-polymer friction assembly development. *Soviet J. Friction and Wear*, 1987, Vol. 8, No. 2, pp. 71–77.
90. A.P. Krasnov, L.B. Makina, S.Yu. Panov and V.A. Mit. Peculiarities of friction surface formation of aliphatic polymers. *J. Friction and Wear*, 1996, Vol. 17, No. 3, pp. 87–89.
91. A.P. Krasnov, L.B. Makina, L.Ya. Nikiforova, et al. The surface layer structure of Ultra High Molecular Weight Polyethylene at friction against steel counterface. *J. Friction and Wear*, 1993, Vol. 14, No. 5, pp. 53–57.
92. D. Landheer and C.J.M. Meesters. Friction and wear of PTFE compounds. *Kunststoffe*, 1993, Vol. 83, No. 12, pp. 996–1001.
93. I. Sugimoto and S. Miyake. High lubrication performance of tribologically oriented fluore polymer molecules analyzed by polarized infrared microscopy. *J. Appl. Phys.*, 1990, Vol. 67, No. 9, pt. 1, pp. 4083–4089.
94. V.V. Korshak, I.A. Gribova, A.P. Krasnov, et al. Tribochemical processes and wear resistance in Polyphenylene polymers. *Soviet J. Friction and Wear*, 1985, Vol. 6, No. 6, pp. 1–6.
95. V.V. Korshak, I.A. Gribova, A.P. Krasnov, et al. Investigation of high-temperature pressing of prepolyphenylene based on 4, 4′-diacetyl diphenyloxide. *Vysokomolecul. Soedinenia*, 1984, Vol. A26, No. 12, pp. 2625–2631.
96. V.V. Korshak, I.A. Gribova, A.P. Krasnov, et al. Chemical transformation in friction of filled phenolformaldehyde polymer. *Soviet J. Friction and Wear*, 1985, Vol. 6, No. 5, pp. 9–13.
97. A. Pogosian, A. Karapetian and K. Oganesian. Tribochemical processes and wear of composite polymer materials. *Intern. Conf. Wear of Materials*, Denver, Colorado, 1989, Vol. 7, No. 2, pp. 521–528.
98. A.P. Krasnov, I.K. Taratuta and L.B. Makina. Triboengineering aspects of improving wear resistance of commercial polymers. In: *Practical Tribology. World Experience.* Ser. 1: Intern. Eng. Encyclopaedia, Moscow, Center of Science and Technology, 1994, pp. 170–172.
99. A.P. Krasnov, I.K. Taratuta, G.I. Gureeva, et al. The nature of Poly(arylenetherketone) self-lubricity. *J. Friction and Wear*, 1994, Vol. 15, No. 3, pp. 68–72.
100. W. Korniluk and K. Dubicki. Tribological properties of Polyamide-6. *Polymery (Poland)*, 1988, Vol. 33, No. 6, pp. 231–233.

101. V.V. Korshak, I.A. Gribova, G.I. Gureeva, et al. Friction of copolymers of ε-caprolactam with ω-dodecalactam. *Soviet J. Friction and Wear*, 1983, Vol. 4, No. 6, pp. 1–5.
102. V.V. Korshak, S.V. Vinogradova, I.A. Gribova, et al. Friction properties of mixtures of aromatic polyamides obtained by matrix polycondensation. *Soviet J. Friction and Wear*, 1989, Vol. 10, No. 3, pp. 145–146.
103. V.V. Korshak, I.A. Gribova, A.P. Krasnov, et al. Study of surface layers during friction of filled polymer system based on polyphenylquinoxaline (PPQ). *Soviet J. Friction and Wear*, 1981, Vol. 2, No. 2, pp. 22–25.
104. V.V. Korshak, I.A. Gribova and A.P. Krasnov. On the phenomenon of tribochemical lubrication at friction of heterochain polymers. *Vysokomolecul. Soedinenia*, 1985, Vol. B27, No. 5, pp. 323–324.
105. V.V. Korshak, I.A. Gribova, A.P. Krasnov, et al. Antifriction properties and wear resistance of plastics based on cross-linked polyphenylquinoxaline. *Soviet J. Friction and Wear*, 1981, Vol. 2, No. 3, pp. 118–121.
106. A.Yu. Shaulov. On triboengineering effects in dispersion-filled polymers. *Vysokomolecul. Soedinenia*, 1990, Vol. A32, No. 11, pp. 2335–2339.
107. V.V. Korshak, I.A. Gribova, A.P. Krasnov, et al. Effect of thermodynamic flexibility of polyphenyl quinoxalines on their thermofrictional properties. *Vysokomolecul. Soedinenia,* 1979, Vol. A21, No. 10, pp. 2214–2220.
108. A.A. Vojno, W.A. Struk and W.G. Sawkin. Physikalisch-chemische Aspeckte bei Herstellung und Anwendung von verschleissfesten Kompositionswerkstoffen SAM. *Schmierungstechnik*, 1984, No. 5, pp. 148–151.
109. V.A. Struk. Effect of composition and topography of surface compounds of metal counterbody on wear of metal-polymer friction pair. In: *Mechanics and Physics of Contact Interactions*. Kalinin, Kalinin State University, 1979, pp. 69–76.
110. V.A. Struk. On mechanochemical concept of wear of metal-polymer friction pairs. *Proc. Belarus AS*, 1987, Vol. 31, No. 3, pp. 223–226.
111. V.A. Struk and Yu.M. Pleskachevskii. On corrosion-mechanical concept of wear for tribological metal-polymer structures. *Proc. Intern. Conference on Tribology "Friction, Lubrication and Wear"*, London, 1987, pp. 339–344.
112. A.V. Nikolskii, A.T. Korzakov and V.N. Kravchenko. Dynamic change of chemical state of friction surfaces of metal-polymer pair in the frictional interaction process. *Soviet J. Friction and Wear*, 1988, Vol. 9, No. 5, pp. 73–79.
113. V.I. Nefedov. *X-Ray Electronic Spectroscopy of Chemical Compounds.* Ref. Book. Moscow, Khimiya, 1984.
114. N.I. Egorenkov. *Variation regularities of adhesive, frictional, and physical-chemical properties of polymer films on metals under thermal effects.* Dr. Chem. Sci. Thesis, Kiev, 1986.
115. S.S. Pesetskii and O.N. Alexandrova. Effect of dicarbonic acids on adhesive interaction of polyamide films with copper. *Colloidal J.*, 1990, Vol. 52, No. 2, pp. 302–308.
116. L. Pinchuk. Corrosion factor in metal-polymer wear. *Proc. 5th Intern. Sympos "Intertribo-93"*, Bratislava, 1993, Vol. 3, pp. 52–56.
117. V.G. Savkin, V.M. Ken'ko and V.A. Struk. Some physico-chemical aspects of frictional interaction of metal-polymer systems. *Proc. Intern. Conf. Intertribo-81*, Vysoki Tatry, 1981, Vol. 1, pp. 172–179.

118. V.A. Belyi, V.G. Savkin and V.A. Struk. Effect of counterface origin on tribochemical phenomena in metal-polymer contacts. *Proc. Intern. Conf. Eurotrib-85*, Paris, 1985, Vol. 2, part 5–111, pp. 1–5.
119. V.A. Goldade, A.S. Neverov and L.S. Pinchuk. Polymer inhibited materials of structural purposes. *Proc. Belarus AS*, 1981, Vol. 25, No. 12, pp. 1100–1103.
120. V.A. Goldade, A.S. Neverov and L.S. Pinchuk. Engineering materials based on polymer gels. *New in Machine-Building*, 1979, No. 9, pp. 68–69.
121. V.A. Goldade, L.S. Pinchuk and V.A. Struk. Plastics containing corrosion inhibitors. *10th Intern. Congress on Metallic Corrosion,* Madras (India), 1987, Oxford and IBH Publ. Co., Vol. 3, pp. 2861–2870.
122. S.P. Papkov. *Physico-Chemical Bases of Processing Polymer Solutions.* Moscow, Khimiya, 1971.
123. V.A. Goldade, A.S. Neverov and L.S. Pinchuk. *Low-Modular Composite Materials Based on Thermoplastics.* Minsk, Nauka i Tekhnika, 1984.
124. A.V. Efimov, P.V. Kozlov and N.F. Bakeev. Phenomena of plastification and antiplastification in polymer glasses. *Proc. USSR AS,* 1976, Vol. 230, No. 3, pp. 639–641.
125. W.J. Jackson and J.R. Galdwell. Antiplasticization. III. Characteristics and properties of antiplasticizable polymers. *J. Appl. Polym. Sci.,* 1967, Vol. 11, No. 2, pp. 227–244.
126. L.M. Rodeson and J.A. Faucher. Secondary loss transition in antiplasticized polymers. *J. Polym. Sci., B,* 1969, Vol. 7, No. 1, pp. 35–40.
127. A.S. Neverov. Antiplastification in polyethylene-mineral oil systems. *Trans. Belarus AS, ser. Phys.-Techn. Sci.*, 1978, No. 4, pp. 70–73.
128. G.V. Rechits, L.S. Pinchuk, A.S. Neverov and V.A. Goldade. On the kinetics of electrochemical processes in friction of polymer composites containing corrosion inhibitors. *Soviet J. Friction and Wear*, 1982, Vol. 3, No. 2, pp. 93–96.
129. UK Patent 2186882, C 08 L 77/00, C 08 J 5/16. *Self-lubricating antifriction composition.* L.S. Pinchuk, V.A. Struk, V.A. Goldade, et al., 1989.
130. U.S.S.R. Patent 1051915, C 08 L 59/00, C 08 J 77/00. *Self-lubricating antifriction material.* V.A. Struk, G.G. Kolyago, A.F. Manulenko, et al., 1983.
131. U.S.S.R. Patent 1420916, C 08 L 23/26, C 08 K 5/00, C 08 J 5/16. *Compound for antifriction material manufacturing.* V.A. Struk, L.S. Pinchuk, V.A. Goldade, et al., 1988.
132. U.S.S.R. Patent 1607364, C 08 J 5/16, C 08 L 77/02. *Inhibited antifriction material.* L.S. Pinchuk, V.A. Struk, V.A. Goldade, et al., 1990.
133. U.S.S.R. Patent 970870, C 09 K 3/10, C 08 L 77/00. *Hermetizing material.* V.A. Goldade, A.S. Neverov, L.S. Pinchuk, et al., 1982.
134. U.S.S.R. Patent 1722021, C 08 J 5/16, C 08 L 77/06. *Antifriction hermetizing compound.* L.S. Pinchuk, E.N. Aleshkevich, G.V. Rechits, et al., 1991.
135. U.S.S.R. Patent 1075712, C 08 L 75/04, C 08 L 59/04. *Thermoplastic preformed compound.* V.A. Struk, G.G. Kolyago, L.S. Pinchuk, et al., 1983.
136. U.S.S.R. Patent 759570, C 09 K 3/10, F 16 J 15/20. *Hermetizing material.* V.A. Belyi, V.A. Goldade, A.S. Neverov, et al., 1980.
137. L.S. Pinchuk. *Hermetology.* Minsk. Nauka i Tekhnika, 1992.
138. The OILES Bearing. Catalogue No. EC-200-5, OILES Corporation, Tokyo, Japan.

139. Belarus Patent No. 394, C 08 J 3/18. *A method of manufacturing composites based on thermoplastic polyamides or polyolefines.* V.A. Goldade, A.S. Neverov and L.S. Pinchuk, 1994.
140. U.S.S.R. Patent 1603725, C 08 J 5/16, B 29 C 39/38. *A method of manufacturing articles.* L.S. Pinchuk, V.A. Goldade, V.V. Lisovsky, et al., 1990.
141. T.A. Osswald, L.-S. Turng and P.J. Gramman, eds. *Injection Molding Handbook.* Munchen, Hanser, 2001.
142. G. Pötsch and W. Michaeli. *Injection Molding: An Introduction.* Munchen, Hanser, 1995.
143. W. Michaeli. *Plastics Processing: An Introduction.* Munchen, Hanser, 1995.
144. U.S.S.R. Patent 887588, C 08 J 5/02. *Method for producing polymer goods.* V.A. Goldade, A.S. Neverov, L.S. Pinchuk and I.M. Vertyachikh, 1981.
145. U.S.S.R. Patent 1237683, C 08 L 67/06, C 08 J 5/16. *Antifrictional compound.* G.G. Kolyago, A.F. Manulenko, V.A. Struk, et al., 1986.
146. U.S.S.R. Patent 568773, F 16 J 15/00. *Production method of materials for seals.* V.A. Belyi, V.A. Goldade, A.S. Neverov and L.S. Pinchuk, 1977.
147. V.A. Belyi, V.I. Vrublevskaya and B.I. Kupchinov. *Wood-Polymer Structural Materials and Goods.* Minsk, Nauka i Tekhnika, 1980.
148. K.M. Gilman, V.A. Goldade, F.S. Mamedov, et al. Antirust sealing material based on polyethylene. *Corrosion and Rust Prevention in Oil and Gas Industry*, 1980, No. 2, pp. 27–29.
149. Belarus Patent 1954, F 16 J 15/32. *Piston for slush pump.* I.M. Vertyachikh, V.A. Goldade and L.S. Pinchuk, 1997.
150. I.M. Vertyachikh, V.A. Goldade. Reduction of erosion and mechanical wear of drill mud pump seals. *J. Friction and Wear*, 1995, Vol. 16, No. 3, pp. 150–152.

5 Inhibited Engineering Service Materials

Engineering materials are a broad group of auxiliary materials subdivided into two subgroups according to their application. To the first subgroup belong materials intended to raise the productivity of technological processes. The second subgroup embraces engineering materials promoting trouble-free operation and extending the lifetime of machines and equipment.

Inhibited polymer materials of the first group are those used for in-process preservation and rust-inhibition of hardware. They are various kinds of films and coats, discussed previously in Chaps. 2 and 3, coolants and inhibited polymer containers. Engineering polymer materials assisting in operation without failure of machines include numerous kinds of glues, sealants and polymer CI carriers protecting hardware in sealed vessels. A specific place within this versatile group of engineering materials is occupied by polymer microcapsules abundant during both the processing of materials and the maintenance of machines.

The properties, structure and range of application of inhibited engineering service materials on a polymer base are considered below.

5.1 Glues

Glues are either liquid or solid (meltable) polymer compositions whose purpose is to create non-detachable joints of structural materials or their products.

Glue compositions in practice contact metal crystals and oxides formed on the surface layer of metal parts as well as adsorbed molecules of impurities and their transformation products. Thus, adhesive bonds formed in such joints acquire variable strength and water-repellent properties. Adhesive joints operating in aqua electrolytes are weakened by diffusion of the electrolyte into the glued seam (via, first of all, the seam butt) upon which these weak bonds dissociate. Corrosion starts initially on the sites where the bonds have failed and then propagates over the whole surface of the seam. Metal corrosion in the glued joint has a local character since electrolyte diffusion in the metal is restricted and corrosion products are commonly removed from the seam. As has been noted earlier, local corrosion is one of the most dangerous types and a serious goal of antirust techniques is its elimination.

The most obvious way of raising the corrosion resistance of adhesive joints seems to be the introduction of corrosion inhibitors into glue compositions. Let us analyze the possible consequences of this measure.

Filling of the glue base with neutral powder particles is employed rather rarely, mainly to cut the cost of the glue or if the glue seam is uneven in thickness [1]. Impregnation of powder CI in the glue composition is ineffective because the glued seam will be weakened by inclusions of unstable inhibitor particles even when using optimum pasting technologies. What is more, CI particles will be encapsulated in the polymer binder, so the speed at which inhibitor molecules travel to the metal will be conditioned by this diffusion. Liquid-phase CI are already brought into contact with the metal surface at the stage of formation of the adhesive joint. This is the reason in the majority of cases why adhesion between the metal and glue is impaired. Isolation of CI in the gaseous phase from the glue composition in the course of curing deteriorates the strength of the glued splice due to gaseous inclusions. Hence, even a simplified analysis shows a certain ambiguity in the method of modifying glue compositions with CI.

The problem of improving the corrosion resistance of adhesive joints of metal parts is often solved by one of the following approaches.

A. Rust-inhibition of glued steel structural elements by filling the glue composition with particles of an electrically negative metal, e.g. zinc [2].
B. Priming of the surfaces to be glued together by inhibiting polymer compositions. Priming is often used prior to sticking steel and aluminum alloys as there is great danger of forming a galvanic element in such splices [3]. The metal part is, in this case, protected by the CI isolated from the primer layer. In an optimum variant of this method the adhesive joints, such as the primer-metal and primer-glue layers, are equally strong.
C. Chemical modification of the glued surfaces by the formation of passivating layers. The modification technique depends on the nature of the metal. The parts are most often subjected to acid pickling, e.g. aluminum alloys are anodized in sulfuric and chromic acids. It is preferable to anodize aluminum parts in sulfuric acid followed by treatment of the anodic film in a bichromate. There are several methods of pickling carbon and stainless steels, chemical oxidation of magnesium alloys as well as copper and titanium alloys before gluing [4].
D. Finishing, i.e. treatment of the surfaces to be glued by surfactants (finish agents) forming an adsorption monolayer that decreases wetting of the metal part by water. The finishing of the silicone-organic series are commonly employed as water-repellent agents. They enter into chemical interactions with adsorbed moisture and crystalline structures of the surface layer (metals and oxides) and form strong adhesive bonds with the glued splice [3, 4].
E. Development of glue compositions with the properties of metal corrosion inhibitors. Resorcin-thiokol glues possess the required properties. Steel

workpieces stuck together by this glue show no traces of corrosion in the seam after a year of exposure in water and humid atmosphere. This effect is commonly attributed to the properties of polyphenols incorporated in the resorcin resin composition [5]. Epoxy glues can be modified by coal tar, bitumen and other substances containing aromatic compounds. The mechanism of corrosion inhibition by these materials consists of their isolation on the metal part surface from the glue seam via syneresis and fulfilling of the functions of specific CI. A glue composition based on a polymer binder containing a heterocyclic polymer with tertiary nitrogen [6] as an antirust additive has been patented.

It should be underlined in conclusion that the choice of the means of modifying glue compositions with CI is in the process of evolution. Towards this aim, additives such as guanidine chromate, some chromic acid esters and chromates, etc. that adsorb onto the glued surface and maintain the strength of the seam are used. They raise the resistance of the glued metal joint to the effects of seawater.

5.2 Sealants

Compositions based on polymers and intended to seal immovable joints of machine parts tightly are called sealants. In contrast to glues, whose purpose is to form strong junctions between materials, sealants aim to create a barrier to permeation of hostile media into these junctions. This is critical when the working volume of machines needs to be insulated from the environment to avert leakage of the interior media, exclude ingress of corrosive components from the atmosphere into clearances of stationary structures, and son on. Adhesion of sealants to the members of joints is desirable but is not requisite. In a number of cases the ability to neutralize corrosive activity of the matter permeating into the joints turns out to be far more essential. Therefore, the modern generation of sealants is represented by a vast variety of inhibited polymer materials.

Over a dozen characteristic features of sealants, numerous formulations and diversity of application fields serve the base for their subdivision into groups. The simplest classification of sealants, by their deformation properties, divides them into plastic and solid ones. The former operate in a viscous-flow state or close to it and may deform irreversibly under low enough loads. The latter are able to preserve the properties of a solid body within a considerable loading range.

Plastic sealants appeared on the world market in the 1940s and, despite stiff competition from elastic sealants made on the base of vulcanizable rubbers, they are still extensively applicable today.

Typical representatives of this group are non-drying sealants based on buty1 rubber, polyisobutylene, ethylene-propylene rubber in combination with polystyrene, polypropylene and polyethylene. To regulate their rheological characteristics and cut cost they are often impregnated with fibrous or powder mineral fillers (50% to 75%). The requisite components of non-drying sealants are petroleum oils, which add malleability, as well as rosin, thermosetting resins, bitumen and tar. This type of sealant is widely applicable in construction to insulate butts of concrete blocks, profile glass, glass packs and in waterproofing. CI are not usually introduced into such sealants.

A large group of non-drying sealants is employed to seal threaded joints of steel pipes intended for the oil and gas industry. These sealants should necessarily contain CI. For instance, a sealing antirust composition [7] containing heavy fractions of petroleum oil along with a mixture of tall pitch and fuming oxides is an efficient corrosion inhibitor for steel in acidic and salt environments. Another sealant [8] on a rubber or polyurethane base shows high protection efficiency of metals against deep-sea corrosion. The polyurethane base can be modified by a synergetic mixture of CI (benzaldehydes, tall pitch, morpholine) that simultaneously provide a number of functions in the sealing composition. This sealant has successfully passed deep-sea submersion tests in the Indian and Atlantic oceans and has been included in the USSR Sea Register.

Solid sealants is a code name of the totality of initially liquid or paste-like sealants that transform into a solid state (spontaneously or as a result of certain effects) when impregnated into the clearance of the joint to be sealed. This term is conditional because, even after solidification, these sealants occupy an intermediate position between a liquid and solid according to their deformation characteristics.

This group includes *vulcanizing sealants* that contain reagents that transform the rubber-based sealing composition into a cross-linked system. They are impregnated with surfactants made on the base of fatty acids [9] that fulfill two functions: (i) to improve the colloidal stability of the system on vulcanization and (ii) to inhibit corrosion of metals. Thiokols are frequently co-vulcanized with epoxide, phenolic, polysulphide and polyester resins. Vulcanizing sealants incorporate evaporative agents as well as containing volatile CI. The sealants of this class are produced by the Products Research Co. (USA) for the aviation industry. In addition, elastic inhibited sealants based on siloxane rubbers of TVP grade are produced by Dow Corning (USA), Caf by Ron Pulenk, (France), KE by Shinetsu Chemical Co. (Japan) and are extensively used in ship-building, electric engineering, machine building, and other sectors.

Rubber-free inhibited sealants based on plasticized thermoplastics have been developed for oil and gas pipelines. A non-drying plastic material [10] in the form of a gel is based on PE and a tar of mineral and vegetable oils, which is used as a plasticizing and anticorrosion additive. PE gels are filled with fibers and powder particles to improve the strength of sealing layers [11]. For PE-compatible CI a synergetic mixture of potassium and triethanol amine salts of fatty acids of tall oil, sodium sulfonate and the extract of phenolic refinement of transformer oil distillates is most often used [12]. An analogous composition for sealing threaded joints has been patented, whose composition includes a monoxicarboxylic acid serving as an adhesion promoter [13]. A ferrous sealant [14] contains a magnetic fluid in the PE-based gel. The viscosity of the liquid phase of the gel increases in the magnetic field and improves the resistance of the sealing layer to squeezing from the thread clearance. Cobalt particles filling the magnetic fluid acquire a more electronegative electrode potential than that of the steel pipeline and so are able to function as protecting elements and provide antirust functions in threaded joints.

Drying sealants are represented by solutions of rubber blends in organic solvents. As soon as the solution is impregnated into the clearance and the solvent evaporates, the sealant becomes rubbery and gains elasticity. The original compositions contain adhesive additives (coumarone, terpenic and phenolic resins, rosin or its esters), Pl and solvents (toluene, xylol, benzine, etc.), fillers (chalk, titanium dioxide, talc, etc.) and stabilizers. These compositions can be easily impregnated with CI [15].

Anaerobic (vulcanizing without oxygen) *sealants* are initially multi-component liquid compositions. They include polymerizable compounds of either the acrylic or methacrylate series and modifying agents. They can be stored for a long time in the liquid state and are able to solidify at room temperature in clearances of joints with oxygen deficiency. Anaerobic sealants cure in contact with active metals (copper and its alloys, cobalt, manganese, nickel) within a few minutes, in contact with carbon steel (as well as aluminum alloys, titanium and zinc) within 8–24 hours, while curing takes days and even weeks in contact with passive materials (anodic, oxide, cadmium, chrome and zinc-based coatings or stainless high-alloy steels) even in the presence of cure activators. The sealants of this class are highly sensitive to the chemical structure of their anaerobic base, which restricts its modification by CI. Anticorrosion protection of the junctions of hardware is largely ensured by the high chemical resistance and perfect adhesion of anaerobic compositions to metals. Nevertheless, adhesive additives containing amine groups are commonly introduced into the anaerobic compositions as CI [2, 5].

Polymeric compounds are specific sealing materials intended to line or impregnate conducting hardware as well as electric radio circuits for electric insulation. They are based on epoxy and unsaturated polyester resins, liquid organosilicon rubbers and monomers (initial products for synthesizing polymethacrylates and polyurethanes). Compounds based on thermoplastic materials (tar, rosin, cerezin) in the form of solid or wax-like masses, heated for transformation into the liquid state, are confined to this application.

Corrosion of conducting non-ferrous elements of electric circuits insulated by these compounds rarely affects the functioning of these devices. The probability of malfunction can be limited still more by the impregnation of CI into the compounds. Polymeric compounds are, as a rule, perfect dielectrics and do not sorb water. Therefore, their impregnation by water-soluble CI is inefficient. Oil-soluble CI usually contain low-molecular conducting components that impair the dielectric characteristics of the insulation. If the risk of corrosion of metal conductors with this insulation is aggravated it is better to modify the surface layers at the conductor-insulation interface as recommended in Sect. 5.1.

5.3 Plastic Carriers of Inhibitors

Engineering goods transported or stored in closed volumes or metal vessels necessitate rust prevention. For this purpose volatile corrosion inhibitors (VCI) are commonly used whose vapors penetrate into narrow gaps and slots of metal structures and become adsorbed on the surfaces, thus inhibiting corrosion propagation.

Rust-inhibition efficiency of VCI depends largely on their pressure and vapor saturation. The higher the saturated vapor pressure, the more reliably the inhibitor protects the article, especially in the initial period of preservation. To keep the article in a passive state for longer, the ambient VCI concentration should be maintained at the required level. This concentration can be defined accurately only in experiments via prolonged field tests [16].

So, the problem of maintaining the pressure of the saturated vapor at a given level within a required period is a serious engineering task in which the superposition of two major factors can be highlighted: (i) the degree of tightness of the vessel containing the vapor, and (ii) the evaporation velocity of CI. As has been shown in Sect. 2.3, the evaporation rate of the majority of industrial VCI is rather high. CI in their supply state, especially liquid CI, are not convenient for the preservation of engineering goods. In this connection, special plastic containers for carrying CI have been elaborated, from which the inhibitors are liberated at a given velocity specified by the container structure. When placed into a vessel, even if insufficiently tight, the carrier is able to maintain a regulated liberation of inhibitor vapors, compensate for their leakage from the vessel and maintain a rustproof concentration.

The following types of plastic CI carriers will be considered below: (1) cor-tabs, plastabs, granules, (2) vapor capsules and foam sheets and (3) fibrous CI carriers.

Cor-tabs are tablet VCI used to protect ferrous and nonferrous metals. They are produced by compacting powder VCI (above 95% of the tablet mass) together with a polymer binder solution. The binder adds strength and forms a thin-walled matrix in the tablet to regulate the inhibitor evaporation rate.

Plastic VCI carriers can be made as granules or briquettes by processing the mixture described above in a worm unit with further granulation and then cut into strips of the required length.

Plastabs can be efficiently used in a similar way to cor-tabs in the preservation of machines and equipment. They are made in the form of thin plates cut from sheets formed from a mixture of a powder inhibitor and polymer solution. VCI-based plastabs are believed to provide close to ideal protection for ferrous and nonferrous metals against corrosion, particularly in gaps and narrow clearances. One of the well-known manufacturers of such plastic carriers is the Northern Instruments Corp. (USA).

Vapor capsules are made as strips of foam plastics impregnated with a liquid VCI and encapsulated into plastic shells in the form of tubes, boxes or pallets. Their chief field of application is the preservation of electric and electronic equipment during shipment, storage and operation. The capsules are placed in closed tanks containing the equipment. The foam plastics used in vapor capsules should not decompose on contact with CI or their solvents. In addition, they should have large enough adsorptive capacity for prolonged inhibiting protection of hardware without additional preservation.

Foam sheets are a foamed sheet polymer material containing CI in the polymer base and as gaseous inclusions. The Zerust Ferrous Foam Sheet material is able to lower the risk of damage of fragile metal articles in transit and protects them against corrosion.

Fibrous CI carriers represent an alternative to the group of antirust means described above, the difference being in the use of a fibrous instead of porous polymer for the CI vessels.

The anticorrosion element [17] has a melt-blown PE fibrous base with an additive that is simultaneously a PE plasticizer and VCI solvent. The base fibers are cohesively bonded in contact sites and form a porous framework of a certain structural strength. The framework is impregnated with a liquid VCI soluble in a plasticizing additive. A distinctive feature of such a framework is the sufficiently large surface area of its fibers that is transformed on melt-blowing into a gel-like state by a special additive. On contact with the fibers the VCI dissolves in the dispersed gel liquid, which permits the regulation of the VCI evaporation rate on two levels: (i) from vacancies between the framework fibers, and (ii) from PE micropores of the gel matrix. As can be seen, the potential for rust-inhibition of the element are raised significantly and the VCI evaporation rate can be preset in accordance with the preservation conditions by choosing the composition and structural parameters of the framework.

Plastic scrims are related to the same category of inhibited plastics. In a non-woven form, the inhibited film is first extruded. The polymer webbing (usually melt-blown inhibited polymer fibers) is then laminated to the film. Zerust ferrous plastic scrim is used in the manufacture of airtight packages inside which a protective atmosphere is created by VCI vapors. The fibrous scrim layer also protects packed goods from impact damage.

5.4 Inhibited Plastic Containers

Engineering goods are packed to protect them against corrosive and mechanical damage. A packing is formed as a rule by two elements: a master container and an inner package. The master container supports mechanical loads and protects the inner package and the packed article. The inner package is aimed at rust-inhibition by the creation of a protective atmosphere in the package. The most applicable types of master containers to date have been cartons and wooden or wood-plastic boxes. For the inner package grease, oil, VCI paper or inhibited film are commonly used. Inhibited plastic containers combine both functions in one. The following types of polymer materials, modified by CI, are employed during manufacture: structural and laminated plastics, foam plastics and composites based on polymer binders and microspheres.

Inhibited structural plastics are manufactured by the extrusion of blends of structural polymers and CI.

A composition of corrosion-inhibiting thermoplastic alloys has been patented [18]. Master batches of nylon, polycarbonate, and polyethylene terephthalate together with corrosion inhibitors in the solid phase (blend of cyclohexylammonium benzoate, sodium nitrite, benzotriazole, and sodium sebacate) are extruded for use in injection molding operations. Molded containers protect the goods from corrosion and mechanical damage. A material of analogous design [19] has been developed on a base of acrylonitrile butadiene styrene (ABS) resin and powder mixture of inhibitors of a similar class for use in molding operations. It is particularly suited for the preparation of protective enclosures for delicate metals, such as those present in electronic media devices and apparatus.

Injection and blow molded products based on inhibited PE and PP produced by the Northern Instruments Corp. (USA) have become world standard for anticorrosion plastic containers. Similar plastic products modified by VCI are manufactured by the German companies Rose Plastic and Brangs and Meinrich [20]. Another original sort of inner packages for small diameter metal tubes is a tube strip. Tube strips are extruded PE tubings inserted into metal tubes to protect their interior surfaces from corrosion [21].

Laminated plastics are a convenient material for the production of inhibited containers as they allow the distribution of the functions of the master container and inner package between different layers of the plastic.

The most applicable representatives of this class of materials are the Zerust Ferrous Chipboard and Fibreboard. They are produced by applying extruded VCI-modified PE coatings about 1.5 mm thick on either chipboard sheets or Kraft paper. One kind of the fibreboard is made as a three-ply material with a paper-coating-paper structure. Therefore the fiberboard can be used as a conventional cardboard or a lining for layer-wise packing of products for preservation.

Dunnage trays and bins are used to speed up and reduce the necessity for care during the handling of parts on assembly lines. Thermoformed plastic dunnage trays, which were pioneered in the automotive industry, provide significant advantages over the paper and wood pallets that were used before. The trays resist abrasion, absorb shocks to and vibrations of products, are strong and durable, and allow for stacking and nesting, which conserves storage space.

Plastic dunnage trays and bins are produced first by extruding high-density sheets and then thermoforming the sheets to the desired shape. Plastic dunnage extruded from inhibited plastics produces trays and bins that do not require additional antirust means for preserving hardware. Co-extrusion techniques enable the customer to order an A–B (inhibited polymeric top layer – bottom layer from a standard polymer) or an A–B–A construction (top layer from the inhibited polymer, core layer from a conventional polymer and bottom layer from the inhibited polymer) for nesting of trays. Inhibited plastic dunnage trays and bins afford a minimum of five years VCI protection.

Corrugated plastic can have a double-faced or single-faced construction. The double-faced construction has two linerboards separated by a fluted center. The single-faced construction allows one to form tubes or to shape the material around most cylindrical or round products. Profile plastic is produced as a continuous extrusion with two outer linerboards separated by a vertical flute. Inhibited corrugated and profiled plastics are used for boxes, dividers and partitions between metal surfaces.

Inhibited foam plastic is a structural material preliminary foamed by CI vapors or based on foamed inhibited plastics and is used for plastic-based containers. The inhibited vapor phase can be introduced into the polymer base by one of the following methods.

A. Foaming of the polymer in a viscous-flow state by sublimation and thermal decomposition of substances producing solid CI. For this purpose, inhibitors are chosen that decompose most fully and with the required speed within the temperature range of the polymer transition from the solid to the viscous-flow state.
B. Foaming of the viscous-flow polymer mass by vapors of boiling inhibitor liquids that evaporate under the effect of the external heat pulse.
C. Either mechanical or pneumatic foam whipping of the oligomer or polymer solution or emulsion contact of the inhibiting vapor phase in the presence of a structural surfactant. The foam is then fixed by lacing the polymer walls.

Inhibited foam plastics are predominantly of a closed-cell structure, which means that the vapor phase is insulated in each cell from the neighboring cells by polymer walls. Diffusion of CI through the cell walls inside polymer containers made of this inhibited foam plastic diminishes the possibility of corrosive damage of the packed hardware.

One of the most widespread foam materials designed for inhibited containers is expanded polystyrene. A method for its production was first patented in

1935 by Swedish engineers Manterson and Tanbergue and its large-scale manufacture began eight years later at the Dow Chemical Company (USA) [22]. Plastic containers are manufactured from expanded polystyrene with above 75% porosity. A typical compression diagram of this material is shown in Fig. 5.1.

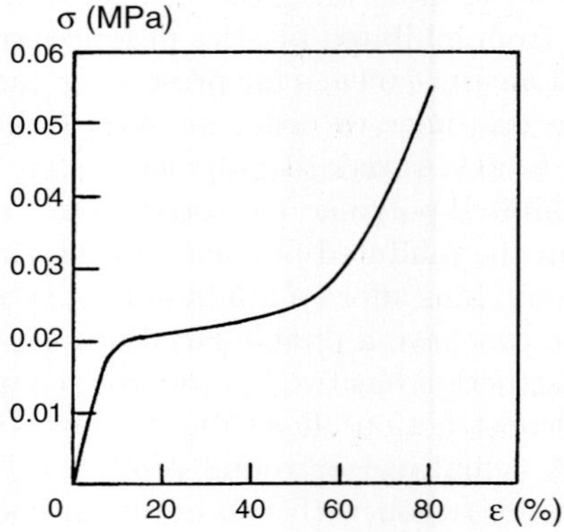

Fig. 5.1. Stress versus relative compressive strain of expanded polystyrene PS (75% porosity)

There are three obvious areas on the curve:

– An initially steep section corresponding to the deformation of cell walls until loss of stability;
– a plateau where resistance to compression increases only slightly while the density of the foam plastic rises, corresponding to loss of stability and partial breakage of the walls;
– stress growth due to the crumpling of the cells with transfer to compression of the polymer base of the foam material.

A typical representative of the inhibited foam plastics used in plastic containers is an isocyanate-derived polymer whose cells contain inhibitor vapors of the cyclohexilylamine chromate, cyclohexylamine M-mononitro benzoate, dicyclohexylamine chromate and dicyclohexylamine nitrite series [23].

Composites based on polymer binders and microspheres are plastics filled by hollow spherical particles. The modern generation of these composites employs spheres containing CI as fillers. The next paragraph of this chapter is devoted to these fillers. Composites based on microspheres have no apparent advantages over foam materials but in certain cases they turn out to be optimum in technical and economic respects for the manufacture of containers.

Microspheres, whether made from glass or, even more so, from plastic, cannot be referred to as strong materials. Therefore, production techniques for composites made from them consists of the stages of mixing microspheres with polyester or other resins, filling of molds and curing under elevated temperatures. The microspheres, which are evenly distributed in the binder, impart the material with a closed-cellular structure that is responsible for their high specific strength and isotropy as well as their chemical resistance.

5.5 Microcapsules

Microcapsules are microparticles of a substance encapsulated into shells of the film-forming material. The encapsulated substance may be in a solid, liquid or gaseous state. Its content makes up 7–85% of the microcapsule mass. Microcapsule shells are about 0.1–10 μm thick, and can be single or multilayered, elastic or tough depending on the film-forming agent. CI are most often encapsulated into shells of high-molecular compounds of animal or vegetable origin, as well as synthetic polymers and oligomers, paraffins and stearines. The inhibitors isolate from the microcapsules on mechanical fracture or solution of the shell, or as a result of diffusion.

There exist more than several dozen methods of encapsulating CI in polymer shells [24,25]. These methods are based on film-forming processes in heterogeneous systems, including at the liquid-liquid, liquid-solid, gas (vapor)-liquid or gas (vapor)-solid interfaces. The technologies of microencapsulating inhibitors can be subdivided according to the film-forming mechanism into three groups:

- film-forming in polymer solutions,
- film-forming in a heated viscous-flow polymer composition,
- film-forming by polymerization or polycondensation of low-molecular substances on inhibitor microparticles.

Film-forming in polymer solutions is based on the coacervation phenomenon, meaning the generation of high-molecular clusters of drops enriched by the dissolved substance. Coacervation in a system made up of inhibitor dispersion in a polymer solution is evoked by temperature or pH variations, or by the evaporation of part of the solvent. As a result, the phase enriched by the polymer deposits on inhibitor particles, forming a solid envelope. To choose the process parameters phase diagrams of polymer-solvent or polymer-solvent-precipitator systems are employed [24, 26]. Typical combinations of components of the film-forming solutions intended for microencapsulating CI are presented in Table 5.1.

The required conditions for shell formation on inhibitor particles in these systems are:

(i) insolubility of the inhibitor in the solvent and precipitator;
(ii) greater surface tension at the solvent-inhibitor interface than at the solvent-polymer-enriched interface;
(iii) negligible surface tension at the inhibitor-polymer-enriched interface.

Table 5.1. Components of polymer systems used for microencapsulation [24]

Polymer	Solvent	Precipitator
Polyvinyl acetate	Acetone Trichloroethylene Methanol	Water Hexane Butyl or isoamyl alcohol
Polyvinyl stearate	Benzol Chlorophorm Kerosene	Acetone Methyl ethyl ketone Mineral oil
Polyvinyl chloride	Cyclohexane	Glycol
Nitrocellulose and cellulose acetates	Acetone	Water
Cellulose acetobutyrate	Methyl ketone	Isopropyl ether
Ethyl cellulose	Xylol Ethyl alcohol Benzol	Hexane, heptane Water Mineral oil
Copolymer of styrene and maleic acid	Ethyl or methyl alcohol	Ethyl acetate, methyl ethyl ketone, isopropyl alcohol

This method is used to microencapsulate oil solutions of inhibitors into shells formed from a water-gelatin gel. The process terminates by the separation of microcapsules from the solution via centrifuging, filtration or decanting followed by flushing and drying of the microcapsules.

Microcapsule shells formed in polymer solutions or suspensions can be solidified by drying. A non-volatile inhibitor is dispersed in an easily volatile liquid, which is the polymer carrier. This mixture is then sprayed in a heated chamber and the liquid evaporates quickly. As a result, film shells are formed on the microparticles or microdrops of the inhibitor during evaporation.

Solid inhibitors are often microencapsulated by spraying a liquid polymer phase onto the particles moving in a fluidized bed, the surface of which is sprinkled with the polymer solution in a volatile solvent. The thermal regimes of the process are regulated by the temperature of the fluidized gas.

Film-forming in polymer compositions heated to a viscous-flow state can be achieved as follows.

The simplest variant presupposes that the CI powder is mixed with a colloidal solution of the binder and the oil that is heated above the binder (PE, wax, polystyrene) melting point. The powder particles are enveloped by the solution, which is cooled further during mixture spraying and then hardens, forming shells on the inhibitor particles.

Some methods adhere to the principle of closing the incident drop by fusion. A thin film of a polymer material is formed on the surface of a spinneret with a small orifice size. Microdrops of a liquid CI are forced periodically through the orifices. When squeezed the drop makes contact with the film and is enveloped by it. The shells formed on the drops solidify during cooling or on treatment with some cross-linking agents.

The method of electrostatic coagulation includes the charging of CI aerosols differently to the molten polymer film-forming agent using a glowing discharge. After this, flows of unlike-charged microparticles are joined together. The film-forming particles precipitate under the action of electrostatic forces on the inhibitor particles and coalesce. The inhibitor content in the obtained microcapsules is 50–90% of the microcapsule mass.

Film-forming by polymerization or polycondensation proceeds in either the liquid or gaseous phase.

The simplest procedure of microencapsulating CI by polycondensation consists of the following operations. One of the monomers is dissolved in an organic solvent and the other in water. A CI dispersion is introduced into one of the phases. The organic and water phases should not mix and the solvent should be compatible with the polymer formed at polycondensation. Polycondensation is initiated slowly (accelerated by dissolved catalysts) on the surface of CI microparticles by joining the phases in the dispersion nozzle. The microcapsules are separated from the reaction blend by either floating or dipping.

Microcapsules having polyamide or polyester shells are produced by polycondensation of functional chloranhydrides with diamines or glycols, correspondingly.

Inhibitors in the solid or liquid phase are microencapsulated by polymerization. A typical process of encapsulation in polyethylene shells follows this sequence of stages:

- dispersion of inhibitors in toluene,
- impregnation of polymerization catalysts into the dispersion (of the Ziegler-Natta type),
- blowing of gaseous ethylene through the obtained mixture.

As a result, 0.5–60-μm thick ethylene shells polymerize on the surface of the inhibitor particles, and the thickness of these shells depends on the amount of injected ethylene.

Microcapsules smaller than 10 μm can be formed by the following procedure. The inhibitor is dispersed in an inert gas medium, after which the aerosol is mixed with vapors of a monomer capable of catalytic polymerization of diolefins, triolefins, vinyl esters. Gaseous compounds like NO_2, BF_3 are used as catalysts. Polymerization runs for about 1–2 min as the suspended inhibitor particles are kept in the reaction zone [27].

Microcapsules of CI are considered as a specific type of inhibited plastics possessing a series of remarkable features, namely:

- chemically unstable volatile inhibitors can be kept reliably in microcapsules without any deterioration of their properties;

- inhibiting compositions displaying a synergetic effect can be easily obtained by mixing microcapsules even of relatively incompatible inhibitors;
- the toxicity of inhibitors related to the second and third classes of danger is substantially lowered after encapsulation;
- encapsulated liquid inhibitors gain the form of a friable matter that is easily bagged, batched and transported to corrosion hotbeds;
- polymer shells of microcapsules improve the regulation of the inhibitor liberation rate.

Microencapsulated CI fixed on the surfaces of metal parts forming immovable joints have found application in various industrial sectors. The interior of the shells is liberated when the parts are brought into contact and the capsules become squashed. This technique is used for antirust protection of rivet joints of the skin of airplanes. Microcapsules with CI are carried, with minimum outlay, to the demanding corrosion sites of casing and drilling pipe strings at excavation sites of oil and gas deposits. The effective use of inhibited microcapsules has not been fully realized due to a number of reasons, including technical and economic ones.

The data cited above have shown the vast range and potential of modern inhibited plastics. Inhibited technical service materials on a polymer base are extremely numerous and are developing rapidly. The conceptual ideas lying at the base of their elaboration use knowledge from several scientific directions and influence the progress in a number of engineering spheres. Even if these achievements are not directly utilized, the state of any of today's branches is somehow connected with evolution in the field of inhibited plastics. The examples of microencapsulation of CI reveals the fundamental developments in the domains of physical chemistry, chemical technology, materials science, electrochemical kinetics and so on, were used to realize the merits of these materials. This shows the dominant tendency for all classes of inhibited plastics to become engineering service materials that are increasingly reliant on science.

References

1. R.J. Hussey and J. Wilson. *Structural Adhesives Directory and Databook.* London, Chapman and Hall, 1996.
2. D.A. Kardashov. *Engineering Glues.* Moscow, Khimiya, 1980.
3. A.S. Freidin and R.A. Turusov. *Properties and Design of Adhesive Joints.* Moscow, Khimiya, 1990.
4. A.V. Pocins. *Adhesion and Adhesives Technology.* An Introduction, 2nd edn. New York, Hanser, 2002.
5. I. Skeist. *Handbook of Adhesives.* Fairfield, Bergano Book Co., 1995.
6. E.P. 471336, C 09 J 181/02. *Polymeric adhesive and anticorrosion agents.* A. Iglal, J.T. Walder and H.H. Lien, 1992.

7. U.S.S.R. Patent 1371035, C 09 K 3/10, C 09 D 5/08. *Sealing antirust composition.* L.S. Pinchuk, S.V. Bezrukov, V.A. Goldade, et al., 1987.
8. U.S.S.R. Patent 1653335, C 09 D 5/08, C 09 K 3/10. *Sealing antirust composition.* L.S. Pinchuk, S.V. Bezrukov, V.A. Goldade, et al., 1991.
9. M. Ash and I. Ash, eds. *Handbook of Adhesive Chemical and Compounding Ingredients.* New York, Synapse Information Resources, Inc., 1999.
10. U.S.S.R. Patent 566458, C 08 L 23/06, C 08 K 5/09. *Polyethylne-based composition.* V.A. Belyi, K.M. Gilman, V.A. Goldade, et al., 1978.
11. U.S.S.R. Patent 352065, F 16 J 15/20. *Plastic packing.* V.A. Belyi, B.I. Kupchinov, A.S. Mikhnevich and L.S. Pinchuk, 1972.
12. U.S.S.R. Patent 761537, C 09 K 3/10, C 08 L 23/06. *A composition for sealing material.* L.S. Pinchuk, K.M. Gilman, V.A. Goldade, et al., 1980.
13. U.S.S.R. Patent 1378343, C 09 K 3/10. *A composition for sealing threaded joints.* L.S. Pinchuk and E.N. Sokolov, 1987.
14. U.S.S.R. Patent 925996, C 10 M 5/02, 5/10, 7/02, 7/04. *Sealing lubricant for threaded joints.* V.A. Goldade, A.S. Neverov, L.S. Pinchuk, et al., 1982.
15. L.S. Pinchuk and A.S. Neverov. *Sealing Polymer Materials.* Moscow, Mashinostroenie, 1995.
16. P.A. Vinogradov. *Preservation of Engineering Products.* Leningrad, Mashinostroenie, 1986.
17. Belarus Patent 5805, B 29 D 81/26, C 23 F 11/02, D 01 F 6/04. *Antirusting element and a method of its manufacture.* L.S. Pinchuk, S.V. Frolov, A.V. Makarevich, et al., 2003.
18. U.S. Patent 6555600, C 08 K 3/24, 3/28, 5/098, 5/17. *Corrosion inhibiting thermoplastic alloys.* A. Sobkin, M. Hobday, B.A. Miksis, 2001.
19. U.S. Patent 6054512, C 08 K 5/17, 3/24,3/28, 5/098. *Corrosion inhibiting thermoplastic alloys.* S. Bradley, T. Nelson, B. Miksis, M. Kharshan, 2000.
20. Der Rost Bleibt Dranpen. Kunststoffverpackungen mit Extra. *Production,* 1996, No. 25, pp. 21–25.
21. Verpackung mit eingebautem Inhibitor. *Werkst. Und Korros.*, 1995, Vol. 46, No. 10, pp. 603–606.
22. A.H. Landrock, ed. *Handbook of Plastic Foam.* New York, Noyes Publications, 1995.
23. U.S. Patent 4275835, C 23 F 11/02, B 23 B 03/26, 05/18. *Corrosion inhibiting articles* B.A. Miksis and R.H. Miller, 1980.
24. V.D. Solodovnik. *Microencapsulation.* Moscow, Khimiya, 1980.
25. M. Ash and I. Ash, eds. *Handbook of Corrosion Inhibitors.* New York, Synapse Information Resources, Inc., 2000.
26. A. Kondo. *Microcapsule Processing and Technology.* Basel, J. Wade and Van Valkenburg, 1979.
27. J.-H. Park, ed. *Chemical Vapor Deposition.* New York, ASM International, 2000.

6 Economic and Social Aspects of Adoption of Inhibited Plastics

Several criteria are commonly used to define the expediency of application of inhibited plastics, including their effectiveness, safety, production and operation peculiarities of this class of materials. The authors understand that there is not any unambiguous interpretation of these problems and the effectiveness of each kind of inhibited plastics should be treated individually in both methodological and technical as well as economic and social respects. What is more, it should be born in mind that these estimates change with time as the rust-inhibition technology of metals gains strength. When considering economical and social aspects of using inhibited plastics we have tried to single out the rules that are most typical for this class of materials and underline their distinctive differences from other means of protecting metals against corrosion.

6.1 Cost-Effectiveness

The estimation of cost-effectiveness of inhibited plastics is a multifactor problem and its solution is unambiguous. Let us consider the main aspects involved in the estimation of the expediency of using any anticorrosive material to justify this opinion:

(A) corrosion-induced losses,
(B) choice of corrosion-resistant materials,
(C) combined means of protection against corrosion,
(D) factors defining the effectiveness of inhibited plastics,
(E) estimation of alternatives for protection,
(F) the specifics of inhibited plastics.

A. **Corrosion-induced damage** has inevitably arisen since man learnt to extract metals. The spheres and scale of metal assimilation have widened continuously and the problem of losses from corrosion have been aggravated and become more acute. Machines and metal constructions today constitute the most valuable part of basic production assets in the world. By the end of the 20th century economic losses from corrosion had reached a magnitude commensurate with worldwide investments into the development of the main branches of industry. Experts estimate annual damage by corrosion of up to

3.5% of the EC's gross national product [1]. As for the USA, this number is even higher – 4.2% [2].

Damage by corrosion can be subdivided into three main categories:

(i) economical expenditure composed of running repairs of machines and equipment, pipelines, reservoirs, cars, ships, bridges, marine constructions and so on;
(ii) catastrophic aftereffects from unexpected failure of high-pressure vessels, boilers, containers for toxic agents, turbines, airplanes and so on;
(iii) social, i.e. additional expenses for metal melting, design and reconstruction and so on to replace objects damaged by corrosion.

Losses from corrosion are subdivided into direct and indirect costs. *Direct losses* are estimated by the amount of financial investment aimed at maintaining serviceability of machinery operating in corrosive environment, which includes the cost of substitution and repair of corroded machines and equipment, expenses for the preservation of hardware, substitution of carbon steel by rustproof grades, and so on. In developed countries these losses make up 5% of the national income.

Indirect losses are composed of the costs arising from drop of labor productivity, rising initial cost of machinery, operation expenses, and impaired production quality due to corrosion.

Indirect losses include [2]:

- downtime due to the corrosion of critical members of equipment leading to the failure of machines and lines;
- losses of processed or manufactured goods (oil, gas, water, antifreeze, fuel leakage, etc.);
- power loss of machines and equipment due to rust deposition on pipeline or heat exchanger walls, corrosion of internal combustion engines and other energy converting systems;
- deterioration of the quality of industrial goods contaminated by corrosion products of metal parts (chemicals, water, foodstuff, etc.);
- raised initial cost of machines and equipment as a result of loosened tolerances of damaged hardware.

It is evident that indirect losses are far higher than direct ones and their evaluation represents a serious problem even when greatly simplified. The catastrophic outcomes of corrosion damage that lead to health injuries and loss of life cannot be expressed in monetary units at all.

Recommendations for the evaluation of damage by corrosion in different branches of industry are published in the European Federation of Corrosion series. Two of the recent publications in this series have been devoted to corrosion in the oil and gas industry [4] and in nuclear waste system [5].

B. **The choice of rustproof materials** is recognized as a key problem that governs the technical and economic efficiency of protective measures against corrosion. At first sight it seems better to increase the consumption of stainless steels and ceramics instead of using coatings, corrosion inhibitors, films and other auxiliary means of anticorrosion protection of ferrous hardware. This is however not always possible due to, first of all, design and process criteria. Secondly, industrial structures formed in the course of evolution presuppose the prevalent use of traditional engineering materials. Thirdly, increasing production of rustproof metal alloys turns out to be five times as expensive as fighting corrosion by two methods at once, namely metal modification and the protection of hardware. In the list of products of the metallurgical industry of the USA rustproof alloys make up just 1–2%, while the rest are conventional ferrous metals [2]. Substitution of metals for polymers is not always beneficial since four tons of oil have to be spent per ton of polymer material under the current fuel and energy balance conditions. Note that inhibited plastics are even more costly. This is why we should consider a number of factors in order to choose a most expedient set of rust-inhibiting measures.

C. **The combined character** of rust-inhibiting measures can also be attributed to global contamination of the atmosphere.

As far back as the middle of the 20th century trunk pipelines were protected from corrosion mainly by bitumen and paper and operated reliably with such insulation for 20 to 30 years. Modern systems of protecting pipelines are much more complex and expensive (see Sect. 3.5.3) but their useful operating life has fallen. The main reason is contamination of the environment caused by man, composed of a number of factors, namely:

- overloading of the soil by inorganic matter during the excavation of deposits, streaming of ores and mineral ore-dressing;
- pickling and swamping of soils, spilling of subterranean waters when drilling oil and gas wells;
- contamination of the land by chemical waste from oil refining plants, plastics processing enterprises, gas and chemical recovery factories, wood-processing plants, etc.;
- local accumulation of chemical and agricultural fertilizers in the soil and subsoil waters;
- pollution of the world's oceans due to uncontrolled carry-over of industrial wastes, discharge of ballast water from vessels, shipwrecks of oil tankers and so on; atmospheric contamination by fuel combustion products;
- contamination of the biosphere by domestic waste round urban areas.

At the beginning of the 21st century it was unanimously accepted that it was preferable to solve the problems of environmental protection and corrosion inhibition of metals in close interrelation.

Any form of production presents an integral system where the opposed processes of creation (production) and consumption (usage of materials and energy) are interdependent. If these extremes do not match up, the equilibrium between the production and environment can be violated. The way out of this situation for mankind consists of the development of industrial technologies in high-priority directions, oriented first of all at ecological safety:

- transfer to waste-less processes, relying on the manufacture of a final product without waste or salvaging of waste within the same or other processes;
- extensive realization of low-waste processes where the product or its part is manufactured with low or not fully recovered wastes;
- development of processes based on new principles for, first of all, ferrous metallurgy oriented initially at ecological safety (e.g. iron reduction from iron-ore concentrates by hydrogen);
- correlation of production volumes with the criteria for social and economic expediency, which in fact always leads to production shrinkage;
- use of biofilters in the final stages of neutralizing industrial waste and other engineering methods that perform microbiological neutralization of pollutants in a similar way to nature.

As can be seen, the growing expense of environmental protection from technogenic contamination cuts the cost of anticorrosion means. At the same time, protection of metals from corrosion contributes to the preservation and optimum use of natural resources.

D. **The factors defining the efficiency of inhibited plastics** are a powerful argument in favor of this class of materials.

Above all, this type of anticorrosion material realizes at least two mechanisms for the suppression of metal corrosion, i.e. the barrier and inhibition mechanisms. Therefore, any additional means of protecting metals from corrosion are not commonly employed together with inhibited plastics. In addition, as inhibited plastics are both structural triboengineering materials and volatile CI carriers, they preserve all members of the sealed system from corrosion during its operation.

Among the obvious drawbacks of inhibited plastics is the high cost of the chief ingredient – the CI. The cost of CI used today fluctuates in the range 10–50 $/kg. Even small amounts of inhibitors in plastics (5 wt%) noticeably increase the cost of polymer films, coatings and structural articles. There is only a slight possibility of abating the cost of some inhibited plastics by the realization of several mechanisms of corrosion inhibiting. Thanks to this, the requirements on polymer components become less strict and it appears beneficial to employ cheaper secondary polymer materials as the base for inhibited plastics.

One of most efficient application spheres of inhibited plastics is coatings. Data cited in Chap. 3 proves that numerous procedures for raising the protection effectiveness of polymer inhibited plastics have not been fully realized to date. Above all, this concerns the application of a few kinds of CI that are impregnated layer-wise through the thickness of the material. Great potential for improving the rust-inhibiting ability of substrates lies in the use of inhibited metal-polymer coatings containing protecting fillers or metallized layers.

The effectiveness of plastic protection means containing volatile CI depends much on the liberation kinetics of CI from the polymer matrix. We have already analyzed (see Sect. 1.5) the problem of binding inhibitors within the plastic structure. The delivery velocity of the required amount of CI to the metal part surface turned out to be a reliable criterion for estimating the efficiency of inhibited plastics. Some plastic grades do not need to take account of the CI delivery factor if the inhibiting component is located in the most probable place of corrosion.

An important aspect of the efficiency of inhibited plastics is the stock of CI in the polymer matrix. If the anticorrosion system is designated for long-term operation, a moment when the CI stock appears to be fully exhausted inevitably arrives. Naturally, this substantially lowers the effectiveness of the anticorrosion system, although does not necessarily lead to failure. Types of rust-inhibiting plastics such as coatings and preserving films retain their barrier properties even after volatilization or wash-out of the inhibitors and continue to insulate metal goods from the hostile environment.

E. **Estimates of protection alternatives**. Distribution of capital investment in different sectors of corrosion protection of hardware provides information on the efficiency of these domains. In the twilight of the existence of the USSR, the ratio of various antirust methods for hardware protection was as follows: [3]

– polymer films and coatings	39.5%
– rustproof steels and alloys	20.5%
– metal coatings	15.6%
– electrochemical protection	11.3%
– chemical protective means	8.6%
– metal substitution for nonmetals	4.5%

Almost half of all assets are spent on polymer and inhibiting means of protection. In our opinion, a correlation of expenditure on rust-inhibiting techniques close to this has been retained in developed countries to the present day.

In fact, adoption of any inhibited plastics in a process presumes the estimation of several design and technological alternatives for rustproofing methods.

In principle, the following variants of protecting industrial equipment can be envisaged:

I. Embody the whole hardware stock in a rustproof variant. In the case of the correct choice of steels and alloys this variant will almost fully exclude corrosion damage and corresponding losses from corrosion. Unfortunately, in most cases it is unrealistic because of high cost.
II. Embody part of the hardware stock in a rustproof modification and use inhibited plastics to protect the remaining part of the hardware.

III. Maintenance of equipment in operational status via repairs only. In this case the expense will exceed by many times the value of the main production funds.

A systematic approach to the choice of an optimal variant of anticorrosion protection should include at least

(1) a retrospective economic analysis of preceding protective measures and an extrapolation of its results to forecast the parameters of new protective means, assisting in the initial choice of variants;
(2) construction of economical and mathematical models describing the application of the compared protection means.

The models consider such parameters as the cost of scheduled protective means, predictions for raw materials and production facilities, initial expenses, efficiency of planned measures, ecological safety, and so on. On this base, the most effective terms of use of these means and their functioning time are defined (up to liquidation) with and without the various protective means.

To make an economic evaluation of the adoption of inhibited plastics it is necessary to compare the accompanying additional capital investments and economy of current inputs:

$$E = 1/\tau = (B_1 - B_2)/(P_2 - P_1) ,$$

where E is the comparative efficiency factor of additional capital investments; τ is the payback period for capital investments, i.e. the time during which additional capital investments are reimbursed via the economy of lower initial cost; B is the product initial cost; P is the capital investment; indices 1 and 2 correspond to the variants of anticorrosion protection.

It is sometimes sufficient to compare the basic and developed variants of protection. The basic variant commonly represents the protective means to be substituted or those to be purchased in the required amounts. In the former USSR, additional capital investments were considered to be efficient if $\tau < \tau_{\mathrm{n}} = 6.6\,\mathrm{years}$, $E > E_{\mathrm{n}} = 0.15$, where τ_{n} is the standard reimbursement term, and E_{n} is the standard coefficient of comparative efficiency of capital investments.

The best of the variants that meets the criterion for efficiency of capital investments corresponds to the minimum reduced expense:

$$B_{\mathrm{i}} + E_{\mathrm{n}} P_{\mathrm{i}} \rightarrow \min ; \qquad P_{\mathrm{i}} + \tau_{\mathrm{n}} B_{\mathrm{i}} \rightarrow \min ,$$

where B_{i} and P_{i} are the prime cost and capital investments for variant 1. The first equation is the initial cost of the annual production and the total sum of capital investments. The second equation is the initial cost of a unit product and specific (per unit product) capital investments.

When comparing the variants, some particular technical and economic parameters are used that reflect the magnitude of separate expenses, including material capacity, labor intensiveness and capital investments of the protective means and the products produced. A systematic approach to the choice of variants before computations of their efficiency is required for the determination of data for calculations that is statistically reliable.

The National Association of Corrosion Engineers (NACE) in Houston (Texas, USA) has elaborated a standard, RP-02-72, on the direct calculation of the efficiency of measures for protection against corrosion, which is based on the dynamic discounting method of the balance of incomes and expenses. The standard makes use of the economic terminology adopted in business accounting and management. It helps to compute up-to-date costs quickly after levying taxes, equivalent annual costs and alleviates the retrieval of unknown economic parameters for various protection variants. The method takes into account different rates of taxes and depreciation of equipment. The analytic scheme contained in the standard can be used in various branches of the economy. Examples of the practical use of the methods of technical and economical analysis of antirust systems are given elsewhere [6].

F. **The specifics of inhibited plastics** as an object of technical and economic analysis consists in the following:

(1) ameliorated rust-inhibiting effect of the hardware fitted with inhibited plastics thanks to the actuation of several mechanisms of corrosion suppression;
(2) raised initial cost of protective means due to costly CI and complicated technology for their combination with the polymer base.

The main spheres where the use of inhibited plastics is effective are the following.

- The cost effectiveness of the adoption of inhibited polymer films for preserving hardware is estimated by prolonged storage times and reduced damage by corrosion. The income is formed mainly of lowering costs of preservation and depreservation operations.
- The majority of inhibited polymer coatings serve as structural elements of metal parts. The cost-effectiveness of coatings is a result of the extended time of failure-free operation of the equipment, the raised quality of manufactured goods and the reduction of expense for the protection of equipment against corrosion.
- Friction parts made from structural inhibited plastics retard mechanochemical wear of conjugated metal counterbodies. Liberation of volatile CI from plastic units located in a closed volume hampers corrosion of all metal elements of the design found in this volume. The cost effectiveness of structural plastics is mainly composed of the reduction of repair expenses of the equipment.

It should be noted that the estimation of cost-effectiveness of each concrete type of inhibited plastics is an uncommon problem necessitating a creative approach to its solution. To match up all the factors that influence effectiveness and take their contradictory effects on the spheres of production and consumption into account, as well as the environment, demands a certain compromise.

6.2 Safety Measures

Polymers used in inhibited plastics (polyolefins, polyamides, fluoroplastics and others) are in their majority harmless and friendly to man [7]. Toxicity of this kind of plastics can arise from additives that are impregnated for special purpose or their decomposition products. The extent of the danger to the human organism of inhibited plastics during their production and application (independently of the purpose) depends on the toxicity of the low-molecular components that isolate into the environment.

The present chapter gives a brief account of the main safety measures that should be observed when handling inhibited polymer materials. A toxicological assessment of the ingredients of inhibited plastics (polymers, CI, Pl) and the requirements for the protection of the attending staff and fire safety for production areas are given.

6.2.1 Toxicology

Toxicology is a domain of medicine that studies the physical and chemical properties of toxic substances, their mechanisms and aftereffects on living organisms, and prophylactic means. The aim of toxicological investigations of plastics is to discover the toxic effects of substances liberating from plastics on the human organism.

Below we present the results of toxicological investigations into the ingredients of inhibiting plastics.

Polymers. A two-step scheme of toxicological investigations of polymers is currently adopted. First, the toxic properties of individual ingredients that make up a polymer during synthesis or are introduced in later stages to achieve certain properties are studied. As a result of this step, the norms for isolation of low-molecular components into the ambient air, water or model media are established. These toxicological assessments are not imposed on non-toxic additives. Furthermore, the presence of such additives is admitted in plastics of any designation without obligatory control.

The second step of the toxicological estimate consists of the investigation of the toxic effect of a polymer material as a whole in the anticipated field of application. Toxicological investigations of this type are conducted in two cases:

(i) the liberation of some components is close to the limit but there is a danger from the combined effect of several components;
(ii) integrated sanitary and chemical evaluation methods in the environment indicate measurable isolation of low-molecular components, but the analysis of separate toxic substances cannot define them.

Both cases occur when unidentified substances are liberated from the polymer material. The second step of the toxicological estimate is used in conditions approaching as closely as possible the real operation of polymer materials.

Different polymers form low-molecular matter with varying toxic effect on destruction [8].

During heating to 150–220°C *polyolefins* isolate thermooxidative products of macromolecules with pronounced toxic properties, namely organic acids, ethers, unsaturated hydrocarbons, peroxide and carbonyl compounds (formaldehyde and acetaldehyde), carbon oxide and dioxide, etc. A mixture of such products of the thermooxidative destruction of polyolefins may result in acute chronic poisoning on inhalation.

Ethylene and propylene copolymers are able to isolate aldehydes and carbon oxide on heating to $T = 240°C$. Decomposition with the formation of aldehydes and acids already starts at 150°C, while carbon oxide isolates at 180–200°C.

Polyamides may form caprolactam and aldehydes during destruction. Their vapors evoke headaches on inhalation, sleep disorders and other dysfunctions. The concentration of aldehydes in the operation zone must not exceed 0.5 mg/m^3 and that of caprolactam must not exceed 10 mg/m^3.

Fluoroplastics are harmless hygienically and nontoxic below 200°C. When heated above 200–250°C they liberate hydrogen fluoride and fluoric hydrocarbons. Above 300–350°C they form perfluoroisobutylene, fluorine monomers, fluoric carbonyl and carbon oxide. Inhalation of highly dispersed particles of the polymer and volatile thermodestruction products of fluoroplastics irritates the respiratory tract and may cause coughing, fever and high temperature. Sometimes, the kidney, liver and brain can even be affected. The maximum permissible concentration (MPC) of fluoroplast-4 in aerosol form should be below 10 mg/m^3 in the working zone.

Below 180°C the decomposition products of *pentaplast* contain phosgene, at 200–220°C they include carbon oxide, acidic chloranhydrides, formaldehyde and hydrogen chloride. The MPC of these volatiles in air is rather low: 0.5 mg/m^3 for phosgene, 1.0 mg/m^3 for formaldehyde, 5.0 mg/m^{3}for hydrogen chloride and 20 mg/m^3 for carbon oxide. As experiments with animals have shown, the products of thermal decomposition of pentaplast multiply their toxic effect when combined.

Polyvinyl chloride isolates hydrogen chloride, chlorine compounds and carbon oxide above 150°C. These substances irritate the retina and mucous membrane of the nose. During processing and especially coating application of PVC a considerable amount of plasticizers and their oxidation products are isolated into the air. Meanwhile, the MPC of, for instance, the most applicable plasticizer (dibutyl phthalate) is below 1 mg/m^3 indoors.

Polyurethanes irritate the retina and respiratory tract due to the presence of di-isocyanates. The most toxic of these are hexamethylene diisocyanate (HMDI) and toluylene di-isocyanate (TDI). Chronic intoxication by TDI and HMDI causes bronchitis, affects the liver and vegetative system. They lead to inflammation in contact with skin. The MPC of TDI vapors in the working area air is 0.5 mg/m^3, while that of HMDI is 0.05 mg/m^3. Heating of the elastic polyurethane foam (foam rubber) to 40–44°C for 20 days is accompanied by the isolation of higher alcohols and carbon oxide. At 60°C it also isolates ethylene glycol, and further heating to 200°C for 30 minutes leads to the isolation of TDI, hydrogen chloride, ethylene glycol and aldehydes.

As a result of thermooxidative destruction, *polyethylene terephthalate* liberates the diethyl ether of terephthalic acid, acetaldehyde, methanol and carbon oxide. As investigations have shown, manufacture of PETP films or fibers has brought no change to the state of health of maintenance workers. Moreover, PETP films are used to preserve sterilized foodstuffs and warm up prepared dishes.

Phenol formaldehyde polymers can isolate toxic matter as they contain unreacted residual monomers, including cresol, phenol and formaldehyde. Heating of the polymers to 55–65°C results in the isolation of considerable amounts of these components. The production of articles of phenoplast press powders by hot pressing is also accompanied by the isolation of phenol, ammonium, carbon oxide, methanol and some other materials. The products of thermooxidative destruction of phenoplasts are far more toxic in combination than individually. Workers handling press powders may complain of skin disorders (dermatitis), headaches, irritability and asthenia. This is largely because of the effect of phenols and methanol, which are strong poisons able to affect the nervous system when vapors or dust are breathed in, or on contact with the skin. The MPC of these substances within the working zone air is 5 mg/m^3.

Metal corrosion inhibitors. Harmful substances are subdivided into four classes of danger according to their affect on human organism [9]:

1st – extremely dangerous (MPS is below 0.1 mg/m^3)
2nd – highly dangerous (0.1–1.0)
3rd – moderately dangerous (1.1–10.0)
4th – low danger (above 10.0)

The majority of CI used as components of rust-inhibiting plastics are, according to special literature [9], materials that present a moderate or low danger, except for NDA, MSDA-1 and IFHANGAZ-1, which are considered highly dangerous (see Table 6.1). Most CI present little danger to man on inhalation according to the criteria for labor hygiene [10].

Table 6.1. Toxicological characteristics of metal CI

Inhibitor	Degree of Danger	MPC, mg/m^3
NDA	Highly dangerous	0.5
CAC	Moderately dangerous	2.0
KCA	Moderately dangerous	10.0
MSDA-1	Highly dangerous	1.0
M-1	Moderately dangerous	10.0
G-2	Moderately dangerous	3.0
VNHL-49	Moderately dangerous	10.0
VNHL-20	Low-dangerous	17.0
IFHAN-1	Moderately dangerous	5.0
IFHANGAZ-1	Highly dangerous	1.0
BMEA	Moderately dangerous	5.0
BTA	Moderately dangerous	5.0
Imidazole	Moderately dangerous	3.0
5-phenyltetrazole	Moderately dangerous	5.0

*Harmful substances are subdivided into four classes of risk according to their affect on human organism [9]:

Under production conditions, the transcutaneous penetration mechanism into the organism, i.e. through undamaged skin, follows inhalation in degree of danger. The toxicological experience of the transcutaneous effects of CI has proved to have a local character, characterized by inflammation (skin reddening) that may be complicated by a suppurative process with deep sores. This local affect is common to the majority of CI.

When permeating through undamaged skin, CI may cause intoxication similar to that following injection through the gastrointestinal tract. This fact confirms that CI possess skin-resorptive properties. These properties are peculiar to CAC, NDA, KCA, MSDA, VNHL-49, other amines of the polymethylene series and azoles like BTA, tetrazoles amongst others. Note that contact CI used as oil solutes may deposit when penetrating through skin and gradually soak up blood.

As experiments have shown [11], the polytropic affect of CI on the human organism exhibited, first of all, changes to the blood. Amines of the polymethylene series and their salts affect blood selectively. Large concentrations of inhibitors such as NDA, CAC and KCA in the air may raise methemoglobin in erythrocytes (up to 4.8%, in contrast to the physiological norm of 1–1.5%), and reduce the total amount of hemoglobin and erythrocytes. The derivatives of carbonic acid deplete and ruin erythrocytes in blood vessels leading to their regeneration. Azole derivatives may hamper the regeneration of erythrocytes by impairing their resistance.

Intoxication by CI is accompanied by violation of redox reactions in the organism, in the first place by hindered activity of oxidative enzymes in the organs. The liver is the organ that is most subject to the effect of amine salts from the polymethylene series; kidneys are more strongly affected by the derivatives of carbonic acid.

Experiments with animals have shown [12] that brain and liver cells undergo inflammation under toxic doses of CI. As a result, RNA granules in cytoplasm of neurocytes and hepatocytes contract greatly in size. Along with functional disorders, toxic doses of CI intensify the deep pathological processes exhibited in hemodynamic violations and dystrophy of the brain and spinal cord, liver, kidneys, heart muscle. Structural disorders detected in the brain also prove the high sensitivity of the nervous system to CI.

Plasticizers. Information on the toxicological properties of Pl is necessary to maintain hygienic labor norms, to exclude poisoning or occupational diseases, and to create harmless inhibited plastics.

Comprehensive data on the physiological effects and toxicity of Pl are available today in special literature [13]. The character of the toxicological effect of ester-based Pl has been established, their MPC in the working areas, dwellings and water has been defined and recommendations have been devised for the application of various types of Pl in different spheres.

Nontoxic Pl used in inhibited plastics include (Table 6.2) mineral oils, polyatomic alcohols (glycerin), higher fatty acids (oleic acid), esters of aliphatic carbonic acids (sebacates) and trimellitic acids (mellitates).

Table 6.2. Toxicological characteristics of some plasticizers used in inhibited plastics

Plasticizer	Degree of Danger	MPC, mg/m^3
Mineral oil (petroleum)	Moderately dangerous	5.0
Glycerin	Non-toxic	–
Dioctyl phthalate	Highly toxic	1.0
Dibutyl phthalate	Highly dangerous	0.5
Dimethyl phthalate	Highly dangerous	0.3
D(2-ethylhexyl)sebacate	Moderately dangerous	10
Mellitates	Low danger	>10
Oleic acid	Non-toxic	–

Phthalic acid esters (phthalates) are also related to low-toxicity compounds on single exposure according to the parameters of acute toxicity [14]. Nevertheless, the capability of phthalates to permeate through undamaged skin and their evident cumulative effect leading to intoxication of the organism in a chronic state has classified them as highly toxic substances. Due to the biological character of their effects, phthalates are referred to as polytropic poisons that selectively strike the nervous system. Notice that observations of generally accepted concentration norms of phthalic acid esters in the ambient air and water exclude any poisoning by this Pl.

6.2.2 Labour Protection and Fire Safety

The contamination level of the air in production areas by chemical matter differs depending on the purpose of the production, degree of automation and mechanization as well as the peculiarities of the technological equipment, amongst other factors.

The results of hygienic observations have shown [15] that the determining factors for creating acceptable operation conditions for staff handling CI are the chemical used and the physical, including meteorological and noise, conditions.

CI and other chemical substances isolate into the air during all technological stages of the manufacture of inhibited plastics, as well as during their storage and operation. This includes the preparation of polymer compositions, the application of inhibited coats, the manufacture of inhibited films and structural products, the preservation and packaging of hardware and so on. As a matter of fact, the protected equipment, as well as the inhibited materials themselves, can be a source of isolation of harmful matter into the environment.

To improve conditions when using with inhibited plastics a series of precautionary measures should be taken. When manufacturing any type of inhibited plastics, the following measures are to be envisaged:

- substitution of highly poisonous compounds for less harmful ones (e.g. CI);
- adoption of technological processes that exclude manual labor;
- presence of exchange ventilation in fabrication facilities and local ventilation in working places, technical means of control of the ambient air state; MPC of harmful matter in the operation zone, temperature, umidity and air motion should not exceed the accepted norms;
- elimination of ingress of inhibitor vapors and aerosols of in the adjoining areas;
- systematic cleaning of operation areas (pneumatic and wet);
- individual means of protection and overalls for the maintenance staff; instruction of the attendants about the level of toxicity and fire risk, first aid measures in case of accidents;
- regular medical inspections, including initial examinations and periodic ones thereafter, to disclose early symptoms of harmful effects of CI to prevent diseases.

Along with these general norms there are a number of specific requirements imposed on the handling of certain types of substances during particular process stages.

Application of inhibited paint coats is connected with the usage of inflammable and harmful substances.

Many organic solvents and CI are related to highly flammable fluids whose vapors can form explosive mixtures on contact with air. The lowest concentration inflammability limit (LCIL) of highly inflammable solvents (acetone, benzene, mixed solvents) is about 45–52 g/m^3 or 1.2–1.8% by volume, the flash point is between –18°C and +5°C, and the inflammability limit is between –20°C and +20°C [16].

The following demands are imposed on fire safety when operating with inhibited paint coats:

- painting areas or shops should be in single-storey buildings or on the upper floors of multi-storey buildings;
- finishing operations should be isolated from painting shops; paints should be stored in a separate building or shop with only a daily stock of paint;
- painting shops, paint supply sections and storage facilities should be fitted with fire extinguishing methods;
- shops should be equipped with forced exhaust ventilation;
- painting shops and paint supply sections should be fitted with water, (low-pressure) vapor or air heating at heater temperatures below 95°C.

The air removed from the painting shops (spraying, drying chambers, local ventilation, etc.) is contaminated with vapors of inhibitors, solvents and frequently with paint aerosols. The concentration of these substances in air outlets from spraying chambers surpass the acceptable norms for atmospheric air in inhabited areas. Therefore, definite measures must be undertaken to reduce atmospheric contamination by ventilation outlets.

Powder inhibited paints present no less fire and explosive danger than liquid ones. Although the energy needed for their inflammation is ten times as high as that sufficient for inflammable solvents, the LCIL of inhibited powder paints is rather low (in the main, 17–35 g/m^3). Their self-ignition temperature (325–500°C) is close to that of the majority of organic solvents [17].

Production areas with dust sources and adjoining harmful substances are most hazardous in terms of sanitation and fire. These are the stations for the preparation and charging of loose raw materials, dry grinding and mixing, application of powder paints and coating deposition. Note that coating spraying is accompanied by the generation of volatile decomposition products of unstable paint and vapor components of low-molecular substances.

Dust from any polymer, including harmless ones, may in fact cause diseases of the respiratory tract when inhaled. So far, abating dust load is an important element of prophylactics of occupational diseases. It is critical to ensure that the dust concentration of harmful matter is below the accepted MPC (Table 6.3) [17, 18].

The amount of volatiles formed during coating application is, as a rule, not very high and for most paints does not surpass 3–5 g per kg of paint used. This is hundreds of times less than the amount of solvents evaporating during liquid paint application.

From the hygienic viewpoint, the best coating application procedure is mechanized electrostatic deposition in a chamber fitted with powder recuperation followed by fusing (setting) of the coating in a chamber with an automatic temperature regime.

Table 6.3. Threshold concentrations of dust of some polymers and pigments in operating area air

Powder	MPC, mg/m^3
Polyethylene	10
Polyvinyl butyral	10
Polyvinyl chloride	6
Polyfluoroolefins	10
Epoxy oligomers	10
Titanium dioxide	10
Microbarium sulfate	6
Zinc oxide	5
Commercial carbon	4
Leaden crown	0.01

In spite of the fact that powder inhibited coatings (PIC) present a lower sanitary and fire risk compared to liquid ones, the following requisite demands should be observed during their application:

- the fluidizing bed baths should have exhausts;
- air velocity during paint spraying in a chamber should be above 0.6 m/s;
- spark discharges above the minimal ignition energy of the powder paint are inadmissible;
- powder recuperation chambers must be fitted with safety valves;
- the coating formation furnace should be equipped with a ventilating cap and self-blocking system so as not to let the temperature rise above the set norm.

Fabrication of *polymer inhibited films* (PIF) and hardware packaging must be accomplished by observance of these generally accepted requirements and hygienic norms together with the following specific demands.

1. The equipment for manufacture of PIF (hose-blowing machines, extruders and so on) and hardware packaging should be located in isolated shops.
2. For the introduction of CI into the blown hose, the hose-blowing machine should have an automatic blocking device to cease CI supply in case of violated continuity of the hose (puncture, rupture, etc.)
3. CI whose decomposition temperature is below that of the polymer processing (including all extruder zones and die head) must not be used for extrusion of inhibited polymer compositions into films.
4. The attendants employed in the preservation and packaging into PIF must have individual protection means, such as cotton overalls and rubber gloves. In emergency situations, respirators and gas masks should be used. Analogous measures are to be observed during depreservation processes.

5. Storage and warehousing norms should indicate that PIF is not explosive but catches fire without burst and burns with a smoky flame, forming a melt and thermal decomposition products.
6. Products packed into PIF should be stored in ventilated storage facilities at temperatures of 10–40°C. Direct sunlight is inadmissible. Continuity of the package should be observed during storage and transportation.

The manufacture of INHAM plastic goods as well as the processing of sealants, lubricants and other inhibited materials employs standard safety measures typical of any plastics production process. Since the materials concerned contain CI and some toxic components, specific additional demands should be observed:

- exchange and local ventilation should be installed to maintain the concentration of harmful matter below accepted limits;
- production shops must be equipped with fire extinguishing means;
- attendants should be provided with working clothes, improved hygiene and sanitary conditions, and undergo periodical medical examination.

Taking into account the above, it seems quite feasible to optimize labor of staff operating with inhibited plastics, and make it safer via strict observation of rational hygienic, sanitary, process and prophylactic norms.

6.3 Production and Application Ecology

Ecological problems arising during manufacture and consumption of inhibited plastics can be formulated as follows.

First, the processing of inhibited plastics into products is accompanied by the liberation of harmful and toxic matter into the environment. In particular, these are volatile fractions of VCI, other engineering additives and thermal destruction products of the plastics. Second, the majority of inhibited plastic goods are expendable and their disposal and recovery pose a serious technological and organizational problem. The expediency and efficiency of using inhibited plastics as a rustproof method relies on a successful solution of this problem.

The ecological safety of the manufacture and consumption of inhibited plastics is connected first of all with the presence of toxic CI. These CI are commonly combined with the polymer base under elevated temperatures, so the VCI may leak and be released into the environment due to either the malfunction of the equipment or the permeability of the plastic itself. Moreover, CI may ingress into the engineering and ambient media at the stages of article usage or recycling.

Therefore, many countries in the world have banned the application of highly toxic CI in inhibited polymer materials, including coatings, films, paper-film laminates, etc. For instance, sodium nitrite has been forbidden for use as a CI since November 1994 in the USA, Canada and the European Community in spite of its good rustproofing properties. In addition, attempts in this direction have been undertaken for the addition of chromate CI in inhibited paint coats and plastics. Instead, the employment of less toxic though less efficient organic phosphates is recommended [19]. Environmentally friendly inhibited films as a rule are devoid of amine, chromate and phosphate CI, and the search for still more efficient and safe CI is an urgent problem for modern science and technology. Assimilation of low-toxicity VCI of the PHC series (imidazole, 3-R-5R′-1,2,4-triazole and 5-R-tetrazole derivatives) as modifiers of inhibited PE has been substantiated in some works [20, 21].

Apart from the CI, some low-molecular components of inhibited plastics can be a source of contamination. For example, materials plasticized by a toxic phosphate of phthalate can present a danger to the environment [13]. An alternative to these Pl can be mineral oils, polar Pl of the esters of aliphatic dicarboxylic acids and alcohols, amogst others. These esters represent perfect Pl of a number of thermoplastics, can dissolve CI and are, besides, contact inhibitors and water-repellent agents for metal surfaces [22] (e.g. dibutyl sebacates).

Economic use and elimination of thermal destruction of CI, as well as the restriction of their release into the environment can also be solved by technological means. For this purpose, multilayered inhibited films are manufactured in which the CI carrier can be a polymer interlayer facing the package inner side, while the outer layer plays a barrier function [23]. It has been proposed to use either solid substances or pastes as VCI carriers with a melting point below 40–110°C to minimize their isolation during molding [24]. Different procedures for the combination of CI with the film base that exclude high temperatures and CI losses on evaporation have been developed. They mainly consist of co-extrusion of CI jointly with the polymer melt. Polymer films are modified by CI in the moment the melt leaves the extrusion head and the film base starts to solidify (see Sect. 2.2).

Ecologically safe methods devoid of any organic solvents are given preference in the application of inhibited polymer coatings. The creation of conducting polymers acting as perfect antirust materials has originated a new epoch of environmentally friendly anticorrosion polymer coatings (see Sect. 3.2).

Of no less importance for ecology is disposal and recycling of refuse based on inhibited plastics that decomposes slowly under natural conditions.

Technologies concerned with inhibited plastic wastes are developing according to two main trends (Fig. 6.1).

The first of them embraces reusable products and their recovery techniques, during which energy and new substances or materials can be produced. This group of methods includes incineration, pyrolysis and recycling. The second trend consists of the methods based on the ability of disposable plastics to degrade with time under the effect of natural factors, including storage, burial, composting and self-destruction.

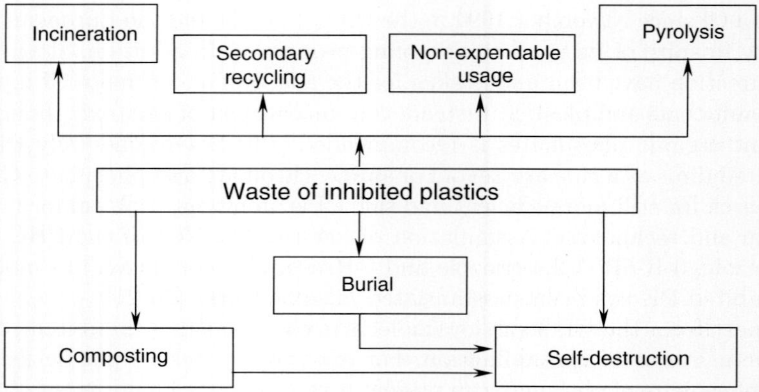

Fig. 6.1. Disposal of inhibited plastic wastes

Non-expendable goods based on inhibited plastics are rarely made and exhaust their anticorrosion resource during their first use.

Incineration of engineering and domestic wastes of inhibited plastics is inefficient in terms of energy production and leads to contamination of the atmosphere by harmful combustion products. In addition, incineration requires thorough sorting of waste, which is uneconomic. According to statistical information [25], incineration of a ton of polymer scrap is twice as expensive as recycling and five times expensive as burial. To obtain official permission to build an incineration plant is in fact improbable in many countries in the world today.

Pyrolysis (high-temperature destruction without oxygen) as a method of polymer refuse recovery is intricate in its technological implementation, very costly and is not in fact used for inhibited plastics.

Recycling presupposes the return of the used-up products into a production cycle and secondary processing into the polymer raw materials fit for molding new goods. Many types of products of inhibited plastics, including film materials, are subject to recycling. In particularly, multilayered films are easily recycled if their CI-containing layer and those carrying the barrier and mechanical load are made of the same polymer [26]. There are examples where one of the layers of such films is made of a first polymer while the others are made of a secondary, similar polymer [27]. Raw materials of secondary cellulose have found application in the production of inhibited combined materials based on paper and cardboard [28]. Apart from these processes there are resource-saving waste-oriented technologies using recovered CI as the components of inhibited plastics. These can be by-products of oil, coal, peat or slate refining, or recovered animal or raw vegetable stock [29, 30].

In spite of the apparent economic advantages of recycling polymer inhibited plastics, their volumes are rather low due to numerous problems in collecting, sorting, cleaning, etc. In addition, additional treatment raises the price of secondary polymers, while their quality deteriorates. In this regard, recycling does not exceed two or three cycles [31].

Burial is the cheapest and most applicable method of destroying wastes of inhibited plastics. It is however far from being safe for the environment. Locations for dumps are often allocated without proper ecological assessment and dumped refuse most often contaminates the adjoining earth, air and subsoil waters. It is becoming more and more difficult to find new places for dumps, even for harmless waste. As for the storage of plastic wastes, it is highly expensive and the cost grows with time. The European Community is prepared to introduce still more rigid standards, methods of burial and dump projects. Ecological laws that restrict the variants of these methods of refuse disposal have been passed in a number of countries.

The most advanced and ecologically friendly method of disposal of inhibited refuse is considered to be the use of self-destructive inhibited plastics with regulated lifetime. For example, Cortec Corp. has developed biodegradable resin products that are inhibited by VCI consisting essentially of a polymeric resin of PE, starch and polyesters, such as polylactic acid or other suitable polyesters [32].

Biodegradable inhibited plastics hold their properties constant during their service life and then undergo fast physico-chemical and biological degradation on burial or composting in a natural medium. Their destruction products can easily enter metabolic processes of either natural or man-made biosystems. This type of recovery does not necessitate any additional areas, while the products of biodestruction of such polymers do not harm the environment.

As a matter of fact, modern industry is armed with a profound arsenal of ecologically friendly methods of manufacture and recovery of products based on inhibited plastics.

References

1. N. Evans. The case for corrosion surveillance. *Corrosion Prevention & Control*, 1993, Vol. 40, pp. 32–35.
2. H.H. Uhlig and R.W. Revie. *Corrosion and Corrosion Control. An Introduction to Corrosion Science and Engineering*. 3rd edn. New York, Chichester, Brisbane, Toronto, Singapore, John Wiley & Sons, 1985.
3. G.S. Keselman. *Efficiency of Preventing Corrosion in Oil Industry*. Moscow, Nedra, 1988.
4. P.S. Jackman, ed. *A Working Party Report on the Life Cycle Costing of Corrosion in the Oil and Gas Industry: a Guideline* (European Federation of Corrosion Series, No. 32). Leeds, Maney Publishing, 2003.
5. D. Feron and D.D. Macdonald, eds. *Prediction of Long Term Corrosion Behaviour in Nuclear Waste Systems* (European Federation of Corrosion Series, No. 36). Leeds, Maney Publishing, 2003.

6. V.R. Pludek. *Design and Corrosion Control.* London, Basinstoke, Macmillan Press, 1977.
7. E.A. Peregud. *Sanitary Chemistry of Polymers.* Moscow, Khimiya, 1967.
8. S.L. Danishevsky (ed.) *Hygiene and Toxicology of High-Molecular Compounds and Chemical Raw Stock Used for Their Synthesis.* Leningrad, Khimiya, 1969.
9. Russian State Standard 12.1.007–76. Detrimental substances. Classification and general safety criteria.
10. V.V. Paustovskaya. Some conclusions of research on the problem of toxicology and labor hygiene; metal corrosion inhibitors. *Protection of Metals*, 2000, Vol. 36, No. 1, pp. 96–101.
11. V.V. Paustovskaya. Main conclusions of experimental investigations of toxicology of atmospheric corrosion inhibitors. *Hygiene of labor and professional diseases*, 1990, No. 3, pp. 7–9.
12. V.V. Paustovskaya, Ya.I. Morgunova, I.S. Brit and V.F. Torbin. On the quantitative aspect of genotoxic effect of chemical compounds. Proc. All-Union Symposium *"Quantitative aspects of chemical effects in oncology"*. Leningrad, 1985, pp. 14–15.
13. R.S. Barshtein, V.I. Kirilovich and Yu.E. Nosovsky. *Plasticizers for Polymers.* Moscow, Khimiya, 1982.
14. N.V. Lazarev and E.N. Levina, eds. *Harmful Substances in Industry.* 7th ed., in 2 Vols., Vol. 2, Moscow, Khimiya, 1976.
15. V.V. Paustovskaya. Actual problems of labor hygiene in manufacture and application of atmospheric corrosion inhibitors of metals. *Medical problems*, 1994, No. 7–8, pp. 93–97.
16. A.D. Yakovlev. *Chemistry and Technology of Paint Coatings.* 2nd edn., Leningrad, Khimiya, 1989.
17. A.D. Yakovlev. *Powder Paints.* Leningrad, Khimiya, 1987.
18. I.P. Maslova (ed). *Chemical Additives to Polymers.* Moscow, Khimiya, 1981.
19. I.L. Rosenfeld and F.I. Rubinshtein. *Anticorrosion Primers and Inhibited Paint Coatings.* Moscow, Khimiya, 1980.
20. I.V. Tsarenko. *Polymer materials modified by polynitrous heterocycles and their protective mechanism.* Ph.D. Thesis, Gomel, 1996.
21. A.V. Makarevich. *Physico-chemical and technological principles of developing active film and fibrous materials based on thermoplastics.* Dr. Chem. Sci. Thesis, 2000.
22. A.I. Altsybeeva, and S.Z. Levin. *Corrosion Inhibitors of Metals.* L.I. Antropov (ed). Leningrad, Khimiya, 1968.
23. L.S. Pinchuk and A.S. Neverov. *Polymer Films Containing Corrosion Inhibitors.* Moscow, Khimiya, 1993.
24. France Patent 2774995, C 08 K 5/521, 5/10. *Compositions containing volatile corrosion inhibitors, articles based on these compositions and their production.* J. Schaptra, P. Droniow, L. Gukllon, et al., 1999.
25. M. Belinsky. Recylcing of polymer packages. *Recycling and packaging technologies,* 2001, No. 8, pp. 45–46.
26. U.S. Patent 5863642, B 32 B 7/02, 5/08, 27/32. *Water resistant and vapor phase corrosion inhibitor composite material.* E. Henderson, L.J. Reed, K.D. Jagielski, 1999.
27. U.S. Patent 5855975, B 32 B 5/16. *Anti-corrosion plastic film containing recycled resin.* B.A. Miksic, J.M. Foley, 1999.

28. U.S. Patent 5894040, B 32 B 5/16. *Vapor Phase Corrosion Inhibitors on Post-Consumer used or Recycled Paper.* J.M. Foley, B. Rudman, 1999.
29. Ukraine Patent 25038, C 09 D 5/08. *Composition of anticorrosion coating.* V.I. Suschenko, 1998.
30. L.P. Maiko, T.Ya. Tsaryuk, V.G. Konstantinov, et al. Alternative protective components for inhibiting mineral oils. *Materials, technologies, tools*, 2002, Vol. 7, No. 3, pp. 45–49.
31. V.A. Fomin and V.V. Guzeev. Biodegradable polymers. State of the art and vistas. *Plastic Masses*, 2001, No. 2., pp. 42–48.
32. U.S. Patent 6028160, C 08 J 63/00. *Biodegradable Vapor Corrosion Inhibitor Products.* C. Chandler, S.J. Bradley, B.A. Miksic, 2000.

Conclusions

Despite their various assessments in production and ecological terms, non-traditional materials combined under the term inhibited plastics have already been assimilated within the sphere of engineering materials that form the base of modern industry. As long as inhibited plastics play a specific role in the global problem of protecting metals from corrosion damage, they fulfill a social function in counterbalancing the cost of metal damaged by corrosion and the repair of industrial equipment.

Inhibited plastics are science-intensive products that emerged from the fundamental achievements of electrochemistry, materials science, polymer technology, tribology and a number of other disciplines. Many aspects of their development and adoption in the sphere of preserving metal stocks are related to strategic activities guarded by the state.

It is believed that the following directions in the development and perfection of inhibited plastics will gain popularity in the near future:

- A class of polymer binders intended specifically for regulated thermodynamic compatibility with CI of different types will be created. Their joint extrusion will not suffer from the destruction or deterioration of the protective ability of inhibitors. Along with this, the problem of developing thermostable CI that can withstand co-extrusion with plastics awaits solution.
- The technology of inhibited plastics will progress in the direction of alternative methods of processing that do not subject raw stocks to high temperatures. Non-traditional technologies for inhibited plastics that are typical of composite materials, such as reactive extrusion, combinations of solution procedures, melt-blowing and others will gain ground. This will widen the range of structural characteristics of plastics for molding for solid as well as porous or fibrous products consistent with antirusting criteria.
- Based on advanced equipment for processing inhibited plastics it will become possible to impregnate CI into the polymer matrix at any stage of molding and in any phase state. A breakthrough can be anticipated in the development of specialized equipment for multilayered polymer films and coatings with a layer-by-layer distribution of inhibitors of various classes to form an integral optimum protective complex for given service conditions.

- As long as inhibited plastics have been classed as risky materials, growth of their production and application has been an issue for engineering ecology. The problem of obtaining CI that are harmless to the environment is of paramount importance today. On a par with this stands the task of biodegradability of inhibited plastics in order to refine their disposal and make it cheaper. Recycling of inhibited plastics has been transformed into a contradictory two-level problem, since they have to be resistant to microorganisms during operation, on one hand, and biodegradable on burial, on the other. Notice that inhibited plastics were initially created for use in engineering.
- Modern trends have taken the path of substitution of biocides, fungicides, insecticides, drugs and other biological active matter by CI. The latter impart novel properties to the polymer materials and expand their usage as far as biotechnology, engineering ecology, medicine and other spheres of activities lying at the junction between the animate and inanimate.

Computer models of inhibited plastics based on taxonomic, heat and mass exchange, diffusion and adsorption representations will add vitality to these materials. Their new functions as carriers of electric and magnetic fields, active biological components, reinforcing elements and so on will limit biased opinions of their application range. Since inhibited plastics are typically active materials, they can be a base for the creation of smart materials; in fact, the majority of today's smart materials are referred to as corrosion resistant. Together with traditional notions on inhibited plastics the authors draw the reader's attention to the concepts that not yet won recognition but are considered as pressing modern scientific tendencies. These reasons underpin our hope that this book will be a stimulus for new investigation in the field of polymer materials.

Index

Printing: Mercedes-Druck, Berlin
Binding: Stein+Lehmann, Berlin